S Westlake
Stewart, BC
Dec 18/72

100 Years of Popular Science

A Century of Wonders

100 Years of Popular Science

A CENTURY OF WONDERS

By ERNEST V. HEYN

in collaboration with past and present editors of *Popular Science Monthly*

1972

DOUBLEDAY & COMPANY, INC. GARDEN CITY, NEW YORK

AN ACKNOWLEDGMENT TO
POPULAR SCIENCE MONTHLY

Preparation of this book has been an independent project, conceived on its own merits and not sponsored by *Popular Science Monthly*. However, the project could not have been undertaken without the fullest co-operation of that magazine, to which (and especially to its President, Eugene S. Duffield) acknowledgment is gratefully made for generously placing its entire files of information and pictures and its bound volumes of past issues at the unlimited disposal of the author and his collaborators.

The material which originally appeared in *Popular Science Monthly*, and was copyrighted at the time, is reproduced here by special permission.

LIBRARY OF CONGRESS CATALOG CARD NUMBER 72–76168

PRINTED IN THE UNITED STATES OF AMERICA

DEDICATION

This book I dedicate with deep appreciation and admiration to my collaborators who also worked with me during my tenure as Editor-in-Chief and Associate Publisher of *Popular Science*.

More specifically, in alphabetical order:

Alden P. Armagnac, recently retired from the magazine after forty-three years as Associate Editor and later Senior Editor, who was creator and master of the huge index of *Popular Science* since May 1872, from which all material, illustration and text, for this book were drawn. Also, he was responsible for the chapters on atomic energy, space, military science, and electric light, besides giving me and the other editors invaluable advice and assistance on all illustrations and text.

Ronald Benrey, formerly Electronics Editor of the magazine, for the chapters on electronics and on radio and television.

Arthur Fisher, currently Science and Engineering Editor of the magazine, for the phonograph and photography chapters.

Devon Francis, formerly Automotive Editor, aviation expert, and assistant to the Editor-in-Chief on the magazine, for the automobile, aviation, and miscellany ("So What Else Was New?") chapters.

Harry Samuels, Art Director of the magazine, ably assisted by Joseph P. Ascherl, senior designer at Doubleday, for design and layout of this book.

Gardner Soule, expert in oceanography and author of many books and articles on that subject, with frequent appearances in the magazine, for the undersea chapter.

Robert P. Stevenson, currently Projects and Activities Editor of the magazine, for the do-it-yourself chapter.

Ruth Westphal, long-time faithful Associate Editor on the magazine and Georgiana Remer of Doubleday's staff for their able and conscientious copy editing of all text and captions.

ERNEST V. HEYN

Contents

for toothpaste and shaving cream – A 1920 hydrofoil boat and a modern successor – The new-found treasures of King Tut – A planet is discovered – Streamline trains, plastics, wonder drugs – The fabulous Dead Sea scrolls – Biggest reflector telescope in the world – Vaccines conquer polio – Climbers scale Mt. Everest at last – Ships ride on air – The fastest trains – World's longest span caps era of great suspension bridges – Human hearts are transplanted – First commercial ship through Northwest Passage – Watches with no moving parts – A new world's tallest building on the way – *and many other milestones of a century of popular science.*

Chapter 16 AFTERWORD

Viewing a hundred years' science as a whole – A technological speed-up unprecedented in history – Alarmists wage a hysterical war on science – Historical examples of anti-science excesses: the persecution of Galileo, machine-smashing rioters of the 1800s, the Scopes trial of 1925 and anti-evolution laws that persisted until 1968 – The real remedy for society's ills: developing even more advanced technology to solve current problems – A 1972 federal program to do it – Seven basic challenges for technology to tackle, and an optimistic look ahead at the new Century of Wonders that it well may bring. 314

ILLUSTRATIONS IN THIS BOOK

have been chosen mostly from pictures in past issues of *Popular Science Monthly,* in keeping with the book's theme of a century of developments as seen through the eyes of that magazine.

All pictures not otherwise credited are from *Popular Science Monthly* and are reproduced here by its special permission.

In the few instances of illustrations obtained elsewhere, the source of each picture is indicated by an accompanying credit line.

COLOR ILLUSTRATIONS:

The Horseless Carriage Grows Up *Following page* 64

Conquest of the Air *Following page* 128

Conquest of Space *Following page* 192

Adventure Under the Sea *Following page* 256
Popular Science Covers 100 Years of Do-It-Yourself

CHAPTER 1

A Century of Wonders: 1872–1972

THAT year Ulysses S. Grant was re-elected President of the United States by a 760,000 majority—the largest ever polled up to that time by a presidential candidate. Luther Burbank was at work on the Burbank potato. New York City erected its first elevated railroad, the "el."

The legendary *Marie Celeste,* which had sailed from New York with a cargo of alcohol, was found adrift in the Atlantic under full sail with no man aboard; what happened to the crew is as much of a mystery today as it was that year.

The summer of that year, fifty-year-old Louis Pasteur perfected a method for purifying and preserving beer. Extended to milk and other products, his process was to make "pasteurized" a household word.

That year twenty-five-year-old Thomas Edison perfected the stock ticker and was working on 45 other inventions. Alexander Graham Bell, also twenty-five, was teaching deaf students in Boston, gaining the experience in acoustics that led to his invention of the telephone.

The year was 1872, and it saw the beginning of a magazine called *Popular Science Monthly* whose first issue appeared in May.

Edward Livingston Youmans, author, professor, and lecturer, blind for much of his life, was its founder and first editor. His new magazine aimed to keep its readers up to date on current developments in science, both theoretical and practical. And in its succeeding ten decades what happenings it saw and reported!

It is my privilege as Youman's twelfth successor to review them in this book. Its purpose is not to present an exhaustive (nor exhausting) history of science in the past hundred years. Rather, it offers the highlights of what we editors brought to our readers during that period, in text and pictures—concentrating on the fascinating and often nostalgic hardware that grew out of practical invention and technology.

The accompanying table on pages 14 and 15 shows at a glance the myriad of great inventions during *Popular Science*'s lifetime. Based on a favorite reference handbook—the *World Almanac*—it has been appropriately abridged to consist of a selection of items dated 1872 or later. Impressively, nearly two-thirds of the original all-time list falls within the magazine's hundred years. Included are many of the epochal innovations that make up the bulk of this book.

Popular Science arrived on the scene just in time to witness a trio of outstanding inventions. Bell's telephone was first exhibited in 1876. His patent for it has been called the most valuable ever issued.

Edison's phonograph of 1877 made history by repeating his words, "Mary had a little lamb . . ." Then this champion inventor (with an astounding 1,093 patents in his lifetime) outdid himself in 1879 with his crowning achievement, the incandescent electric lamp.

With the 1880s came Waterman's fountain pen, Eastman's Kodak, rayon, the clanging electric trolley, and prototypes of "horseless carriages."

The 1890s brought successful automobiles—the Duryeas' and others. Edison, Jenkins, and Lumière machines ushered in the movies. Marconi gave the world wireless. The safety razor, the diesel engine, and halftone engraving were invented. (So were primitive forerunners of the tape recorder and the zipper.)

The next decade saw man take wing. The pioneer zeppelin of 1900 was a curtain-raiser for aviation's great event, the first airplane flight, by the Wrights' historic craft of 1903. Momentous, too, was De Forest's Audion tube, the radio amplifier that inaugurated modern electronics in 1907.

Military tanks and depth bombs of the World War I decade were followed in the twenties by talking movies, automatic elevators, and the coaxial telephone cable. Crude television experiments began. God-

Adapted by permission from *The World Almanac*, published annually by Newspaper Enterprise Association, Inc.

Great Inventions and Scientific

Invention	Date	Inventor	Nation
Adding machine	**1885**	Burroughs	U.S.
Addressograph	**1892**	Duncan	U.S.
Aerosol spray	**1941**	Goodhue	U.S.
Air conditioning	**1911**	Carrier	U.S.
Airplane, experimental	**1896**	Langley	U.S.
Airplane, jet engine	**1937**	Whittle	British
Airplane with motor	**1903**	O. and W. Wright	U.S.
Airplane, hydro	**1911**	Curtiss	U.S.
Airship, rigid dirigible	**1900**	Zeppelin	German
Arc tube	**1923**	Alexanderson	U.S.
Autogyro	**1920**	de la Cierva	Spanish
Automobile, differential gear	**1885**	Benz	German
Automobile, electric	**1892**	Morrison	U.S.
Automobile, experimental	**1875**	Marcus	Austrian
Automobile, gasoline	**1887**	Daimler	German
Automobile, gasoline	**1892–1893**	Duryea, C.E.	U.S.
Automobile, magneto	**1899**	Daimler	German
Automobile, self-starter	**1911**	Kettering	U.S.
Automobile, steam	**1889**	Roper	U.S.
Bakelite	**1907**	Baekeland	U.S.
Bicycle, modern	**1884**	Starley	English
Bomb, depth	**1916**	Tait	U.S.
Bottle machine	**1903**	Owens	U.S.
Camera, Polaroid Land	**1948**	Land	U.S.
Car coupler	**1873**	Janney	U.S.
Carburetor, gasoline	**1876**	Daimler	German
Card time recorder	**1894**	Cooper	U.S.
Carpet sweeper	**1876**	Bissell	U.S.
Cash register	**1879**	Ritty	U.S.
Cellophane	**1911**	Brandenberger	Swiss
Circuit breaker	**1925**	Hilliard	U.S.
Coaxial cable system	**1929**	Affel and Espensched	U.S.
Coke oven	**1893**	Hoffman	Austrian
Comptometer	**1887**	Felt	U.S.
Computer, automatic sequence	**1939**	Aiken et al.	U.S.
Condenser microphone (telephone)	**1920**	Wente	U.S.
Cream separator	**1880**	DeLaval	Swedish
Cultivator, disk	**1878**	Mallon	U.S.
Cystoscope	**1877**	Nitze	German
Diesel engine	**1895**	Diesel	German
Dynamo, hydrogen cooled	**1915**	Schuler	U.S.
Electric fan	**1882**	Wheeler	U.S.
Electrocardiograph	**1903**	Einthoven	Dutch
Electroencephalograph	**1929**	Berger	German
Electron tube multigrid	**1913**	Langmuir	U.S.
Elevator, push button	**1922**	Larson	U.S.
Engine, automobile	**1879**	Benz	German
Engine, gasoline	**1872**	Brayton	U.S.
Engine, gasoline	**1886**	Daimler	German
Engine, gas, compound	**1926**	Eickemeyer	U.S.

Invention	Date	Inventor	Nation
Engine, coal-gas, 4-cycle	**1877**	Otto	German
Engine, compression ignition	**1883**	Daimler	German
Engine, electric ignition	**1880**	Benz	German
Engraving, half-tone	**1893**	Ives	U.S.
Filament, tungsten	**1915**	Langmuir	U.S.
Flatiron, electric	**1882**	Seeley	U.S.
Gas discharge tube	**1922**	Hull	U.S.
Gas mantle	**1885**	Welsbach	Austrian
Gasoline, lead ethyl	**1922**	Midgely	U.S.
Gasoline, cracked	**1913**	Burton	U.S.
Gasoline, high octane	**1930**	Ipatieff	Russian
Geiger counter	**1913**	Geiger	German
Glass, laminated	**1909**	Benedictus	French
Gun, Browning	**1916**	Browning	U.S.
Gun, magazine	**1875**	Hotchkiss	U.S.
Gun sight, telescopic	**1891**	Fiske	U.S.
Gun silencer	**1909**	Maxim, H. P.	U.S.
Gyrocompass	**1911**	Sperry	U.S.
Harvester-thresher	**1888**	Matteson	U.S.
Helicopter	**1939**	Sikorsky	U.S.
Iron lung	**1928**	Drinker, Shaw	U.S.
Kinetoscope	**1887**	Edison	U.S.
Kodak	**1888**	Eastman and Walker	U.S.
Lacquer, nitrocellulose	**1921**	Flaherty	U.S.
Lamp, arc	**1879**	Brush	U.S.
Lamp, incandescent	**1879**	Edison	U.S.
Lamp, incandescent, frosted	**1924**	Pipkin	U.S.
Lamp, incandescent, gas	**1916**	Langmuir	U.S.
Lamp, Klieg	**1911**	Kliegl, A. and J.	U.S.
Lamp, mercury vapor	**1912**	Hewitt	U.S.
Lamp, neon	**1915**	Claude	French
Launderette	**1934**	Cantrell	U.S.
Lens, fused bifocal	**1908**	Borsch	U.S.
Linotype	**1885**	Mergenthaler	U.S.
Loudspeaker, dynamic	**1924**	Rice-Kellogg	U.S.
Machine gun, improved	**1872**	Hotchkiss	U.S.
Machine gun (Maxim)	**1883**	Maxim, H. S.	English
Meter, induction	**1888**	Shallenberger	U.S.
Meter, parking	**1935**	Magee	U.S.
Microphone	**1877**	Berliner	U.S.
Microscope, electron	**1931**	Knoll and Ruska	German
Monotype	**1887**	Lanston	U.S.
Motor, AC	**1892**	Tesla	U.S.
Motor, induction	**1887**	Tesla	U.S.
Motorcycle	**1885**	Daimler	German
Movie machine	**1893**	Edison	U.S.
Movie machine	**1894**	Jenkins	U.S.
Movie machine	**1895**	Lumière	French
Movie, panoramic	**1952**	Waller	U.S.
Movie, talking	**1927**	Warner Bros.	U.S.

dard launched the first liquid-fuel rocket. The principles of radar were discovered.

The thirties introduced nylon, Kodachrome, electric razors, and the parking meter. Whittle designed the jet engine that was to revolutionize military planes and airliners. Sikorsky put the finishing touches on the helicopter he would fly in 1940.

The Atomic Age arrived in the forties—and the Space Age in the fifties.

Demonstrated in 1960 was the first laser, a revolutionary device whose name was short for "Light Amplification by Stimulated Emission of Radiation." Originally from a ruby crystal, a laser emitted a pencil-thin light beam of extraordinary intensity—with such dramatic effects that it would enter into many aspects of our lives. Already lasers have served for eye surgery, welding tiny electronic circuits, 3-D photo-

Discoveries 1872-1972

Invention	Date	Inventor	Nation
Neoprene	1930	Carothers	U.S.
Nylon synthetic	1930	Carothers	U.S.
Nylon	1937	Du Pont Lab.	U.S.
Oil cracking furnace	1891	Gavrilov	Russian
Oil filled power cable	1921	Emanueli	Italian
Pen, ballpoint	1888	Loud	U.S.
Pen, fountain	1884	Waterman	U.S.
Phonograph	1877	Edison	U.S.
Photo, color	1892	Ives	U.S.
Photo, color, controlled penetration	1928	Mannes and Godowsky	U.S.
Photo film, celluloid	1887	Goodwin	U.S.
Photo film, transparent	1888	Eastman and Goodwin	U.S.
Photoelectric cell	1895	Elster	German
Photographic paper	1898	Baekeland	U.S.
Photophone	1880	Bell	U.S.
Phototelegraphy	1925	Bell Telephone Labs.	U.S.
Plow, disk	1896	Hardy	U.S.
Pneumatic hammer	1890	King	U.S.
Punch card accounting	1884	Hollerith	U.S.
Radar	1922	Taylor and Young	U.S.
Radio amplifier	1907	De Forest	U.S.
Radio beacon	1928	Donovan	U.S.
Radio crystal oscillator	1918	Nicolson	U.S.
Radio receiver, cascade tuning	1913	Alexanderson	U.S.
Radio receiver, heterodyne	1913	Fessenden	U.S.
Radio transmitter triode modulation	1914	Alexanderson	U.S.
Radio tube diode	1905	Fleming	English
Radio tube oscillator	1915	De Forest	U.S.
Radio tube triode	1907	De Forest	U.S.
Radio, signals	1895	Marconi	Italian
Radio, magnetic detector	1902	Marconi	Italian
Radio, FM 2-path	1929	Armstrong	U.S.
Rayon	1883	Swan	English
Razor, electric	1931	Schick	U.S.
Razor, safety	1895	Gillette	U.S.
Record, cylinder	1887	Bell and Tainter	U.S.
Record, disc	1887	Berliner	U.S.
Record, long-playing	1948	Goldmark	U.S.
Record, wax cylinder	1888	Edison	U.S.
Refrigerants, low-boiling fluorine compound	1930	Midgely and co-workers	U.S.
Resin, synthetic	1931	Hill	English
Rocket engine	1929	Goddard	U.S.
Searchlight, arc	1915	Sperry	U.S.
Soap, hardwater	1928	Bertsch	German
Spectroscope (mass)	1918	Dempster	U.S.
Steam turbine	1884	Parsons	English
Steel alloy	1891	Harvey	U.S.
Steel alloy, high-speed	1901	Taylor and White	U.S.
Steel, electric	1900	Héroult	French
Steel, manganese	1884	Hadfield	English
Steel, stainless	1916	Brearley	English
Stove, electric	1896	Hadaway	U.S.
Submarine	1891	Holland	U.S.
Submarine, even keel	1894	Lake	U.S.
Tank, military	1914	Swinton	English
Tape recorder, magnetic	1899	Poulsen	Danish
Telegraph, quadruplex	1874	Edison	U.S.
Telegraph, wireless, high-frequency	1896	Marconi	Italian
Telephone	1876	Bell	U.S.
Telephone amplifier	1912	De Forest	U.S.
Telephone, automatic	1891	Strowger	U.S.
Telephone, radio	1902	Poulsen and Fessenden	Danish/ U.S.
Telephone, radio	1960	De Forest	U.S.
Telephone, radio, long distance	1915	AT&T	U.S.
Telephone, recording	1898	Poulsen	Danish
Telephone, wireless	1899	Collins	U.S.
Teletype	1928	Morkrum and Kleinschmidt	U.S.
Television	1926	Baird	Scottish
Television	1934	Zworykin	U.S.
Time recorder	1890	Bundy	U.S.
Time, self-regulator	1918	Bryce	U.S.
Tire, pneumatic	1888	Dunlop	Irish
Toaster, automatic	1918	Strite	U.S.
Tractor, crawler	1900	Holt	U.S.
Transformer, A.C.	1885	Stanley	U.S.
Transistor	1947	Shockley, Brattain, and Bardeen	U.S.
Trolley car, electric	1884–87	Van Depoel and Sprague	U.S.
Tungsten, ductile	1912	Collidge	U.S.
Turbine, gas	1899	Curtis	U.S.
Turbine, steam	1896	Curtis	U.S.
Vacuum cleaner, electric	1907	Spangler	U.S.
Washer, electric	1907	Hurley Co.	U.S.
Welding, atomic hydrogen	1924	Langmuir and Palmer	U.S.
Welding, electric	1877	Thomson	U.S.
Wind tunnel	1923	Munk	U.S.
Wire, barbed	1874	Glidden	U.S.
Wire, barbed	1875	Haisn	U.S.
X-ray	1895	Roentgen	German
X-ray tube	1916	Coolidge	U.S.
Zipper	1891	Judson	U.S.

graphs (by holography), rangefinding; their use is foreseen for large-screen TV, for optical telephony, and even as military weapons.

Events of subsequent years are recent memory.

Few of these things-to-come were dreamed of when, in 1876, the Centennial Exposition in Philadelphia celebrated the hundredth birthday of the republic. But, as Mitchell Wilson observes in his scholarly book *American Science and Invention,* "America in 1876 was in love with machinery and everything connected with it . . . feeling toward machinery was almost religious in fervor."

Highlights of the Centennial included Bell's new telephone; pre-Edison electric lamps, "blue, flickering, and unstable" but "auguring a new era in the history of electricity," in *Popular Science*'s words; and the typewriter, invented in 1868 but only lately perfected and marketed. A famous early user of the typewriter, Samuel Clemens—better remembered by his pen name, Mark Twain—had dashed off this amusing "testimonial" to it only the year before:

> "Gentlemen: Please do not use my name in any way. Please do not even divulge the fact that I own a machine. I have entirely stopped using the Type-Writer, for the reason that I never could write a letter with it to anybody without receiving a request by return mail that I would not only describe the machine, but state what progress I had made in the use of it, etc., etc. I don't like to write letters and so I don't want people to know I own this curiosity-breeding little joker."

First to edit *The Popular Science Monthly*, as it was then called, in a second-floor office in downtown New York, was a stocky man with curly brown hair, mutton-chop whiskers, and a fondness for playing the violin. Founding a magazine that would "break the bread of science for the multitude" had long been the dream of Edward Livingston Youmans. To realize it, he had triumphed over almost hopeless odds.

In his teens, acute eyestrain from a passion for reading was turned to blindness by harsh medications of an ignorant "eye doctor." His consuming interest in science had to be satisfied by absorbing books read to him, and scientific lectures to which he was led. His devoted sister studied chemistry in a New York lab and passed on to him all she learned. So well did he profit from this informal education that an 1851 chemistry book he authored became a standard textbook; Harvard ordered 200 copies.

At thirty, Youmans recovered his sight sufficiently to travel about the country, giving scientific lectures. He received the title of professor from Antioch College, where he lectured on chemistry for a year.

A sensation of this mid-1800s period was the theory of evolution. It had a more profound effect on the sciences than anything since Newton discovered the law of gravity. It aroused such bitter controversy that men even dueled over it. And in a way it played a part in the birth of *The Popular Science Monthly.*

What put "evolution" on the world's lips was the publication in 1859 of Charles Darwin's *Origin of Species.* But a doctrine of evolution had been fathered some years earlier by a distinguished British philosopher, Herbert Spencer. It was Spencer who, for Darwin's principle of natural selection, coined the better-remembered phrase, "survival of the fittest." So grand in scope was Spencer's view of evolution that he saw its laws applying, even to things as diverse as art, language, the celestial system, mental processes, and such social structures as governments and institutions.

Spencer's cosmic-scale picture of evolution intrigued leading scientists of his day—and thrilled Youmans. To Youmans and many others, Spencer was the intellectual colossus of the century, and when Youmans journeyed to England in 1862 to meet his idol, a lifelong friendship sprang up.

Through Spencer, Youmans met and hobnobbed with Huxley, Tyndall, and other scientists. Youmans arranged to have books by them published in America.

In 1872 Spencer began writing a new book, *Principles of Sociology,* incorporating his evolutionary views. To see if the contents could be published serially in an American periodical, he sent the opening chapter, "The Study of Sociology," to his friend Youmans. Two minutes after getting it, Youmans decided the time had come to launch the popular scientific magazine of his dreams. To Spencer he wrote:

"I determined to have a monthly at once, and in time to open with this article . . . We have started a monthly of 128 pages.

No. I.] MAY, 1872. [PRICE 50 CTS.

THE

POPULAR SCIENCE MONTHLY.

CONDUCTED BY E. L. YOUMANS.

CONTENTS

NEW YORK:
D. APPLETON AND COMPANY,
549 & 551 BROADWAY.
1872.

Reproduction of the cover of the first issue of *The Popular Science Monthly*—as it was then called—shows its staid styling. "Smash feature" was leading article of prestigious, though ponderous, series by British philosopher and evolutionist Herbert Spencer. Table of contents remained on magazine's covers until late 1915, when conversion to modern dress brought pictorial covers in full color. At right, above, the first editor and founder, Edward Livingston Youmans, as seen in an old engraving.

The first part of it is now printing . . . and we will have it out in five days more." Articles were hastily assembled, "but with yours . . . and a translation by my sister from the French, a short article by myself, and fragments by my brother, we shall make a very fair show, and as we shall print in better style than any other magazine, the thing will do."

The thing did "do." In eighteen months the circulation of *The Popular Science Monthly* climbed to the dizzy figure—for a time when distributing periodicals was anything but efficient—of 12,000. As the magazine neared its one-hundredth year in 1972, its circulation exceeded 1,600,000.

In many fields of science besides invention—medicine, atomic science, engineering, astronomy, exploration, for example—*Popular Science* saw history being made.

Medical advances within its time included discovery of the typhoid bacillus, of filterable viruses, and of X-rays. Memorable triumphs of research put the finger on mosquitoes as the carriers of malaria and yellow fever. With the twentieth century came vitamins, insulin, the "iron lung," wonder drugs like the sulfa compounds and penicillin, polio vaccines, and the surgical feat of transplanting human hearts.

The world's builders vied in engineering feats that were wonders of their day. Skyscrapers rose to ever-increasing heights; bridges leaped unprecedented spans. The Panama Canal was completed.

The sensational discovery of radium and radioactivity—nature's version of atomic energy, Einstein showed—led atomic scientists ultimately to the unlocking of the stupendous power within the atom and the awesome consequences of atomic and hydrogen bombs. On the constructive side, harnessing the atom's energy brought atomic power plants, artificially radioactive materials for medicine and research, and the prospect of limitless thermonuclear power from the hydrogen in the sea.

Astronomers discovered the planet Pluto, completed telescopes of unprecedented 100-inch and 200-inch size, and received the first close-up pictures of Mars from spacecraft with television cameras.

Explorers reached the earth's North Pole in 1909 and its South Pole in 1911; climbed its highest mountain, Everest, in 1953; and descended virtually to the ocean's greatest depth, in a deep-sea craft, in 1960. In 1969 man set foot on the moon.

Thumbing through the magazine's issues from its earliest days to date, we could see the parade of these events re-enacted.

Often we were amazed by the prescience of early editors. Just fifty years before man's first step on the moon—a highlight of my own editorship—a remarkable vision of it appeared in the April 1919 issue of *Popular Science*. It was written by Waldemar Kaempffert, the magazine's fourth editor (and, by curious coincidence, my uncle by marriage).

"Hurling a Man to the Moon," he titled his article. "How could a lunar Columbus break the grip of gravitation," reads the subtitle, "and reach the nearest heavenly body? What kind of motor would he use? How much power would it take?"

"If I could shoot a projectile from a gun with greater and greater velocity," he writes, "a point would be reached when, instead of returning to earth, the projectile would continue to travel on an independent course through space." With a starting velocity of 25,000 feet per second, "I would keep on traveling in a circle around the globe. Indeed, I would become a satellite of the earth—a living moon . . . But I do not want to travel perpetually around the earth. I want to reach the moon . . . I must increase my velocity to about 37,000 feet (seven miles) a second if I am to leave the earth in my projectile for good and all . . ."

He visualizes a "skyrocket car" built for his lunar voyage. "And when I reach the moon," he asks, "dare I leave the car and walk about on the airless, dead body—that planetary cinder? Even with a tank of oxygen strapped to my back, and with a breathing-mask on my face, I might court death. The moon is cold—bitter cold—cold with the coldness of interplanetary space . . . I am walking in a land of dazzling highlights and black shadows. It seems like some impossible nightmare.

"One minute on the moon would be so harrowing an experience that I would surely rush back to my car, pull the starting lever of my radium engine [a prescience of tomorrow's nuclear rockets?] and hurl myself back to earth, where I belong."

And a full-page drawing, reproduced on the facing page, shows our man floating

A 1919 vision of man-to-the-moon, floating weightless in his spacecraft, fifty years before it really happened.

weightless in his spacecraft—just as our Apollo astronauts do in theirs!

The century of scientific history since 1872, as reported in the pages of just one magazine, produced such a rich treasure of text and pictures that I can hope to bring you only a carefully selected fraction.

An index of all that material took months to assemble. With its aid, my collaborators in the following chapters, experts in their fields, and I were able to review the whole pageant of stories and illustrations and recapture the high spots that would make this both an accurate and entertaining record of the ten decades.

Hopefully, the appeal of *A Century of Wonders—100 Years of Popular Science* will extend beyond the audience of that particular magazine. All are welcome—men and boys, women and girls—and perhaps this book will attract even those like myself to whom the arts and literature were of primary concern, until the fascination and excitement of the world of science revealed themselves by chance to our inquiring minds.

CHAPTER 2

The Atomic Era Arrives

NATURE came up with atomic energy first.

So discovered a French scientist in 1896, when he chanced to put some uranium ore in the same drawer with sealed photographic plates—and found the plates fogged, as if by light. The savant's name was Henri Becquerel, and he thus learned that uranium was radioactive—a property that had gone unsuspected since Martin Klaproth discovered uranium in 1789.

Pierre and Marie Curie, a husband-and-wife team, succeeded in extracting intensely radioactive concentrates from uranium ore. Found in one such preparation, in 1898, was a new element that would make the Curies famous: radium.

Discovery of radioactivity and radium made sensational news in the scientific world—and further revelations of their extraordinary nature kept it agog. Radium endlessly gave off heat, as if it contained some limitless source of energy. Equally mysteriously, it turned spontaneously into a whole succession of other elements.

All this was new, strange, and enormously exciting. A flurry of major *Popular Science* articles of 1900–05 reviewed latest findings about radioactivity and debated what to make of them.

In 1905 a towering genius of intellectual science, Albert Einstein, solved the mystery of the inexhaustible *atomic energy* (for such it was) streaming from radium. Matter, he said, was turning into energy. His dictum contradicted one of the most basic beliefs of physical science—that matter was indestructible. But Einstein proved to be right. So did his famous formula for the colossal amount of energy equivalent to a given quantity of matter: $E=MC^2$, where E stands for energy, M for matter, and C for the speed of light. Just forty years later, the first atomic explosion would reiterate that formula with a thunderous exclamation point.

Its staggering implications already were being seen as early as the 1920s.

"Dare We Use This Power?" asked the headline of a *Popular Science* article in 1920. It quoted an eminent British atomic scientist, Sir Oliver Lodge: "The time will come when atomic energy will take the place of coal as a source of power." He added, "It may take a century." The magazine in 1921 foresaw atomic bombs (though it was wide of the mark as to what they would be made of).

Skeptics there were aplenty among prominent figures of the day. No less distinguished a physicist than Dr. Robert A. Millikan, noted cosmic-ray researcher, called harnessing atomic energy "a childish Utopian dream." Electrical wizard Nikola Tesla told *Popular Science*, "The scheme is worse than a perpetual-motion machine! It would take far more energy to break up an atom's structure than can be recovered in useful work." So indeed it seemed from experiments thus far.

Public imagination, unconcerned then about atomic energy, was captured by the new wonder of radium.

It became a boon for treating cancer. Gold capsules or "seeds" containing its radioactive daughter, radon gas, were implanted to destroy cancerous tissue.

Mixing radium with fluorescent zinc sulfide yielded a luminous paint for watch dials, to tell time in the dark.

Quacks peddled "radium water" as a tonic and restorer of youth. Several who drank it died of radium poisoning and federal authorities cracked down. Deaths occurred, too, among girl workers who painted luminous watch dials in a New

Jersey factory and inadvertently swallowed radium by shaping the point of a paintbrush with their lips. Belatedly the hazards of radioactive materials, and the precautions that were needed to handle them safely, came to be recognized.

Meanwhile, the labs of atomic researchers saw a succession of novel happenings.

A British scientist, Ernest Rutherford, became the world's first successful alchemist in 1919 by artificially transmuting one element into another. With alpha rays from natural radioactive material, he turned nitrogen into oxygen.

The first large "atom smasher"—the cyclotron, a sort of magnetic slingshot for atomic particles—was invented in 1930 by an American physicist, Ernest O. Lawrence of the University of California at Berkeley. It proved an excellent source of newly discovered penetrating particles called neutrons, especially effective in causing atomic transmutations.

Artificial radioactivity was discovered in 1934 by Irène Curie, daughter of the discoverers of radium, and her physicist-husband, Frédéric Joliot. Aluminum and other elements, which they bombarded with alpha rays, became radioactive.

But no one had found a hint of how to release atomic energy and harness it, until 1938 brought a surprise breakthrough: discovery of atomic fission.

Natural uranium consists of two chemical twins, called isotopes, differing in weight—uranium 238 and much-scarcer (1 part in 140) uranium 235. When struck by a neutron, an atom of uranium 235 was found to break almost squarely in two, forming a pair of lighter elements—a phenomenon new to science, which received the name of "fission." The break-up released a flash of energy—and, equally significant, several *more* speeding neutrons. Scientists were quick to see that the new neutrons in turn could break up other uranium 235 atoms, in a spreading "chain reaction" that could release enormous energy—and in fact might yield either atomic power or an atom bomb.

The portentous discovery of uranium fission was made in Nazi Germany—at the Kaiser Wilhelm Institute in Berlin, by Otto Hahn and Fritz Strassman.

War clouds were gathering in Europe in 1939. In seized Czechoslovakia, Germany had taken over the uranium mines, and a

Film picture of key by rays from uranium bar, made by Atomic Energy Commission for *Popular Science* in 1950, re-enacts Becquerel's 1896 discovery of radioactivity.

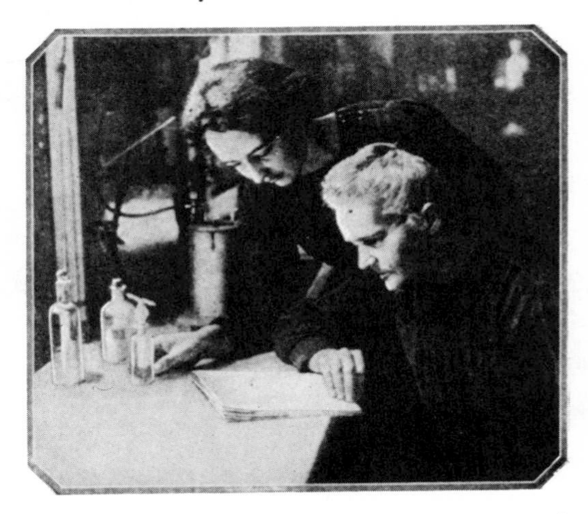

In Paris lab: Mme. Curie, discoverer (with her husband) of radium, and her daughter Irene, co-discoverer of artificial radioactivity.

Striking sight is bright blue glow from radium in crucible below, highly purified after long succession of treatments.

CANADIAN RADIUM & URANIUM CORP. PHOTO

NEW SOURCES OF LIGHT AND OF RÖNTGEN RAYS.

BY HENRY CARRINGTON BOLTON, PH. D.

AMONG the general laws of physical science, none seems more firmly established than that of the conservation and correlation of energy; according to this the various forms of energy that constitute the domain of experimental physics, heat, light, electricity, magneti- and chemical action, have reciprocal dependence

A NEW SOURCE OF HEAT: RADIUM.

BY HENRY CARRINGTON BOLTON, PH.D.

AT a meeting of the French Academy of Sciences held in March MM. Curie and Laborde announced a newly discovered property of that extraordinary substance radium—its salts emit heat continuously and to a measurable extent. Readers of the POPULAR

PRESENT PROBLEMS IN RADIOACTIVITY.*

BY PROFESSOR E. RUTHERFORD,
MCGILL UNIVERSITY.

SINCE the initial discovery by Becquerel of the spontaneous emission of new types of radiation from uranium, our knowledge of the phenomena exhibited by uranium and the other radioactive bodies has grown with great and ever increasing rapidity, and a very large

Sample *Popular Science* headlines, 1900–05, show scientists agog over startling discoveries of radioactivity and radium, then new.

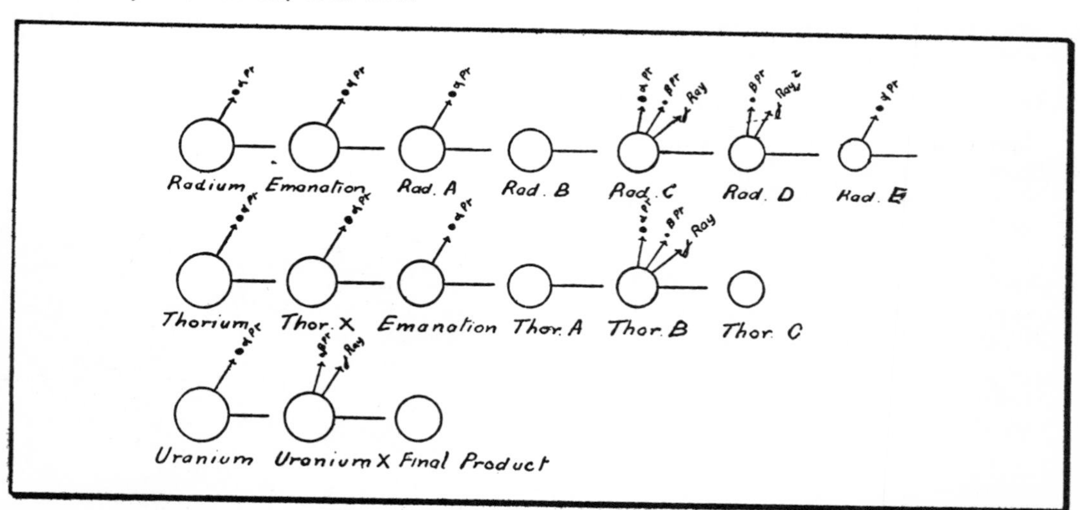

Radium, thorium, and uranium mysteriously changed into a succession of other substances —usually emitting rays in the process. Rutherford's 1905 article contained pioneer attempt, above, to chart what turned into what. Time brought many revisions; later, for instance, uranium and radium proved to belong to same series or "radioactive family."

Albert Einstein, in 1905, rocked the scientific world with his answer to where the energy of radioactive substances was coming from: matter, long supposed indestructible, was turning into energy. He proved right—and his formula bared the colossal amount of energy that atoms of matter could yield.

large section of the Kaiser Wilhelm Institute reportedly was set to work on uranium research—ominous signs that the Nazis planned to exploit the discovery of uranium fission for military purposes.

U.S. atomic scientists saw the nightmare prospect that an unprepared America, along with the rest of the world, might be confronted with an irresistible Nazi weapon. On August 2, 1939, only a month before German invasion of Poland began World War II, a momentous letter went to President Roosevelt from the scientists' distinguished spokesman, Albert Einstein. It warned the President that uranium fission might yield "extremely powerful bombs of a new type." It recommended that the U.S. organize a crash program to develop atomic weapons—and procure ample uranium ore for the purpose.

The warning brought swift action.

Launched was the supersecret U.S. atom-bomb project, so hush-hush that even mentioning the word uranium was taboo. ("Tuballoy" became its code name.)

To carry on the project, the Army's Corps of Engineers created its now-famed "Manhattan District."

For an attempt to bring off a uranium chain reaction, a primitive atomic reactor was hidden from sight in a squash court beneath the west stand of a University of Chicago athletic stadium. Square at the bottom and with a dome-shaped top, it was built up of what looked like black bricks—actually, blocks of purest graphite. Into holes drilled in the blocks went cylinders of uranium—a silver-gray metal whose surface turns black in the air. Since there was not quite enough of the pure metal in the whole country, oxide of uranium had to do for the outermost holes.

Even before the builders of this "atomic pile" had completed the uppermost layers, their instruments showed it was ready for the crucial test. Under the direction of Nobel prizewinner Enrico Fermi, control rods holding the atomic reaction in check were withdrawn, a few inches at a time.

Lest there be a catastrophic runaway reaction, an assistant stood by with an axe—ready to sever a rope and let a weighted control rod slam back into the reactor. But all went well.

Watching a recording pen climb a chart, Fermi announced, "The reaction is self-sustaining." Only half a watt of atomic power was being developed—but history was made. The world's first chain reaction had been achieved, on December 2, 1942.

A succession of major developments followed. A man-made element, plutonium, first created in 1940 by bombarding ordinary uranium (uranium 238) with neutrons in a cyclotron, proved to be fissionable like uranium 235. It could be made on a large scale with neutrons from an atomic pile. Here was an alternate material for an atom bomb.

The all-out Manhattan District project was overlooking no bets. At Oak Ridge, Tennessee, a huge plant rose to separate uranium 235 from the common kind by the tedious process of passing a gaseous uranium compound time after time through porous barriers—since it could not be isolated by chemical methods. Another giant plant was built to make plutonium at Hanford, Washington, in big-scale atomic reactors; unlike uranium, it could readily be separated by chemical means.

Simply bringing together a solid mass of plutonium the size of a baseball, or a roughly comparable one of pure uranium 235, would be enough to touch off an atomic explosion. Two alternate methods could do it in a bomb: a gunlike device could shoot the missing quantity at an almost sufficient amount; or surrounding explosive charges could compress a hollow mass, not quite compact enough to go off, into a closer packed one that would.

A trial atom bomb was made of plutonium. Before dawn of July 16, 1945, it was exploded atop a high steel tower near Alamagordo, New Mexico. A great ball of fire lit up the desert like a super-sun on earth. The tower vanished in vapor. Acres of surface sand were fused by the heat.

In the far Pacific, America then was girding for what promised to be the bloodiest operation of World War II—the invasion of Japan itself, against the fanatical resistance of last-ditch defenders. Weighing the awful cost against the newly available alternative, President Truman made the fateful decision to use the atom bomb.

On August 6, 1945, a uranium bomb blasted Hiroshima, a Japanese military port and industrial center. Three days later a plutonium bomb razed Nagasaki. Japan surrendered and the war was over.

Actually, German efforts to produce an atom bomb had gotten nowhere. Better at nuclear technology, or espionage, or both, was the Soviet Union. In 1949 the Russians had the atom bomb.

An even more fearful one was to follow—the hydrogen bomb. Its story began with a major astronomical discovery.

Long a mystery was the source of the sun's boundless energy, constantly reaching us in the form of heat and light, but in 1939 the answer was announced by a Cornell physicist, Dr. H. A. Bethe:

The sun drew its energy from a new-found kind of nuclear reaction: the *fusion* of light elements into heavier ones, in contrast with the *fission* of heavy elements into lighter ones. Through a series of intermediate steps, which Bethe and others were able to duplicate with cyclotrons, the net result was to turn hydrogen into helium. Weight for weight of material, far more energy was releasable by this "thermonuclear" process—so called because it could occur only at extreme temperature—than by uranium or plutonium fission.

To produce a thermonuclear, or fusion, reaction on earth would take sunlike heat of millions of degrees. A uranium or plutonium bomb could provide it—and serve as the detonator of a hydrogen bomb.

In 1950 the U.S. made public its decision to attempt to produce a hydrogen bomb. Reported *Popular Science:* "Lead-

May, 1920

Dare We Use This Power?

Sir Oliver Lodge says atomic energy will supplant coal

By **E. F. Richards**

SIR OLIVER LODGE thinks that man is not yet civilized enough to use the energy hidden in ordinary matter.

"The time will come when atomic energy will take the place of coal as a source of power."

The man who spoke thus before the Royal Society of Arts in London was Sir Oliver Lodge—one of the towering figures in modern science, a man who has devoted the better part of his life to the study and interpretation of the atom. This new form of energy, which our great-grandchildren may utilize instead of oil and coal, has possibilities so appalling that Sir Oliver almost rejoices that we do not know how to release it. "I hope that the human race will not discover how to use this energy," he says, "until it has brains and morality enough to use it properly, *because if the discovery is made by the wrong people this planet would be unsafe.* A force utterly disproportionate to the present sources of power would be placed at the disposal of the world."

Sir J. J. Thomson, England's great authority on the atom, gives a picture of this terrible form energy that wins one over to Sir Oliver Lodge's view. He tells us that the atomic energy stored in an ounce of chlorine "is about

iron, copper, wood, or stone. The food we eat is made of atoms, and so are the tables and chairs in our houses. Every one of us, then, locks up within himself immense stores of

© International Film Service

"I hope the human race will not discover how to use this energy until it has brains and morality enough to use it properly," says Sir Oliver Lodge, in explaining the terrible possibilities that lie not only in radium but in a piece of wood or iron

energy. In a little finger there is

of as beta particles or electrons, were subsequently found to be given off also by radium, as one of the products of the breaking up of its atoms. For, though the atom ordinarily remains *undivided*, we know today that it is not *indivisible.* In certain cases the atom breaks up of its own accord, as in the case of radium, shooting off the fragments at speeds which make a rifle bullet appear like a snail in comparison.

Active Radium Particles

There are, indeed, good grounds for suspecting that all matter, not only radium, is thus shooting off particles and giving out energy. But we become conscious of the fact only in the case of radium and a few other radioactive substances. Radium is giving up its atomic energy more rapidly and more violently than limestone, for example, which explains why it seems more explosive than other elements. A radium atom is like a two-ton gun firing a hundred-pound shot. Just like the gun, the rest of the atom recoils after having been fired. This is not merely a speculation, a picturesque guess. *The recoil has actually been observed.* After five such projectiles have been fired, radium settles down into another existence—a quieter

August, 1921

How an Ounce of Iron Might Blow Up a City

If man could release the powerful force that holds atoms together the world's power problems would be solved

SCIENTISTS today are probing the secrets of an explosive power capable of blowing the universe to pieces. When Sir Oliver Lodge startled the world with his declaration that an ounce of

Do you believe that engines may some day be run with the power hidden in paving-stones?

Did you know that the modern theory of matter—the electron theory—tends to

crystals might be similarly revealed. He extended this investigation and found that he was actually able to count the number of electrons in each atom!

The exact positions of the

Harnessing atomic energy was foreseen as early as 1920–21 in *Popular Science* articles. Sir Oliver Lodge's words on its Jekyll-and-Hyde possibilities were strikingly prophetic.

U.S. radium mines, like this New Mexico one for carnotite ore, began springing up in 1912. For nearly a decade, until 1922, America was world's chief supplier of radium.

Radium, being refined and concentrated above, became an article of commerce. One early application: radium-painted luminous dials for watches (inset).

ing scientists say they hope it won't work —but are afraid it will."

It did. The first U.S. hydrogen bomb exploded in 1952, in the Pacific test area. Russia had an H-bomb the following year. For the first time, nuclear weapons' power was measurable in *millions* rather than thousands of tons of TNT—"megatons" instead of "kilotons."

What could possibly call for a bomb of such staggering size? The purpose: A missile with a hydrogen warhead could miss a direct hit by a far greater margin and still destroy its target, because of the wider range of its blast.

The following years saw extensive U.S. and Soviet tests of H-bombs of growing size. Largest of all was a 58-megaton monster exploded by Russia in 1961.

Radioactive debris from H-bomb tests—"fallout"—showered down on the whole northern hemisphere. With a steadily rising level of "background" radioactivity everywhere, the world became concerned. Though still far short of being dangerous, the ray-emitting litter was certainly untidy housekeeping. In 1963 the U.S. and Russia signed a treaty banning further nuclear tests in the open air, under water, or in space. (Still permitted were underground tests, provided no radioactivity escaped beyond a country's borders.)

Ten years after Hiroshima, in 1955, *Popular Science* had commented:

"The H-bomb's scarlet fire has brought the power to destroy on a fantastic scale.

Eerie blue beam of radiation streams from a cyclotron, the first large atom-smashing machine. Devised in 1930, cyclotrons aided the quest for a way to release atomic energy. They readily transmuted elements into others and created man-made radioactive substances—feats achieved before only on an infinitesimal scale. And they provided experimenters with a prolific source of particles called neutrons—newly discovered ingredients of matter, destined to play a key role in the ultimate success of atomic-power seekers.

Yoking the atom to whirring electric generators promises miracles for constructive ends. Ready to do whichever may be man's bidding is the mighty atomic genie he has summoned up."

Happily, in succeeding years, the magazine could report the atomic genie's signal services for peaceful purposes.

From the first chain-reacting atomic "pile" in Chicago have evolved practical atomic reactors for power.

In 1956 the world's first full-scale atomic power plant, initially generating about 50,000 kilowatts, began operating at Calder Hall in England—a country where the scarcity of other fuels made uranium an especially attractive competitor.

By 1957 America too had its first big atomic generating plant—a 90,000-kilowatt station at Shippingport, Pennsylvania.

With the complete success of these pioneer atom-power plants and their rapidly multiplying successors, exhaustion of the world's dwindling reserves of conventional fuels—coal, oil, natural gas—no longer threatens a dire shortage of energy. A tiny 1-inch cube of uranium holds enough energy to provide a 6-room house with electricity and heat for 1,000 years.

What has happened so far is only the prelude. In 1969 just 1 per cent of America's electric power came from nuclear plants; by 1972, about 3 per cent.

But a revolution is on the way—so soon that most of us will see it happen before our eyes. By 1980, experts' forecasts show, the atom will be generating a quarter of all the country's power; by 1990, 40 per cent of it. Some time between then and the year 2000, the phenomenal rise of atomic power is expected to take it past the halfway mark—and make it the nation's predominant source of energy.

Outmoded will be conventional power plants, belching smokestacks and air pollution. Atomic power will be clean power.

In principle, an atomic power plant and a conventional steam-electric plant have more resemblances than differences. Both make steam in boilers, convey the steam to turbines, and spin electric generators with the turbines. The major difference is simply what heats the boiler. An atomic power plant supplies the heat from a nuclear reactor instead of a boiler.

Of course, there must be some flowing medium to carry heat from the reactor's hot fuel to water in the steam boiler. Britain's first plant and most of its successors have employed carbon dioxide gas. America's plant at Shippingport and many others have used plain water, pressurized so it can reach a higher temperature without boiling. Liquefied metals of low melting point are seen as promising.

At sea, nuclear energy has propelled ships since 1955, when the U.S. submarine *Nautilus* signaled the historic message, "Under way on atomic power." Civilian vessels, America's *Savannah* and others, have harnessed the atom, too. Atomic power will

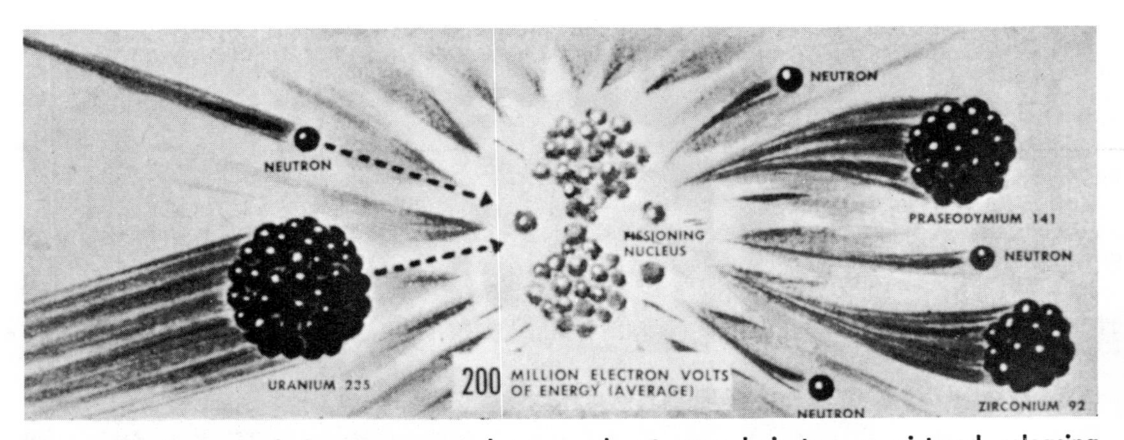

Discovery of atomic fission—a new and unexpected phenomenon, first evidenced in 1938 by experiments in Germany—proved the breakthrough to tapping nuclear energy. Struck by neutrons, atoms of uranium 235 (a minor constituent of natural uranium) broke almost squarely in two, as pictured, releasing stupendous energy. Formed in the break-up were pairs of various lighter-weight atoms, such as zirconium and praseodymium—and more speeding neutrons, capable of continuing the fissioning process in a "chain reaction."

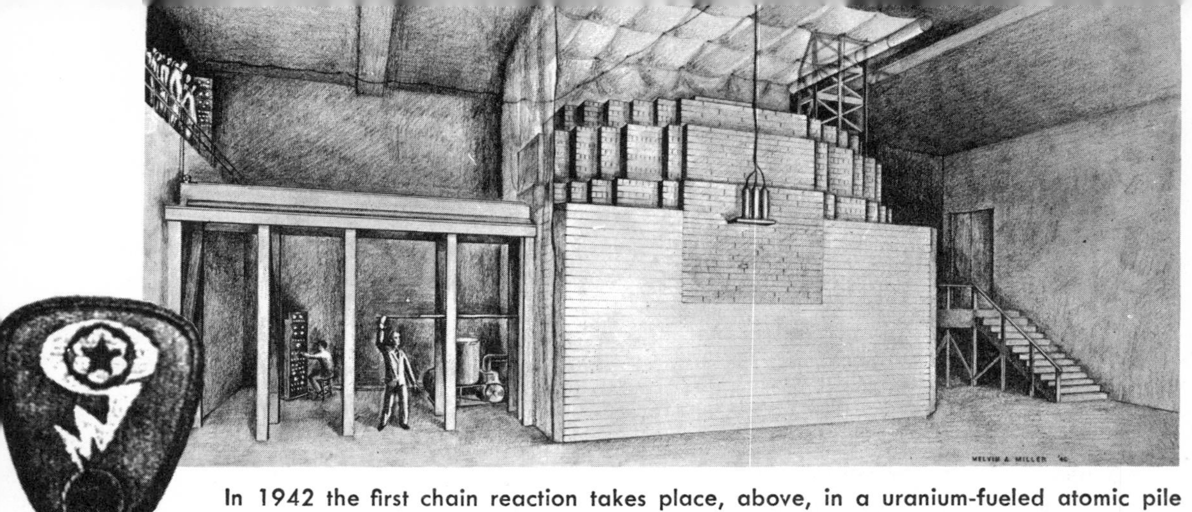

In 1942 the first chain reaction takes place, above, in a uranium-fueled atomic pile hidden in a squash court. From here on, it's an all-out race by America's super-secret Manhattan District project (emblem shown in inset) to get the atom bomb first.
AEC PICTURE

The man-made element plutonium, found fissionable like uranium 235, offers an alternate route to the atom bomb—and this huge plant to produce it rises at Hanford, Wash.

Trial atom bomb explodes, right, in awesome New Mexico test. This first one was made of plutonium—a synthetic, silvery metal, with appearance shown below when first obtained in form of "buttons."
AEC PHOTO

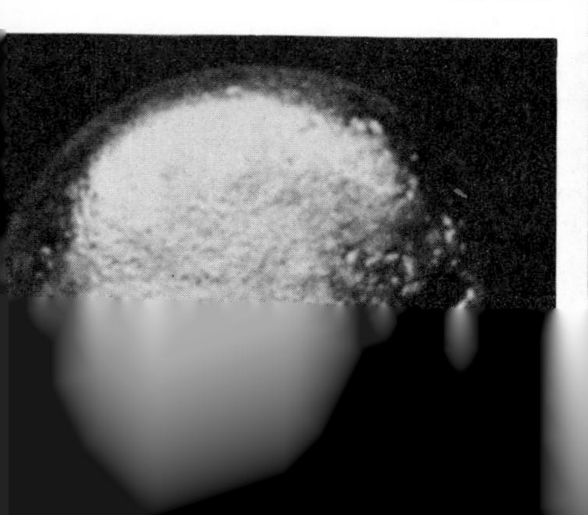

What atom bombs looked like was top secret until release of these official photos fifteen years later. Hiroshima-type bomb at left, 10 feet long, contained uranium; Nagasaki-type bomb at right was of plutonium. Each was equivalent to 20,000 tons of TNT. AEC PHOTOS

Sea is hurled skyward by 1946 atom-bomb trial to test effect on warships, seen silhouetted at water line. Heeding results, fleets will never again travel in close formation.

POPULAR SCIENCE MONTHLY
SEPT. 1946
Fountain of Fury:
This 2,000-foot-wide column of water a mile high shot up when Atom Bomb No. 5 went off at Bikini. Ships are silhouetted against the wave. See "Taking Hell's Measurements," on Page 91.

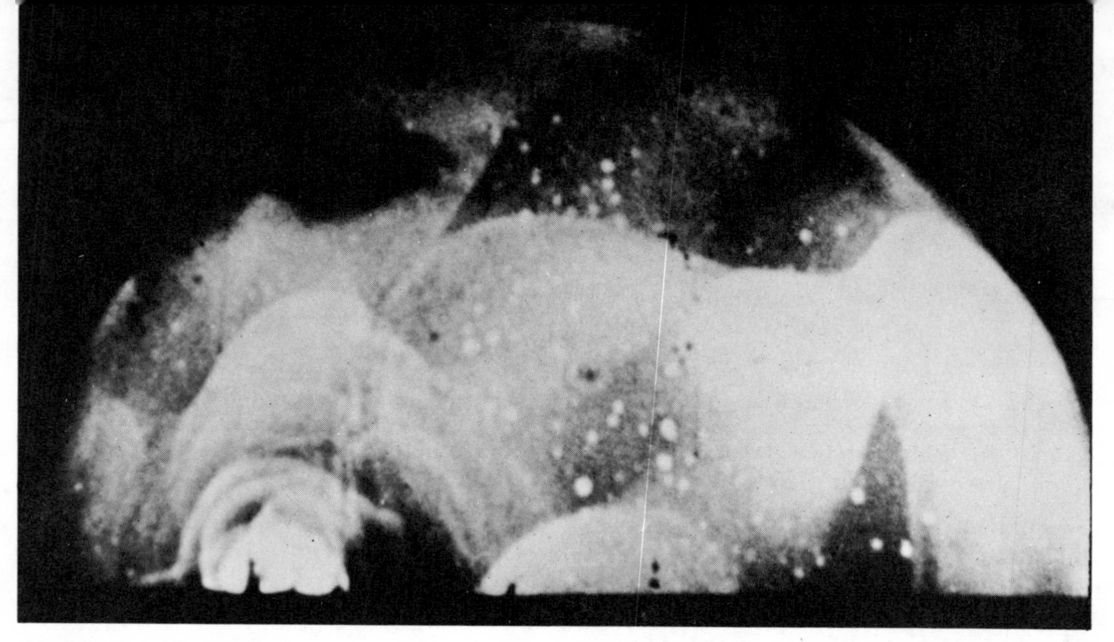

World's first thermonuclear explosion creates spectacular fireball more than 3 miles in diameter, in initial U.S. trial of a hydrogen bomb in 1952. For the first time the power of a single nuclear bomb became measurable in *millions* of tons of TNT.

propel future spaceships on interplanetary voyages.

Another atomic milestone, in 1960, went almost unnoticed by the world:

The mining of ore for radium ended.

Uranium and radium come from the same ores. Once, the radium was extracted and the uranium thrown out in the discarded tailings. From now on, it was the other way round. The day of radium, the wonder of more than half a century, had passed.

In 1946 man-made radioactive preparations known as radioisotopes—"artificial radium," *Popular Science* called them—had become available in quantity. The Joliot-Curies had made infinitesimal amounts before; now they were mass-produced at Oak Ridge in Tennessee by exposing everyday substances to neutrons in the heart of an atomic reactor.

Radioactive fallout from H-bomb tests is visualized by *Popular Science* with non-radioactive but identical-looking form of a major ingredient—fluffy, white strontium carbonate.

Plentiful radioisotopes did everything that could be done by rare and costly radium—once valued at $180,000 a gram, or 1/28 of an ounce—and far more.

For radiation treatment of cancer, radioactive cobalt and cesium superseded radium. The first U.S. plant for making radon seeds, which began operation in a New York hospital in 1917, was dismantled in 1970.

Radioactive isotopes yielded brighter and safer luminous paints than the watch-dial kind containing radium.

Most novel and varied of the uses of radioisotopes was as "tracers."

From a catalogue of available radioisotopes, you could order a radioactive twin of virtually any element you wanted—iodine, phosphorus, iron, you name it. Chemically identical with its everyday counterpart, the radioactive element behaved exactly the same in the human body, in a plant, or in an engine. Because it was "tagged" with detectable radioactivity, its location was readily observable.

In medicine this offers a new aid to diagnosis—say, to gauge uptake of iodine by a thyroid gland. In industry it shows a piston ring's wear by detecting radioactive iron in an engine's oil. In agriculture it reveals what plants do with elements in the soil. Researchers in every field are finding new uses for radioactive tracers almost daily.

Peaceful atomic explosions open vistas of engineering and mining feats on a grand scale. In 1957 the Atomic Energy Commission created Project Plowshare to explore the possibilities. Trials demonstrated that nuclear blasts could excavate a new Panama Canal or create an "instant" artificial harbor—*if* the AEC could develop nuclear explosives virtually free of objectionable radioactive fallout. That endeavor is currently under way.

Underground explosions, releasing no radioactivity to the open air, have meanwhile scored Plowshare's greatest successes. In New Mexico and Colorado, nuclear blasts from 4,000 to 8,000 feet deep have stimulated the flow of natural gas from wells by shattering subterranean rock. Future underground blasts may recover oil from shale; pulverize ore beds so they can be mined by pumping leaching solutions through them; and create caverns in naturally hot rock formations, which would then turn pumped-down water into steam for geothermal power.

Growing use of atomic energy spurred history's greatest uranium hunt. In 1948 the U.S. offered a $10,000 bonus for discovery of high-grade uranium ore. The Great Uranium Rush that followed was successful beyond its dreams. By the mid-1950s the U.S. was the biggest producer of uranium in the free world, and it has remained so since.

By the 1980s a new generation of atomic reactors, called "breeders," will stretch the supply of uranium fuel.

Today's reactors extract their energy from the scarce uranium 235 in natural uranium. Left over like ashes from a fire, the rest is then called "depleted" uranium. In contrast, a breeder reactor surrounds its chain-reacting core with depleted uranium and turns it into plutonium, which can serve as reactor fuel. Thus a breeder reactor produces new fuel at the same time that it generates power. Small-scale trials proved in 1952, *Popular Science* then reported, that *a breeder reactor can actually create more fuel than it consumes!*

To pioneer the practice, President Nixon has called for completing a big-scale breeder power plant, of from 300,000 to 500,000

Lowering uranium-fuel core into first large U.S. atom-power plant, at Shippingport, Pa., harnesses atomic energy for peaceful use.

Great bulb-shaped container is being readied below to house giant million-kilowatt power reactor for operation in 1972 at Alabama site of TVA's new Browns Ferry station.

At sea, atomic power first propelled the submarine *Nautilus,* at top. Then came the N.S. (for "nuclear ship") *Savannah,* above, a civilian cargo vessel that has sailed the globe.
SCIENCE SERVICE PHOTO

A nuclear rocket engine is readied for trial. Such an engine is needed to propel future manned spaceships beyond the moon.
NASA PHOTO

Radioisotopes—liquids and solids made radioactive artificially—have proved a boon to medicine and to research of every sort.

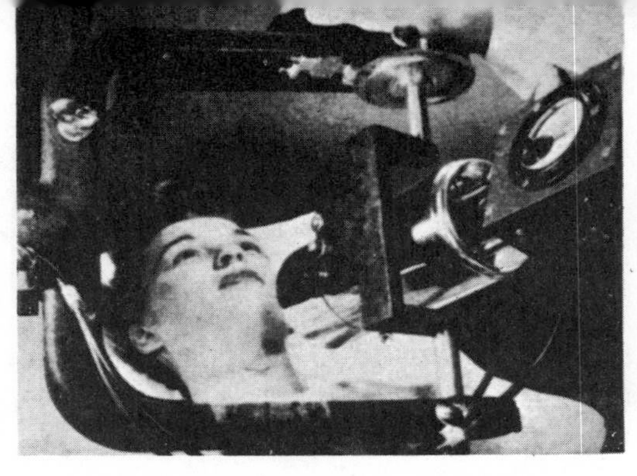

For diagnosis of thyroid-gland disorder, patient drinks "atomic cocktail" of radioactive iodine. Geiger counter detects normal or abnormal rate of iodine uptake by the gland.

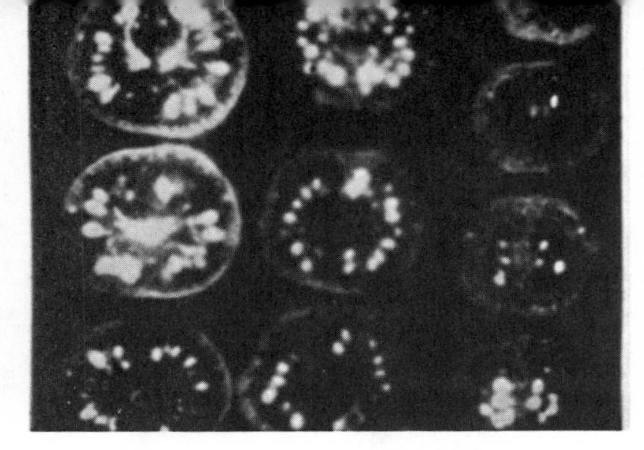

"Tracer" radioisotope reveals what a tomato plant does with zinc in soil. Element turns up in seeds, radioactive zinc shows conclusively—by taking its own picture, above.

A peaceful nuclear explosion: atomic blasting is tested for excavation—like digging a canal.
AEC PHOTO

Atomic explosive, right, goes down 4,000-foot hole to tap natural gas.
AEC PHOTO

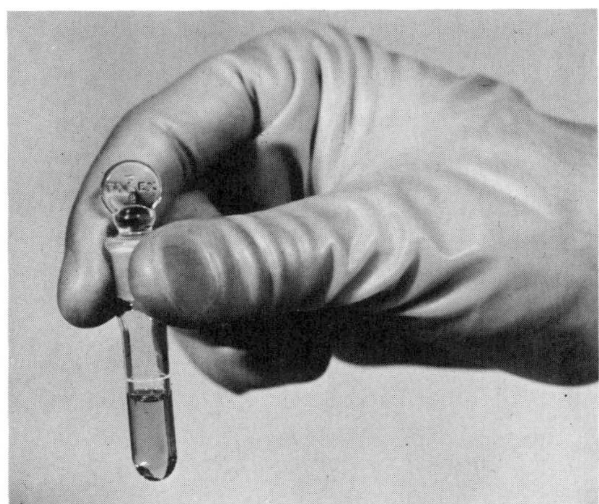

Vial holds plutonium, in yellow solution, made by first experimental breeder reactor—proving feasibility of a power reactor that manufactures more fuel than it consumes.

To help design coming big-scale breeder reactors, this plutonium-fueled one—Argonne National Laboratory's Zero Power Plutonium Reactor—currently serves as a stand-in.
ARGONNE NATIONAL LABORATORY PHOTO

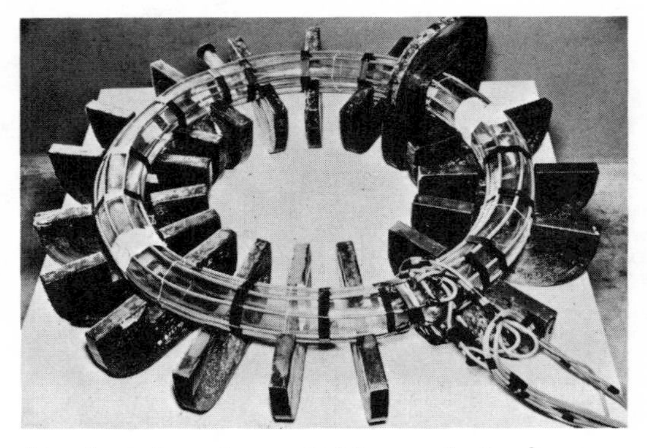

Atomic science's next big goal: to harness *fusion* reaction, as of hydrogen bomb, for power. Table-top device above, whimsically named Perhapsatron, served in early trials.

The promise of the future: limitless energy for coming generations, from the inexhausti-ble fuel in the seas, when—sooner or later—the dream of fusion power comes true.

Trial fusion-power devices have now reached size of Oak Ridge lab's Ormak, above. Com-posite photo shows both inside and outside.
OAK RIDGE NATIONAL LABORATORY PHOTO

kilowatts, by 1980. It is expected to set the style for most uranium-fueled power plants of the future.

Then will come the Atomic Era's greatest triumph—taming the fusion reaction of the hydrogen bomb for peaceful power. That will end dependence on uranium for fuel. Heavy hydrogen, a natural form of hydrogen obtainable in unlimited quantity from the sea, will replace it.

Harnessing fusion power poses staggering problems. First you must heat the fuel, perhaps by sudden compression, to the fantastic temperature required to start a fusion reaction. Then you must somehow confine it. Only a magnetic bottle will do; the heat would vaporize a container of any known material.

Weird-looking machines shaped like doughnuts and pretzels began to be built in the early 1950s to test these far-out ideas —in the U.S., Great Britain, and Soviet Russia. Early secrecy about their designs was soon lifted, enabling researchers to benefit by work abroad. For example, a promising new Soviet machine called the Tokamak has inspired U.S. adaptations, the Ormak and others.

Trials to date encourage the belief that sooner or later fusion power will come—possibly by the year 2000. But they have yet to achieve the first "flash in the pan" signaling release of more power than was put in.

Ultimately it will come, and practical fusion power soon after. It may happen here or abroad. Happily, this time the international race—to present this great gift to the world—is one that can have no losers. Whoever wins it, everybody wins.

CHAPTER 3

The Horseless Carriage Grows Up

ONE of the standard jokes in the early days of the automobile was about the driver who asked his wife to look at the rear tire on her side and see if it was flat.

"No," she reported, "only on the bottom."

Flat tires were, in fact, no joke. They happened too often. Right into the second decade of the twentieth century one manufacturer boastfully advertised a "10,000-mile tire." Fifty years later a tire that wouldn't last four times that long wouldn't pass muster.

Yet the horseless carriage persisted. It survived miserable roads. It survived cut-and-try engineering. It survived prospective famines in fuel. It survived critics. "The ordinary 'horseless carriage,'" said the prestigious *Literary Digest* in 1899, ". . . will never, of course, come into as common use as the bicycle." Woodrow Wilson, later the twenty-eighth President of the United States, said in 1906 that "nothing has spread socialistic feeling more than the use of the automobile . . . a picture of the arrogance of wealth." U.S. Senator Joseph W. Bailey of Texas in 1909 stated: "If I had my way I would make it a crime to use automobiles on the public highways because no man has a right to use a vehicle . . . that is dangerous."

The horseless carriage survived ever-mounting taxes, endless restrictive legislation, appalling casualties from its operation, stylists who shaped it into caricature, designs that were outright bloopers, and NO PARKING signs by the millions. It survived because it was useful. It supplied personal mobility to hundreds of millions of people. It was a status symbol. It satisfied man's lust for power.

Often it was beautiful and always, as it matured, a marvel of ingenuity. At the threshold of its invention, nothing in the archives of humankind had approached it as a transportation device.

The horseless carriage survived because men fell hopelessly in love with it.

Now, after eighty years as of the publication of this book, it stands charged with being the chief contributor to pollution of the air in the great centers of population of the planet earth.

To take a charitable view of history and invention, the "automobile" first appeared in 1769. A French army officer, Nicolas Joseph Cugnot, built a military "steam-carriage." It worked. An American, Nathan Read, patented a steam carriage in 1790. On the testimony of *Popular Science* in 1878, Oliver Evans, another American, inventor of a non-condensing, high-pressure steam engine, said early in the nineteenth century, "I have no doubt that my engine will propel . . . wagons on turnpike roads." The magazine in 1897 made a pitch for "horseless [steam] locomotives." They had been tried abroad with limited success. In England, Parliament finally killed them off by law.

All this was overture. In 1877 a German inventor, Dr. N. A. Otto, received a patent for an internal-combustion engine he had built the year before. Unlike the heavy, inefficient steam engine, which converted energy from a source outside its cylinders into mechanical movement, the light internal-combustion engine manufactured its own power. It burned coal gas on a 4-stroke cycle—suction, compression, burning stroke, and exhaust. This was fine for stationary engines fed with coal gas from mains, but it offered nothing as a perambulating power plant.

Then, in 1886, two other Germans, Gottlieb Daimler and Karl Benz, independently put "petrol" engines in what were, in truth, only carriages without a nag in front. The horse as a means of locomotion on roads began to number his days.

Like horseless carriages for the next fifteen years, Daimler's and Benz's contraptions offered seats for driver and passengers on—not in—the vehicle. Besides springs, they soon had some insulation from road shock—the pneumatic tire was invented in 1888. Daimler's company later produced the world-known Mercedes automobile and, still later, the Mercedes-Benz. As a footnote to the times, Daimler's success occurred only after Dr. Otto fired him as technical manager. Otto found him "indescribably thickheaded."

The first automobile in America is credited to Gottfried Schloemer, who drove a "horseless buggy" of his own design and construction through the streets of Milwaukee, Wisconsin, in 1889.

Historians generally date the origin of today's automobiles from 1892, when serious production of internal-combustion engines started. The Germans and the French—Panhard, Renault, the Comte de Dion, Peugeot, Lepape, and Delahaye among them—were in the vanguard.

The British lagged, hobbled by a law requiring that a horseless carriage be preceded by a man carrying a red flag by day and a red lantern by night. The U.S.

World's first auto was 1769 "steam-carriage" built by Nicolas Cugnot of France. He made this improved version in 1770.

Next self-propelled road vehicle of record was this monstrosity built in 1804 by Oliver Evans for use at Philadelphia's docks. Another steamer, this one was amphibious.

The "road locomotive" was becoming quite the thing in England by 1810. This one had a crewman at the rear blowing a horn. Such steam coaches had problems—great weight and the frequent need for water and fuel.

That incredible conveyance at lower left transported the British jet set in third decade of nineteenth century. It took muscle to steer it.

Fancy went unrestrained in the magnificence of design. Three such vehicles were built to prowl the bumpy roads in the London area.

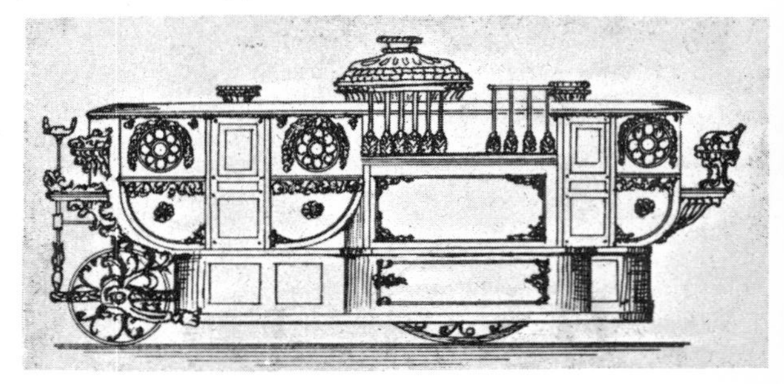

lagged, too, apparently because its inventors and manufacturers couldn't take the horseless carriage seriously.

Charles E. and J. Frank Duryea, brothers, of Springfield, Massachusetts, turned out the first American automobile that would lend itself to "mass production" in 1893–94, and 13 of the vehicles rolled off the Duryea Motor Wagon Company's assembly line in 1896. Elwood Haynes, a metallurgical engineer, was a contemporary. His first car, built by Elmer and Edgar Apperson at Kokomo, Indiana, achieved a speed of 7 miles an hour on July 4, 1894.

Between 1898 and 1903, the U.S. began making up for lost time. Cars bearing the names Olds, Knox, Autocar, Pierce, Franklin, Cadillac, and Winton appeared on the market. The Winton won a special distinction. In 1903 it became the first automobile to cross the American continent. The elapsed time was 64 days, 44 of which were spent on the road. Threescore years later the tiller-steered, "curved dash" Olds, designed by Ransom E. Olds, who subsequently fathered the Reo, was a prized museum piece.

All these self-propelled machines had characteristics in common. They had one lung or two. Their wheels were chain-driven. Their ignition was make-and-break. Ten horsepower was enormous. A 1900 car that could achieve 30 miles an hour for 100 yards was a racer. They were hard-starting. Right into the second decade of the century some engines came with "ether cups" atop each cylinder to encourage starts. A petcock released the highly volatile ether into the combustion chamber. The first of the Duryeas had a single con-

America's first automobile was built by Gottfried Schloemer in Milwaukee in 1889. Its internal-combustion engine had one cylinder.

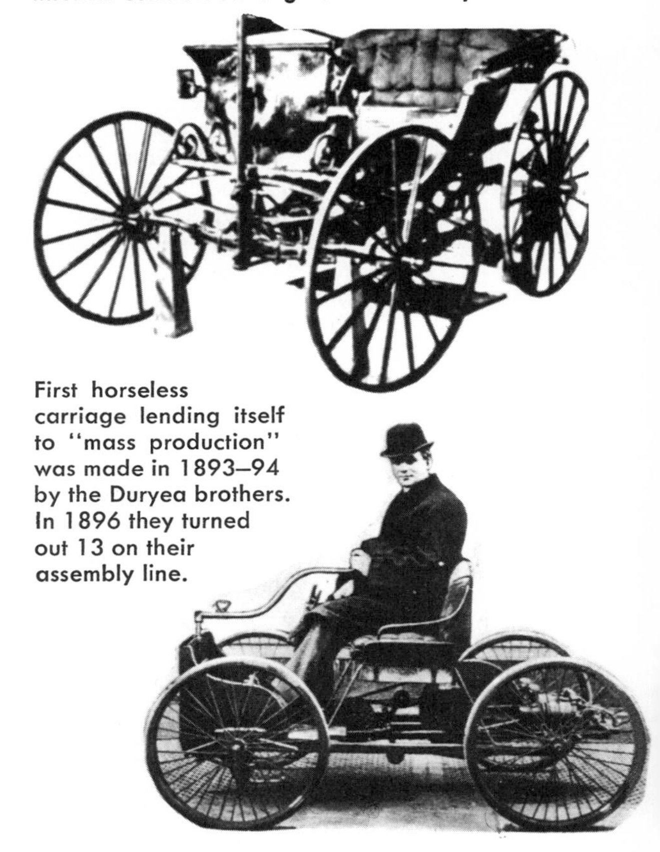

First horseless carriage lending itself to "mass production" was made in 1893–94 by the Duryea brothers. In 1896 they turned out 13 on their assembly line.

Henry Ford (above) produced his first car in 1896. (See color pages.) Elwood Haynes, another pioneer (below) was stopped in 1895 by Chicago bicycle cop for impeding traffic.

First mention of a U.S. gasoline car occurred in 1879 when George Selden, an engineer, applied for a patent on a "road engine" (below). Royalties made him rich. Ford refused to pay. Selden sued. Later, Ford won.

trol lever that took care of steering, gear changing, spark adjustment, the throttle, and braking.

Headlights were flickering, oil-wick carriage lamps. Some manufacturers used camel's hair for brake linings. The "steering wheel" was a tiller. Except for electrics, closed bodies and windshields were unknown, and a trip into the countryside on dirt roads, with passengers arrayed in goggles and "dusters," called for a bath afterward. Since inventors had only the past to guide them, running wheels were the spoked military type or adaptations of bicycle wheels. The entire chassis was borrowed from horse-drawn carriages, bowed leaf springs and all. One car boasted an all-mahogany frame.

The automobile was such a novelty that for a time Barnum & Bailey led its daily circus parade with a Duryea "runabout."

Automobiling became a communicable fever. A wealthy motoring enthusiast, Charles J. Glidden, organized trips, with a trophy as a prize, to demonstrate the reliability of the horseless carriage, and for ten years beginning in 1905, Glidden Tours were as much a fixture in the American scene as the Chautauqua Circuit.

Then, in America, a man named Henry Ford began tinkering with a self-propelled vehicle and, for millions of people all over the world, the course of transportation history was changed. A former farmer, now a machinist, Ford had built a tiller-steered, 4-cycle, 2-cylinder "quadricycle" in a tiny shop in Detroit's Bagley Avenue and exhibited it in 1896. No dunce at merchandising, he built a series of racing cars for publicity. In 1903 the Ford Motor Company offered for public sale its first car, a 2-cylinder job, and three years later a 6-cylinder one that sold for $2,500.

That was too costly. It wasn't what Henry wanted. "The market for a low-priced car," he said, "is unlimited." In 1908 he brought out the first—soon to be famous—Model T for $850. That covered controls limited to a hand brake, 3 foot pedals, and an ignition switch. Standard equipment included a folding windshield, collapsible (2-man) top, bulb horn, and kerosene tail and side lamps. It did not include a spare tire or even a speedometer. The suspension was transverse leaf springs, front and rear.

The engine was a water-jacketed, 4-cylinder, 4-cycle L-head with a 3¾-inch bore and a 4-inch stroke, developing 22½ horsepower at 1,600 revolutions. The fuel was magneto-fired. A planetary transmission provided 2 speeds forward and a reverse. The driver controlled the forward speeds with his left pedal, reverse with the center pedal, and a transmission brake with the right one. A left-hand lever engaged low or high speed. Under the steering wheel were spark-advance and throttle levers. The car ran out of breath at 35 miles per hour.

The Model T came in one color. To a convention of dealers clamoring for multi-hued Fords at the height of the Model T's popularity, Henry said jokingly, "You can have any color you want—so long as it's black."

Thus, the "flivver," the "tin Lizzie," with its brass radiator, of legend. Before production was closed out in 1927, 15 million had been sold at prices that, steadily declining, reached a nadir in 1923—$265 for the runabout. Ford's volume was more than that of all other U.S. car manufacturers combined. More than any other automobile of its time, the Ford put city dweller and farmer alike on wheels.

A secondary industry grew up around the Model T. The engine often "kicked" when it was cranked—Henry supplied no self-starter—sometimes breaking a man's wrist or arm. So a brace of Wisconsin inventors came up with an automatic spark-retarder to prevent it. Another inventor offered a cable-pull ratchet device to rotate the engine from the driver's seat. (When the oil congealed in the winter, some owners jacked up one rear wheel, in gear, and used its inertia to help start the balky engine.)

Other after-market suppliers made tidy fortunes cashing in on the Ford's primitiveness. One offered a "1-man top," others a rear-view mirror, shock absorbers, rubber running boards, a gas-tank gauge (located, incidentally, under the front seat), and a choke coil for brighter lights when, finally, the Ford acquired electric lamps.

Apart from popularizing the automobile, Henry Ford freed the industry from royalty payments exacted by George B. Selden, inventor of a "gasoline road engine."

It was Ford who arbitrarily switched the driver's seat to the left-hand side for better judgment of the distance of an approaching vehicle. The industry followed suit.

Ford's competitors were not idle. If they couldn't compete on price, they could produce fancier cars. As early as 1903 they introduced shock absorbers and sliding-gear transmissions. They replaced acetylene headlights with electric ones. The Locomobile introduced the first electric generator for storage batteries. Demountable rims came in, exorcising that horror of puncture repair, the pinched tube.

But interior heat for passengers in winter? Buy a Clark Auto Warmer "made in 20 styles, $2 to $10." It used a carbon brick good for twelve to sixteen hours—an idea purloined from bed warmers.

A lot of self-starters had been tried, using acetylene, compressed air, or electricity, but it was not until 1912 that a successful one was put on a car, a Cadillac. It was developed by Charles F. Kettering ("Boss Ket," the General Motors genius of later years), when a friend of the company's president was killed cranking a car.

Decorative gadgetry such as the Boyce Moto-Meter, a radiator-mounted engine-water thermometer, was a must for the younger set.

A few manufacturers, notably Franklin, introduced air-cooled engines to do away with radiators. One car, the Carter, carried a spare engine—if one failed, the other got you home. Even Sears, Roebuck got into the act with a motorized delivery wagon in 1911. It sold for $445. (Forty years later Sears popped up again, selling, briefly, an Allstate passenger car made by Kaiser-Frazer.)

Oldsmobile made a wooden-framed closed car, a novelty because right through the decade of the 1920s the label "touring car" clung to anything with a removable top. Olds also brought out the first speedometer. By 1907 the V8 engine had appeared, in both the U.S. and England.

The first effort to effect a fundamental change in the valving of the gasoline engine was made in 1911. Charles Y. Knight developed a substitute for the poppet valve. It consisted of a sleeve interposed between piston and cylinder wall that, riding up and down on cams, exposed ports for intake and exhaust. Known as the "sleeve valve," it appeared on the Stearns, Stoddard-Dayton, and Columbia; and, later, on the Willys.

Another nineteenth-century hopeful in the horseless carriage business was Alexander Winton. This phaeton was the forerunner of the famous Winton Six.

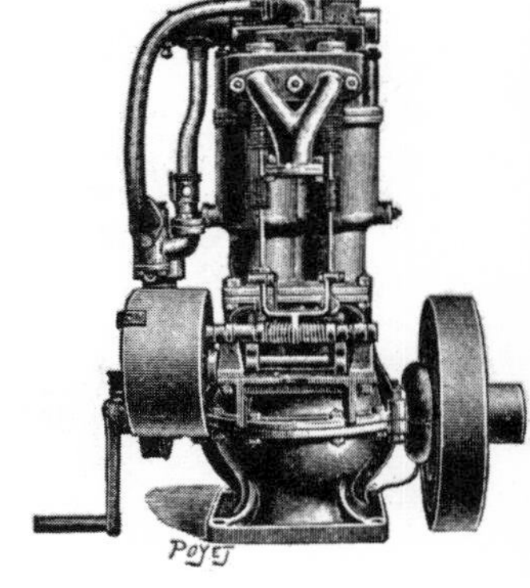

Abroad, the French firm of Panhard and Levassor produced an engine that took part in the world's first automobile race.

Louis Renault, whose name was to be famous, produced his first "voiturette" (light car) in 1898. Unlike other cars of the period, it was driven by geared shafts, not chains.

First car to cross the U.S., in 1903, was a 2-cylinder Winton. The trip took 64 days, only 44 of them used in actual travel.

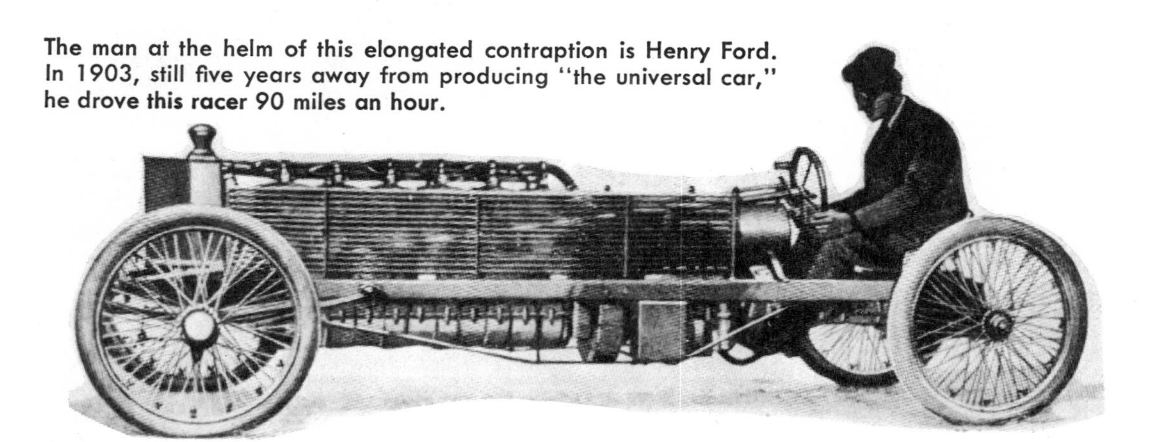

The man at the helm of this elongated contraption is Henry Ford. In 1903, still five years away from producing "the universal car," he drove this racer 90 miles an hour.

But it lacked the snap of the poppet valve.

Advanced styling? A "wrap-around" windshield appeared on the Kissel Kar in 1913, and both Briscoe and Owen displayed convertibles in 1915. The Packard Twin Six —12 cylinders, mind you—was marketed in 1915. Ford was not the only manufacturer reaching for the low-price octaves. The Saxon sold that same year for $395 f.o.b.

Several brands of cars offered hand-operated windshield wipers, rear-view mirrors, and even stoplights as standard equipment. The first engine-source heaters appeared in the year the U.S. entered World War I: 1917. And in 1918 *Popular Science* issued its first "Motor Manual" for do-it-yourself auto mechanics.

In 1920 4-wheel hydraulic brakes appeared. They had external-contracting bands for shoes, and they were forever getting gummed up with oil and dirt.

Car makers were appealing to the sporty set with rakish cars like the Stutz Bearcat and the Jordan Playboy, the latter advertised in a romantic setting—SOMEWHERE WEST OF LARAMIE, said the headline above a winsome damsel and a cowboy galloping in the background.

Unique among the French cars was Peugeot's Decauville, designed by Ettore Bugatti. It was the first European machine in what later would fall in the "mini-compact" category. For sheer elegance the British Rolls-Royce surpassed anything abroad. At the opposite end of the scale, the Morris was a rough counterpart of the Model T. There were others: the Turner, Vauxhall, Napier, Humber, and Sunbeam among them.

Anything with velocity was bound to spawn races. The first international race for horseless carriages was run between Paris and Rouen in 1894. A French Gobron-Brillie as early as 1904 set a world record of more than 100 mph on 110 hp. A Napier averaged more than 65 mph for 24 hours. A Paris-Madrid race was marked by so many gory accidents that it was stopped at Bordeaux.

In the U.S., the goggled gentry, chanting that a mile-a-minute was attainable, were rewarded when Barney Oldfield, most famous of America's early racing drivers, in 1903 drove a Ford racer a mile in 55.8 seconds. Henry Ford, himself, eclipsed that in the same racer, innards refurbished. The next year he ripped off a mile in 39.4 seconds. The most ambitious race was one from New York to Paris—the long way around—underwritten by the Paris newspaper, *Le Matin,* and the New York *Times* in 1908. Starting at New York City's Times Square, 6 cars—3 French, 1 German, 1 Italian, 1 American—drove to San Francisco, went by boat to Alaska, drove to the Bering Sea, boated across the Bering Straits, and drove on across the Asian and European continents. The American entry, a Thomas Flyer, won in an elapsed time of 170 days.

In the beginning the best-publicized U.S. events were the Vanderbilt Cup races on Long Island, N.Y., originated by millionaire William K. Vanderbilt, Jr. Vanderbilt once ran afoul of the constabulary and was charged with "speeding" down New York's Broadway. When he argued that his car could do no more than 15 miles an hour, said the judge: "You may not think that 15 miles an hour is very dangerous, but for the average man, 8 miles an hour is

More "curved dash" Oldsmobiles were sold in 1904 than any other make. The car got 7 hp from 1 cylinder. Wheelbase was 66 inches, weight 1,100 pounds, price $650.

fast enough." That was the pace of a horse at a slow trot.

Indianapolis' first closed-course race was held in 1909, and its speedway oval—the famous "Brick Yard"—was dedicated formally two years later.

The steam engine, the thing that started the mania for autolocomotion, refused to lie down and play dead in the presence of the internal-combustion engine. Steam did have its advantages. It was almost ghostly quiet and, unlike the IC engine—which had to turn at a smart clip to develop power and needed gears or a substitute therefor to get a vehicle moving—a steam engine produced maximum power right from a standstill.

But its debits far outweighed its credits. It was an anomaly. One of the simplest of prime movers, it demanded a horrendous array of appurtenances to make it run. The sheer weight of its component parts was against it. It took time to generate enough steam to get going. The water-supply tank froze in winter. Boilers leaked, and until the day when a way was found to condense the used steam and recycle the water, steamers had to visit a horse trough too often.

A steamer made by the White Sewing Machine Company in Cleveland was one of the best. A product of two brothers, F. E. and F. O. Stanley—who sold out to Locomobile before the turn of the century—was the most famous. As late as September 1923, *Popular Science* published details on a steam car designed by a San Franciscan, Abner Doble, that could get up an operating head of steam in a half-minute. Alas for steam's enthusiasts, it was love's labor lost. As the internal-combustion engine

Recognize this as an ancestor of anything today? It's a 1904 Cadillac with 1-cylinder, 9-hp engine. Bore and stroke were each 5 inches. Note starting crank at side.

This is the forefather of all Buicks as it appeared in 1904. Already getting muscular, it had a 2-cylinder engine of 22 hp.

Demountable rims took terror out of fixing punctures and made L. H. Perlman (above) rich before the day of removable wheels.

Steam cars had their champions. Stanley produced "Gentlemen's Speedy Roadster" (below) in 1906 and set record of 127 mph.

Here, Thomas A. Edison stands beside the Bailey electric auto of 1912. Powered by an Edison storage battery, the car passed endurance test by running 1,000 miles (with recharging) without trouble.

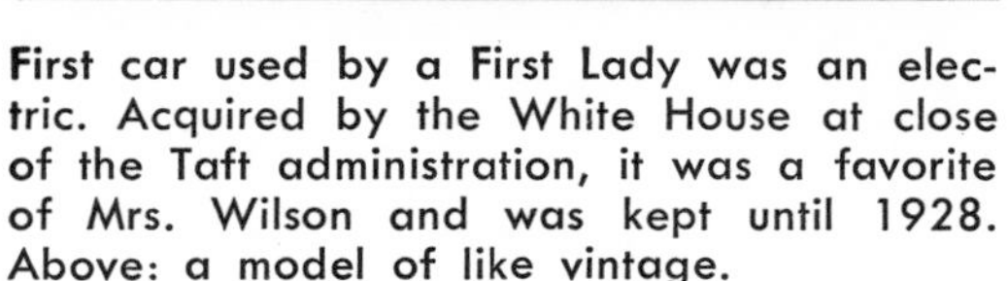

First car used by a First Lady was an electric. Acquired by the White House at close of the Taft administration, it was a favorite of Mrs. Wilson and was kept until 1928. Above: a model of like vintage.

grew in sophistication, a requiem was read over steam. Less than fifty years later it began to look as though the last rites had been premature.

Another early entrant in autolocomotion was the electric car. As early as 1888, vehicles propelled by motor and storage batteries were humming around the streets of Paris. Electrics were even quieter than steamers (their audible approach a low, barely perceptible whine) and, mechanically, much less complex than gasoline buggies. They were cleaner and more dependable. Riker, Waverly, Columbia, and Rauch & Lang electrics sold in the U.S. in modest numbers during the first fifteen years of the century, mostly to sedate elderly gentlemen and timid ladies (crowned by vast, imitation-flower hats) who were quite content with 20 mph. A range of only 40 miles between battery recharging was par for the times.

The electrics did leave a legacy to Gasoline Alley. They were the first closed cars. Their coachwork was exquisite. And, as with the steam car, the electric's funeral may have been a bit premature. That corpse, too, leaped up later.

Still another type of vehicle—the cyclecar—appeared on U.S. streets and highways before the first two decades of the century were closed out. Roughly the counterpart of the French Decauville, it was short-lived—succeeded by the Smith Flyer.

The Flyer, made first by A. O. Smith in 1917 and later by Briggs & Stratton, was a buckboard with five wheels on the ground. The fifth wheel supplied the power. Mounted on the stern, it carried its own one-lung engine bolted directly to it. The engine, the Smith Motor Wheel, had been invented five years before as a power unit for bicycles. A "shift" lever up front raised the motorized wheel clear of the road.

A Flyer driver cranked his engine into life, lowered the wheel, and was off and running. To stop, he stomped on a pedal that pressed a brace of brake-lined fenders against the rear wheels of the car. That also stalled the engine if he had failed to use the shift lever.

True motorcycles, their engines integral, were a boisterous improvement on the Smith Motor Wheel. They were legitimate descendants of a steam velocipede made in the late 1860s by Sylvester H. Roper in Roxbury, Massachusetts. Best-known were the Indian and the Harley-Davidson.

A phenomenal growth in automobile output and ownership was matched by endless rhetoric on motoring in the press. In 1914, P. G. Heinemann, a doctor of philosophy at the University of Chicago, wrote in *Popular Science* that automobiles could contribute to the public health by: (1) bringing improvement of streets and roads, thus reducing the amount of germ-laden dust in the air, (2) eliminating horses and

The self-starter was inspired by death of a man cranking a car. In 1915 *Popular Science* showed starter's four-year weight reduction: 80 to 14 pounds (bottom to top view).

STUTZ Bearcat

Swankiest U.S. car on the road in 1915 was Stutz Bearcat, engineered by man whose name it bore, Harry C. Stutz. In mid-1920s, Stutz introduced one of first safety windshields.

For first decade of century, the highway system, built for horses and buggies, was miserable. On a West Virginia road (above) a car is dug from mud with planks and shovels.

Bumpers were conspicuous by their absence. One inventor devised a person-catcher (left) to drop down and scoop up anyone struck.

Engine sophistication matched the incredible growth of automobile popularity. Cars had to have at least 4 cylinders. Cutaway of a 1915 Model T Ford engine and its power transmission shows what had happened to engines since the carriage became horseless.

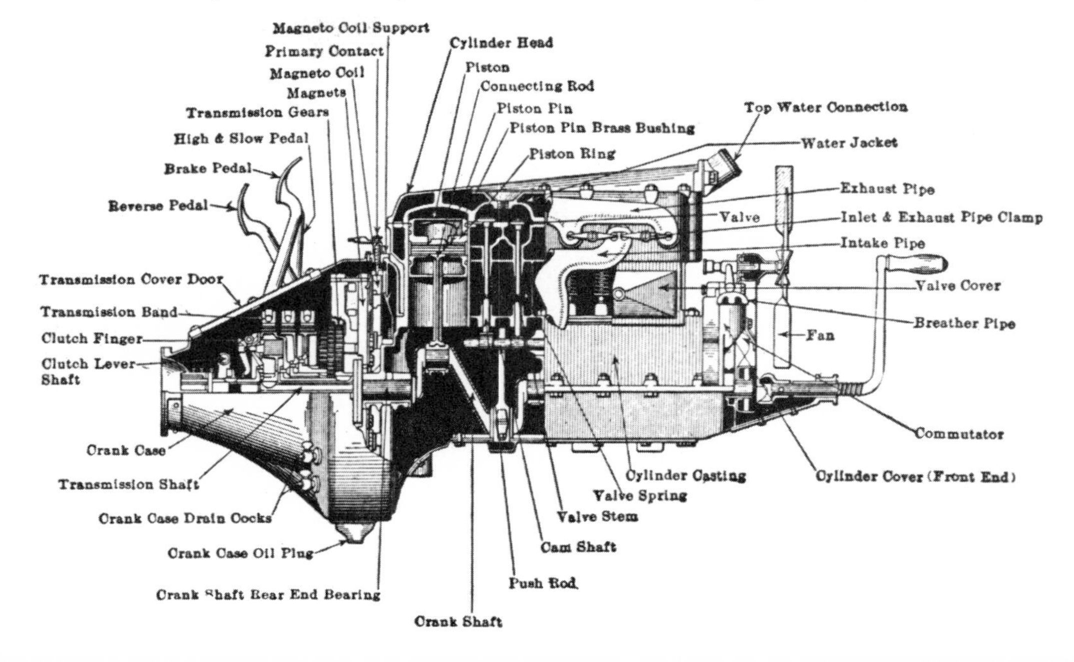

stables from neighborhoods of human habitation, with a consequent reduction of germ-carrying flies.

Boss Ket thought it was not too much to predict that "some day we may have a carburetor that can start an engine smoothly and quickly in the coldest weather." That transpired, but he was far off target when he said we'd be getting 80 miles from a gallon of gas by 1939.

In 1900 the U.S. boasted 8,000 cars. In 1919 this had grown to 6 million, a 750-fold increase. Though the automobile had been born in Europe, America's vast distances—and the fatter paychecks of its workers—soon made the U.S. the land of its development.

In the U.S. the specter of a gasoline shortage plagued motorist and industry. As early as 1913 the car manufacturers offered through the International Association of Recognized Automobile Clubs a prize of $100,000 for a substitute for gasoline that would cost less. By 1916 headlines began appearing in the press: WHAT WILL WE DO FOR GAS? One answer already was at hand —"cracking" crude oil to get more fuel palatable to a car engine. Simple distillation had been producing little gasoline and a great volume of heavier by-products.

Science had other tricks up its sleeve—the hydrogenation of coal to produce synthetic gasoline, polymerization of light gases to make refinable heavier hydrocarbons, and a catalytic process invented by Eugene Houdry of France to improve crude's gasoline yield and its ignition point.

In the end, refiners found rich new pools of oil.

A fuel fright of a more intimate nature swept the U.S. in 1925. Tetraethyl lead had been added to gasoline since 1923 to improve its anti-knock quality. At a Standard Oil of New Jersey refinery, 5 workmen exposed to the stuff died and 36 others were hospitalized, several "showing symptoms of insanity."

The U. S. Bureau of Mines after 10 months of investigation, reported *Popular Science,* said the danger to the public of breathing exhaust gases from "ethyl" gasoline was seemingly remote. In late spring of 1926 the U. S. Public Health Service announced that its own investigation showed the compound was dangerous only in concentrated form and could be used safely in automobiles. Dr. Yandell Henderson of Yale, an authority on the action of gases on the human body, disagreed. Breathing the exhaust of ethyl gas, he claimed, was a public menace. He called the turn, as events forty years later would prove.

By the mid-1920s the elements that launched the Gasoline Age were well in mesh. An incredible 17½ million "pleasure cars" were under registration in the U.S. alone. The legal speed limit in most states was 10 mph; Michigan allowed 25.

The first transcontinental road, the Lincoln Highway, was finished in 1920. In 1921 a Federal Highway Act integrated federal and state systems.

Between 1900 and 1908 no less than 500 U.S. firms turned out automobiles. By 1925 that mighty army of optimists, plus hundreds of others that had joined up, had shrunk to 50. Some had been gobbled up by companies that were better-heeled, some had failed, some had just quit. E. V. Rickenbacker, leading U.S. ace in World War I air combat, tried his hand at producing a car bearing his name. It was short-lived. Where the stakes were astronomical, motor-making was a bruising business.

By the mid-1920s Henry Ford was in trouble. Sole owner of his business, he was as rich as Croesus. He had put America on wheels, but he had clung to the Model T and the Model T had seen its day. Technology was passing him by. His business was being nibbled away. His dealers were being proselytized to sell machines with more sparkle, both in appearance and performance. The Chevrolet in particular, by now a General Motors product, was making inroads on the Ford. Named for Louis Chevrolet, a pioneer in racing and design, it had a respectable gearbox, not planetaries, a 4-cylinder engine of 26 horsepower, a self-starter, and a foot, instead of a hand, throttle. It also looked classy and it cost only $25 more than a Ford.

An upstart engineer, Walter P. Chrysler, had entered car manufacture to do battle with the giants. His cars had a compression ratio of 4.5:1, sensational for their day. This lent them sass when the throttle was opened. They could top 70 mph. All automobiles, Ford excepted, had adopted 4-wheel brakes, and Chrysler added to their safety with hydraulics. He put ball joints on his front suspension.

Clamp-on rear-view mirrors were an oddity rather than the rule in 1917. (This driver's pose suggests hers wasn't so useful.) Such accessories were not supplied by automobile factories but were products of a secondary industry riding Detroit's coattails.

Inside rear-view mirrors were almost the ultimate in classy gadgets in 1920–21. This one attached to windshield's top. Later: Wills Ste. Claire's automatic backup light.

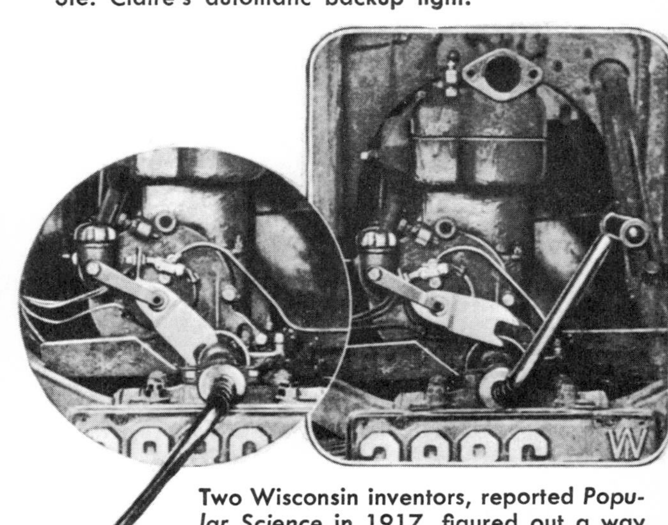

Two Wisconsin inventors, reported *Popular Science* in 1917, figured out a way to prevent a Ford's "kicking." Before the starter crank could engage the crankshaft, an arm had to be pushed out of way of ratchets. This retarded spark and prevented backfiring.

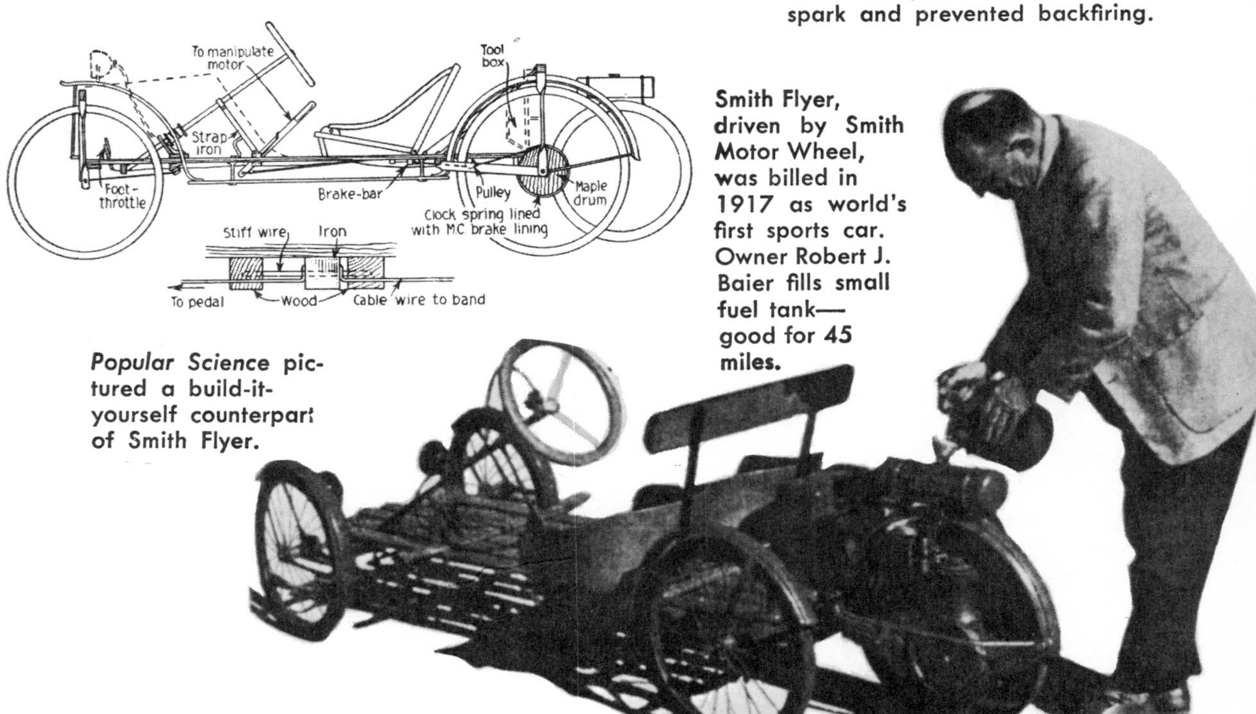

Popular Science pictured a build-it-yourself counterpart of Smith Flyer.

Smith Flyer, driven by Smith Motor Wheel, was billed in 1917 as world's first sports car. Owner Robert J. Baier fills small fuel tank—good for 45 miles.

Ah, at last, the "1-man" Ford touring-car top in 1920! Gone was the center post, gone the unsightly front bow sockets. The conversion was not available from Ford Motor Co.

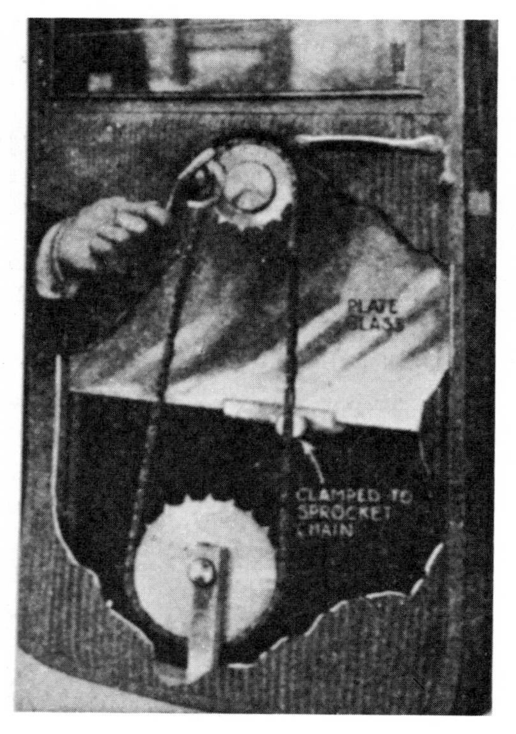

For closed cars there was another tidbit: pulley gears to replace the sash bands used to raise and lower windows. Add a crank, and a driver (circa 1920) was ready for winter.

The air-cooled engine began gaining favor at just about this time. Without radiator, jackets for cylinders, and plumbing, it didn't risk freezing in winter and steaming in summer.

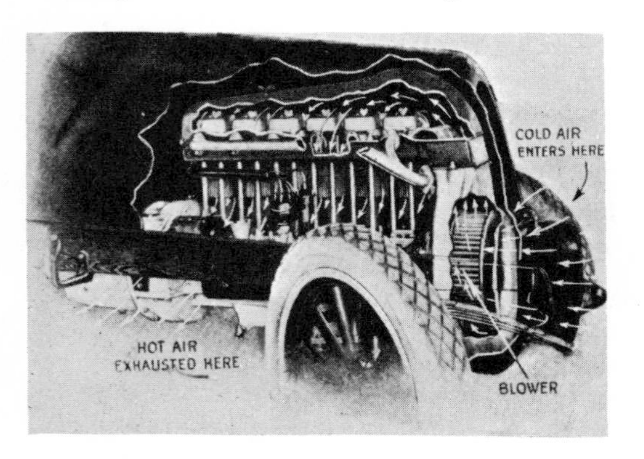

The automobile was becoming a marvelous piece of mobile machinery. Synchromesh transmissions soon obviated the clashing in shifting gears. Balloon tires, reducing the air pressure from 65 to 35 pounds, softened the ride. Pumps had begun to replace gravity fuel feed to engines.

As early as 1918 *Popular Science* ran a contest for devices to improve automobile operation, and first prize went to C. A. Butterworth of Newton Center, Massachusetts, for a gear-shift system actuated by solenoids. A driver merely pushed some buttons on the steering column. Second prize was won by P. C. Haas of Ann Arbor, Michigan, for a solenoid-assisted steering design.

Streamlining was in vogue. Body shells were shaped to imply speed. On closed models, lift straps on the windows were replaced by cranks. The granddaddy of today's power steering was invented. Francis W. Davis, a New England Yankee, toured Detroit's automobile plants in a Pierce-Arrow roadster trying to interest the industry in his system for turning a steering wheel with a fingertip. It was no sale. Hypoid gears were adopted, lowering the drive shaft and permitting lower bodies. One-shot lubrication, automatic

Chevrolet's 5-passenger touring car was only a mild threat to Ford's dominance of low-cost field in 1923. Ford's cheapest, a runabout, was $265, and you could pay $5 a week toward a car for delivery in 53 weeks.

windshield wipers, shatterproof windshields, internal-expanding brakes, and vacuum-assisted braking all arrived in a space of four years.

A beleaguered Henry Ford sat down and took stock. He closed down his production lines. It was six months later, in 1928, before a new Ford car was announced. The Model A, it was an immediate hit. Its 4 cylinders turned out 40 horsepower that could drive the car 65 mph. Three synchromeshed forward speeds gave it snap at takeoff. Hydraulic shock absorbers erased bumps.

But the Model A was only a stopgap. Henry had something else up his sleeve. In 1932, smack in the Great Depression, he gambled with his first V8, a flathead of 65 hp. For the first time it brought within reach of the store clerk, the dirt farmer, and the factory hand a car powered by an engine theretofore reserved to citizens who counted money by the yard. Its lowest f.o.b. price was $500. Ford again was off and running.

In the decade before the U.S. was catapulted into World War II, sophistications in the automobile trod on one another's heels. Front wheels got independently sprung. Steel tops for safety became uni-

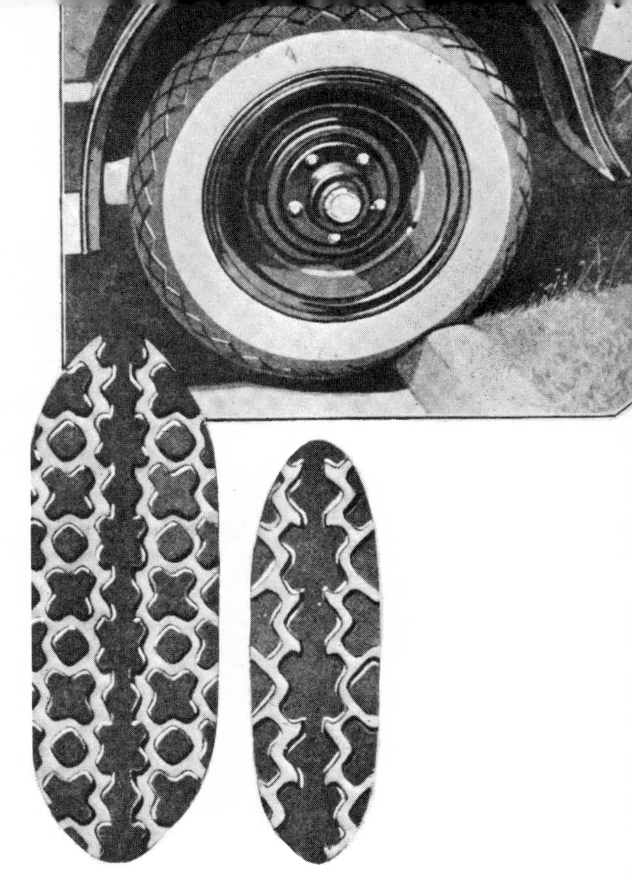

Balloon tires came in 1923, with 35 instead of 65 pounds' pressure. Tread imprint was bigger, tires suffered less damage on curbs.

By 1924 Detroit stylists began to soften boxy lines by streamlining cars. Ford's coupé (lower left) had bigger rear window, deeper cushions; Buick's roadster (right), a rakish silhouette; Dodge's touring car (top) had a higher radiator. All had spoke wheels.

Ford's assembly line was wonder of the industrial world. For benefit of the Prince of Wales, it turned out a complete automobile in 26 minutes. Other companies adopted Henry's manufacturing techniques. At left: inventor in 1928, at age 64, with the Model A.

versal—twenty-five years after Daimler introduced them in Europe. "Overdrive" provided a gear above "high" for the open road. Engines acquired hydraulic tappet clearance adjusters. They were mounted flexibly to insulate their vibration from the car body. Walter Chrysler interposed a fluid coupling in his drive train to soften the transmission of power to the wheels. In an excess of zest Cadillac produced a car with 16 cylinders. Vents on front windows supplied draft-free ventilation.

A good many stabs had been made at a fully automatic transmission. Between 1907 and 1916, the Cartercar sported a clutchless friction drive with an infinite number of gear ratios. In 1917 the clutchless Owen Magnetic used an electric generator to produce a magnetic field. This acted on a shaft connected to the driving axle of the car. Varying the intensity of the field controlled the car speed.

In 1930, Studebaker ballyhooed "freewheeling"—gearshifting without clutching. It was promptly adopted by 16 other brands. In 1935, the Hudson and Terraplane had an "electric hand"—fingertip gearshifting—on the steering column. Reo had a nominal success with an automatic transmission in 1934.

Then, in 1939, General Motors bravely put a combination fluid coupling and planetary gears (shades of the Model T!) into some production Oldsmobiles for its 1940 model. The device jerked in upshifting and downshifting but, wholly automatic, Hydra-Matic, as it was called, was a portent of developments to come.

Other things were happening. Sealed-beam headlights bowed in. Cord put out a front-wheel drive. While FWD was no novelty in Europe, in the U.S. it had, among other drawbacks, too high a price tag, and in eight years the Cord was in limbo. Packard brought out a Twin Six. It also introduced air conditioning, which promptly became a casualty of the war. Davis' power steering did come into use, but only on military vehicles. Auto radios were a common option.

Finally, the car manufacturers had hit on a gimmick to boost sales. This was an annual model change, putting the motorist in hock for a new bauble to park in his driveway almost before he had made the last payment on the old one.

Hardly had the fires of Pearl Harbor been quenched before the soothsayers were in print with what was coming in cars after the war. Their batting averages turned out to be surprisingly good.

One list of forecasts in *Popular Science* saw synthetic cord fabrics for tire bodies, far stronger than the cotton then universally

The Pierce-Arrow, born in 1901, was epitome of elegance in U.S. motordom, like Rolls-Royce in England. It looked like this in 1919, nine years before factory folded.

Another car of transient fame: the sumptuous Peerless, with 6 cylinders, 61 hp. Pictured here, in 1929, it died in 1932.

Streamlining had almost reached point of caricature when car makers exhibited their wares at 1934 National Automobile Show. This "Airflow" Chrysler was a sales disaster; it turned customers away in droves.

General Motors took the plunge into fully automatic drive (right) with Hydra-Matic Olds in 1940 models, as announced in December 1939 ad. Cadillac adopted it in 1941.

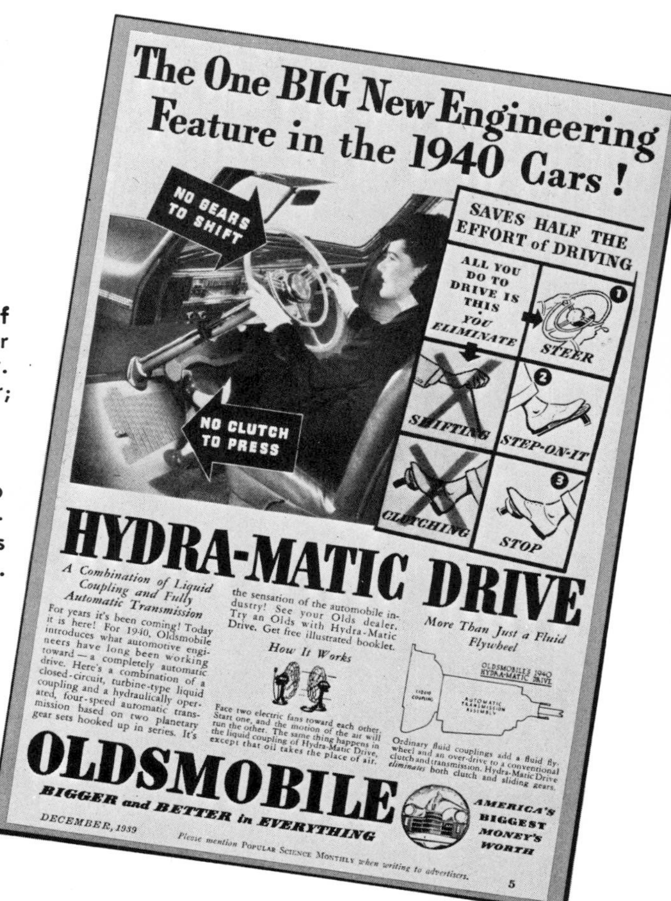

Packard was first, in fall of 1941, with air conditioning, working on the same principle as a home refrigerator. Car trunk housed the condenser (above); at right, a schematic.

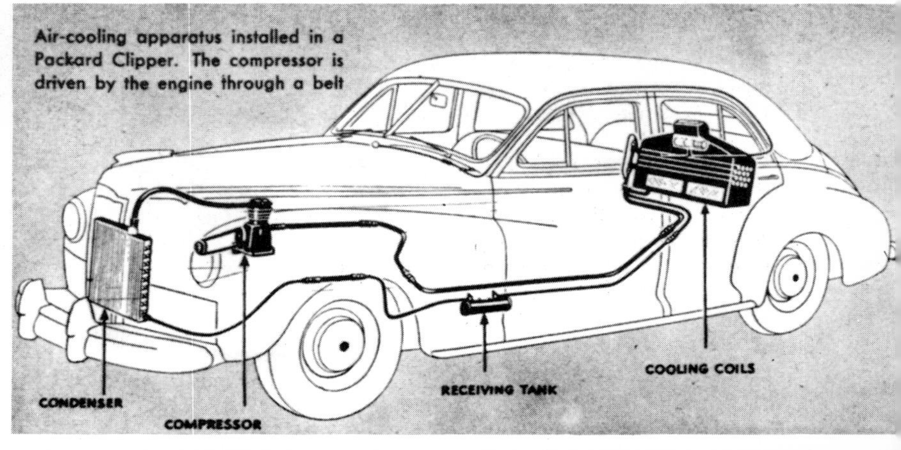

Air-cooling apparatus installed in a Packard Clipper. The compressor is driven by the engine through a belt
COOLING COILS
RECEIVING TANK
CONDENSER
COMPRESSOR

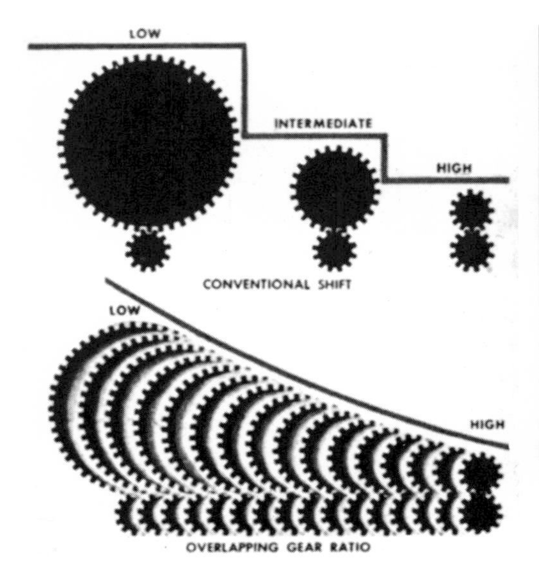

Buick was first with the torque converter, an automatic drive without Hydra-Matic's fidgety up-and-down shifting. Its Dynaflow provided infinite number of overlapping gear ratios.

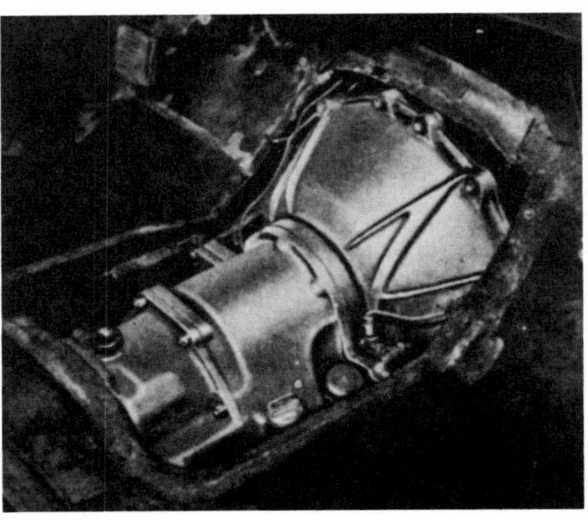

The transmission was bulky and weighed 50 pounds more than a manual, but it opened the door to industry-wide adoption of automatics—by Packard, Borg-Warner (for Studebaker and Lincoln-Mercury), Chevrolet, and the Chrysler Corp., which soon began playing variations on the same theme.

Running boards had begun to disappear—that is, when a car's doors were closed—before W.W. II. The sheet metal hid them. This remnant of the iron step used to get in and out of the high-wheeled horseless carriage was abandoned entirely after the war. In 1949 Nash offered a foretaste of disappearing fenders by shrouding the entire car, wheels included. Only a narrow front-wheel track made it possible to negotiate a turn.

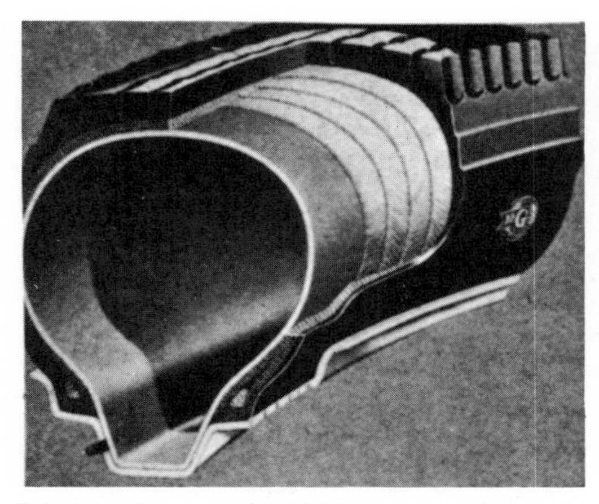

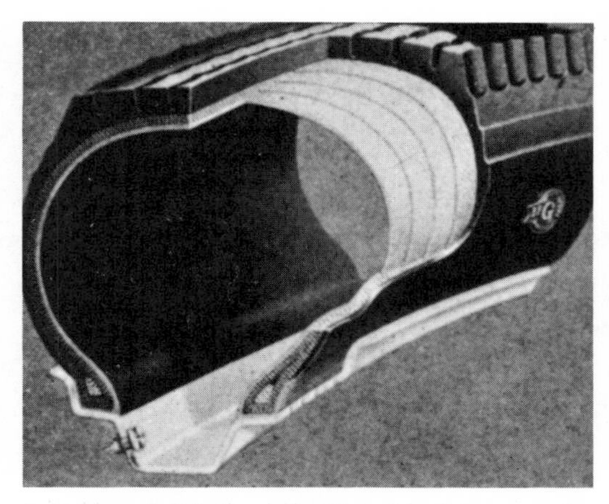

Tubeless tires, out in 1947, appeared on new cars in early fifties. Suspect by car buyers at first, they soon proved themselves. Tubed tire (above, left) insulated from rim the heat buildup of driving, resulting in air expansion and a harder ride. Tubeless shoe (right) exposed the heat to rim, which acted as a radiator, reducing its temperature.

Power steering—invented in 1926—finally arrived in 1951 via the Chrysler Corp. Its two units: power package on steering axis and pump-reservoir.

Tail fins made their debut in 1948 on Cadillacs. By 1957, just look at what had happened to them!

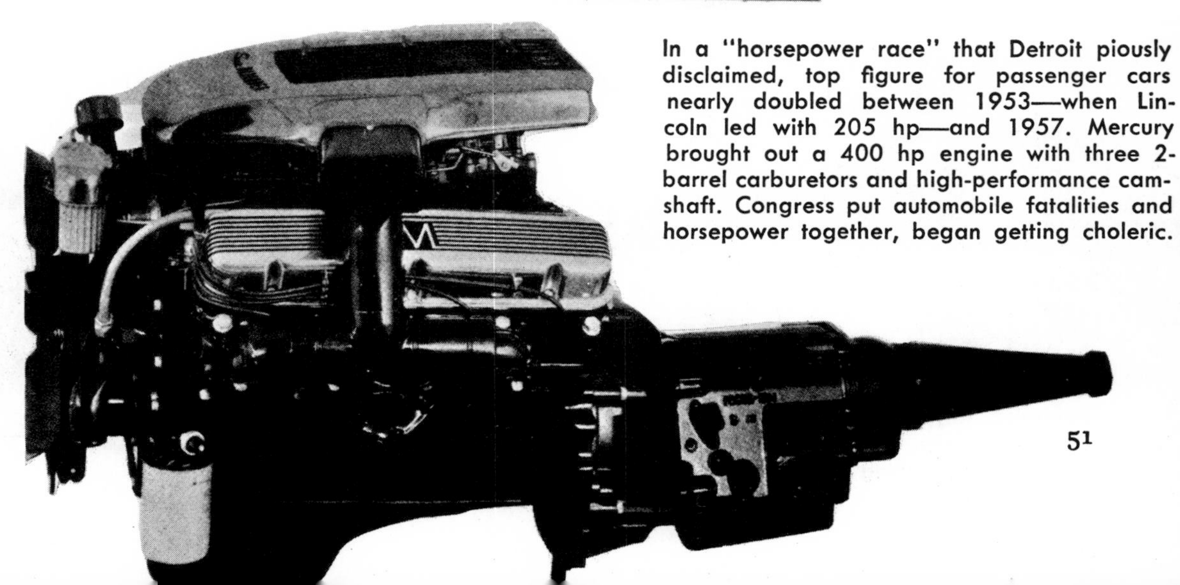

In a "horsepower race" that Detroit piously disclaimed, top figure for passenger cars nearly doubled between 1953—when Lincoln led with 205 hp—and 1957. Mercury brought out a 400 hp engine with three 2-barrel carburetors and high-performance camshaft. Congress put automobile fatalities and horsepower together, began getting choleric.

Meantime, an invasion of foreign cars was sneaking up on the U.S. auto industry. West Germany's homely little Volkswagen, with all of 25 horses, became more and more popular.

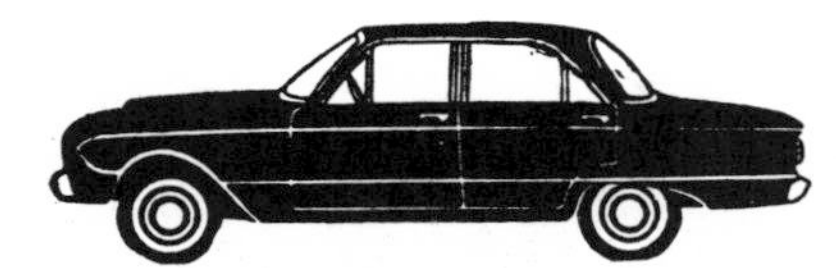

FALCON: 27" shorter than Ford

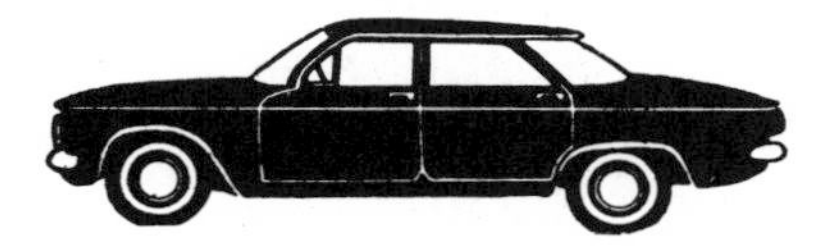

CORVAIR: 31" shorter than Chevrolet

VALIANT: 24" shorter than Plymouth

It took a business recession to do it, but in 1959 out popped 3 low-horsepower, shortened, lower-cost U.S. cars—Ford's Falcon, Chevy's Corvair, Plymouth's Valiant.

used (they were right), synthetic rubber casings (partly right), plastic car tops (wrong), frameless chassis (some), hydraulic drives (right), superchargers (an experimental sprinkling), lighter engines (not with all the garbage that Detroit would hang on them), revolutionary streamlining (wrong), small-car prices ranging from $500 to $1,000 (wrong), higher-octane fuels (right), 30 miles to the gallon (wrong), and highways to permit superspeeds (right—if, in most states, a motorist didn't mind a ticket).

The automobile changed not one whit for more than two years after the war. The car makers hauled their 1941–42 tools and jigs out of storage to hastily satisfy four years of pent-up demand.

But in the decade 1948–58 the engineers turned out some sparkling achievements. The first genuinely smooth automatic transmission, Dynaflow, was pioneered by Buick in 1948. This was a torque converter and a gas-eater. Not until it was backed up by gears could a car equipped with it be forced past a fuel pump.

Power braking and, at last, power steering—the need for the latter an open confession by designers that they were improperly distributing car weight by pushing the engine forward to boost the size of passenger compartments—were oft-bought options. The V8 engine with overhead valves became the standard power plant. Chrysler offered torsion-bar suspension. Compression ratios crept upward, boosting miles-per-gallon, until 9:1 was a commonplace.

Air conditioning escaped the novelty class and, later, General Motors developed

Fade out, fade in, eleven years later. With 1 of every 6 cars sold in the U.S. an import, Detroit began selling its own midgets: Chevy's Vega (left), Ford's Pinto (right), and Plymouth's Cricket. They weren't much on performance, but were cheap to buy and run.

their Climate Control, with sensors scattered about the interior of the car to maintain any fixed temperature selected, summer and winter. Master brake cylinders were divided for front and rear wheels for fail-safe operation. The rubber industry introduced tubeless tires.

At their best, the stylists captured lines that, to a car buff, were sheer poetry. The hardtop convertible was a case in point, though it didn't convert. It only lost its center pillars. At their worst, stylists inflicted some pretty garish plumage on cars. Sheet metal got "sculptured," evoking the comment by one wag that Detroit was "pre-denting its fenders." Hoods steadily lengthened to impart an impression of power. "Anyone for table tennis?" asked another critic, eyeing the expanse of hood on a new model.

Tail fins appeared, first on the Cadillac, and they grew and grew in acceptance and size until the Chrysler Corporation practically killed them off with an outrageous exercise in French curves.

A "horsepower race" got under way. Stock-car racing became a sales tool. Cars kept growing in size. Chevrolets began looking like Cadillacs, Fords like Lincoln Continentals, and Plymouths like Chrysler Imperials.

In the midst of this yeasty brew, events occurred that ultimately had a sobering effect on U.S. car manufacture. In 1930 a British design, the Austin Bantam, was built here under license. It went almost unloved. Powell Crosley, who had made his money in radios, tried marketing a car of somewhat the same dimensions after World War II. It bombed. So did a Kaiser-

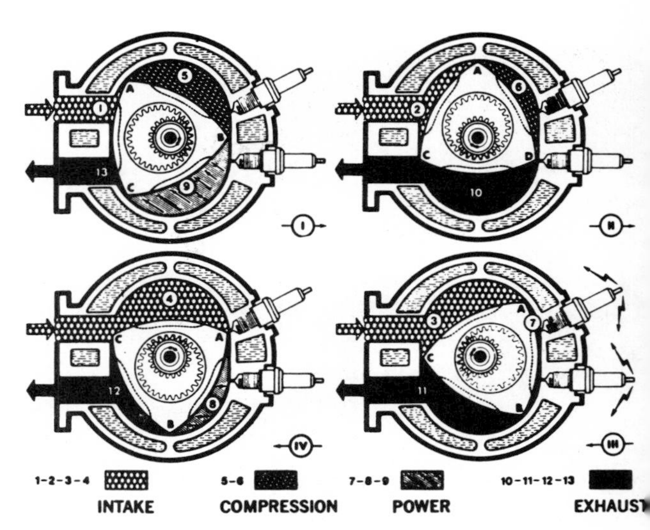

The first acceptable radical change in the internal-combustion engine in eighty-two years came out of Germany—where piston engine was born—in 1954. It substituted a 3-faced rotor for a piston. Like the Otto engine, it worked on the 4-phase cycle. The 2 spark plugs merely lent more heft to the ignition.

Four-rotor engine in Mercedes-Benz C-111 Mark II in 1971 weighed only 396 pounds, delivered 400 hp. Wankels, as type is known, come in any size. Smallest is ½ hp.

Let's not forget American Motors. It, too, had a tiny car, the Gremlin, that the company said was aimed right at the VW market.

Mercedes coupé weighs 2,734 pounds or 6.8 per hp, which gives it top speed of 186 mph and a 0–60 mph time of 4.7 seconds on 5-speed transmission. Axle ratio is 3.77:1.

Europe adopted disk brake long before U.S., where it began appearing in quantity in '67. Instead of shoes and drums, it uses calipers acting like powerful fingers on a disk.

Frazer midget, the Henry J. Nor did a smallish Nash Rambler and a tiny Nash Metropolitan cause much of a stir.

In 1950, unheralded, a tiny, funny-looking—aye, ugly—car of German manufacture began making its appearance in U.S. showrooms. Born before the war as a military reconnaissance vehicle but propagandized by the Hitler regime as (the name said it) a "people's car," the Volkswagen was cleverly merchandised in America. Its ugliness was made a virtue. Its numbers grew.

The VW, or any of its European counterparts, didn't worry U.S. auto makers. "Those things," said a Ford official in 1952, "won't claim more than two per cent of our market. We can live with that."

By 1958 he was eating his words. An economic recession had set in. A surprisingly substantial portion of the buying public wanted less expensive automotive hardware. The small foreign cars remorselessly munched into the U.S. market. The imports had other things going for them. One was strict quality control. And the imports didn't change styling with the annual first frost.

Nash, which became American Motors, had scoffed—as a sales ploy—for years against "gas-guzzling dinosaurs," and sud-

How they work: Fundamental differences between drum and disk brakes

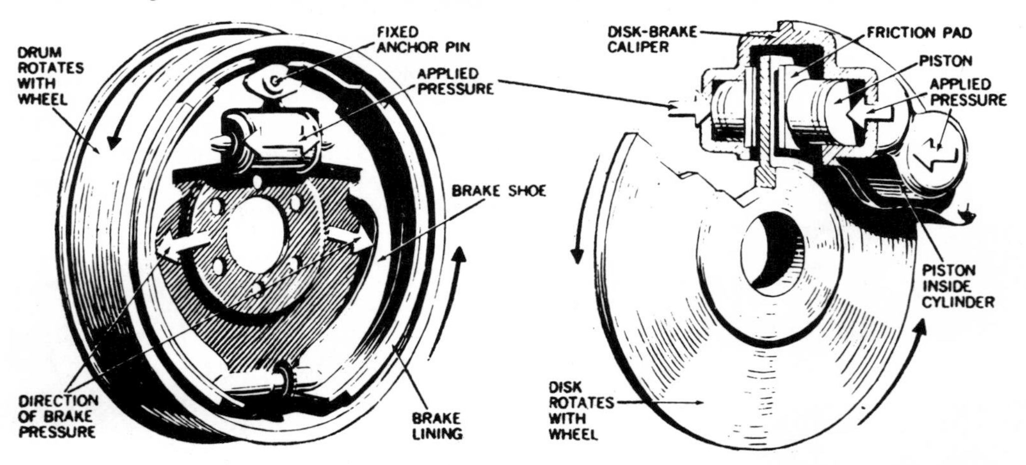

Difference between two braking systems is seen above. Conventional drum brake (left) has two stationary shoes anchored to backing plate fixed to hub. The drum rotates around them. Hydraulic pressure applied by a wheel cylinder forces shoes outward to make contact with drum. Hydraulic pressure is low and the friction area quite large. In the more effective disk, line pressure is high and the friction area small. Brake pads are bonded to caliper held rigidly to wheel hub. Hydraulic pressure forces the pads against the disk.

denly the Big Three plucked some dust-laden ideas off their shelves and the "compacts" appeared. Chevrolet's Corvair, Ford's Falcon, and Plymouth's Valiant turned up in showrooms in 1959. Six other brands of compacts quickly bolstered their number. All were bigger than the imports, and they cost more.

It was a transient victory for smaller-car enthusiasts. As the economic pendulum swung back toward normal, the enthusiasm of U.S. factories for the things chilled. They did remain on the production lines, but they tended to grow in dimension. So German, French, Italian, and British small cars kept arriving in growing numbers at U.S. wharves.

Detroit coppered its bets. To its standard, compact, and luxury cars, it added "intermediates" and sport/specialty cars.

Presently the Japanese discovered the bonanza of the American market for their own postwar automobiles and, for the U.S. industry, the fat was in the fire. Faced with another economic recession, Detroit for 1971 responded with sub-compacts. Aptly enough, Ford's offering, the 4-cylinder Pinto, was referred to in the company's councils as "the Model T."

That was not all the industry had to

Safety-consciousness finally made seat belts—supplemented by shoulder harness—mandatory by law to combat the rising toll of automobile fatalities. Here children are protected against collisions *inside* the car.

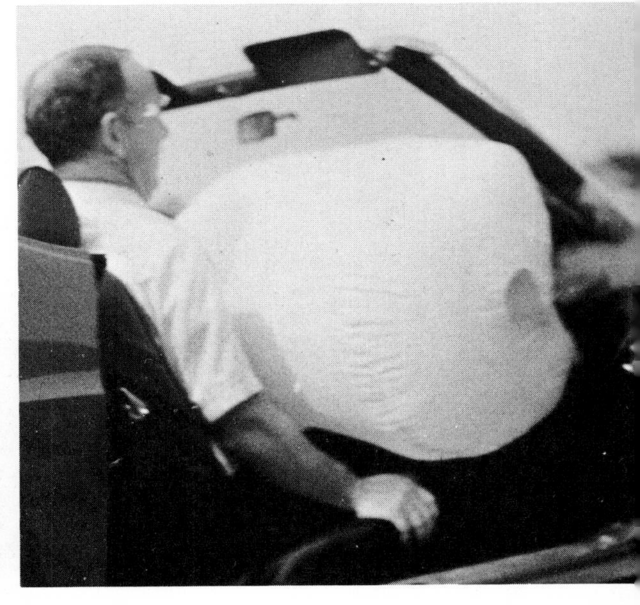

The federal government then went a long step beyond that—requiring car makers to install, by 1976, air bags (above) that would inflate instantly in a collision. This experimental bag deflates only a half-second later.

Finally Washington decreed energy-absorbing bumpers to reduce collision hazards and, incidentally, cut damage and repair costs. Those in use couldn't take an impact of 2 miles an hour. The goal: 5 mph. At left, a GM experiment.

worry about. Washington's Capitol Hill began grumbling about the automobile's pollution of the air.

As early as 1960 *Popular Science*'s "Detroit Report" took cognizance of smog traceable to automobile emissions. Southern California's mountain-girt coastal plain suffered most from it, but it had begun to afflict other densely populated areas.

The Chrysler Corporation and Thompson-Ramo-Wooldridge were working on an exhaust-manifold afterburner to meet part of the problem. A more immediate and simpler fix for another part of it, quickly adopted industry-wide, was the elimination of the "road draft tube" on crankcases and plumbing to funnel their unburned gases, blown past the piston rings, into the intake cycle. Work was being done on catalytic converters to mix fresh air with exhaust gases and filter them through chemicals to produce harmless carbon dioxide and water.

A completely different attack on smog was mounted by Chrysler—experimentally—by substituting a gas turbine for the piston engine. Boeing Aircraft in 1947 (and, later, Ford and GM) had built turbine engines for trucks experimentally. Gas-turbine exhausts did contain oxides of nitrogen but they were almost free of other contaminants. Turbines were not finicky about their diets. They needed no leaded fuels.

The automobile reached a sort of zenith in fun cars, among them all-terrain vehicles, ski scooters, dune buggies, and campers appealing to sportsmen and vacationers. ATV pictured, with power on all 6 wheels, is typical of its breed. Some even swim.

Ski scooters, also known as snowmobiles, have endless treads, like army tanks, in the rear for propulsion. Some can top 45 mph.

Hundreds of scooter clubs in snow areas of U.S. and Canada participate in cross-country rambles, rallies, jumping contests, racing.

One of newest dune buggies on the market (above) offers a useful cargo platform behind its 2 seats. It's popular in racing events.

ATV at right has motorcycle engine, weighs 230 pounds, can tackle almost any terrain.

Unlike many dune buggies built on shortened Volkswagen chassis, this one has its own, specially designed. Small, it can be stowed in station wagon or utility trailer, or trailered partially piggyback on a snowmobile.

They would burn kerosene, or Chanel No. 5, or (almost) stove wood.

Chrysler's turbine test-bed car, *Popular Science* noted in July 1961, topped 60 miles an hour from a standing start in less than 10 seconds. Its top speed was about 115 miles an hour.

After more than ten years of research Chrysler in 1963 turned out 50 turbine-powered cars for selective consumer evaluation over the next two years. They ran pretty well, but turbine cars had two seemingly insurmountable faults. They were expensive to build and their engines lacked the flexibility of pistons. The turbine essentially was a constant-speed engine, more suited to aircraft than to cars.

By 1966 California was enacting some tough laws for smog control, and Congress had given the Department of Health, Education, and Welfare authority to promulgate demands for cleaner engines. The research was centered on the elimination of unburned portions of gasoline, though an attack on carbon monoxide was included for general health reasons.

In May 1967 Dr. John T. Middleton, director of the National Center for Air Pollution Control, let fly at the horseless carriage with both barrels. Writing in *Popular Science*, he quoted John W. Gardner, then HEW secretary, as saying: "The automobile internal-combustion engine and the interests of the American people are on a collision course."

Wrote Dr. Middleton: "We will drown in our polluted air unless we act fast to clean it up. Automobiles account for roughly half of the noxious fumes that are poisoning the air we breathe."

His estimate, as of then, was that cars and trucks each year released 66 million tons of carbon monoxide (a deadly gas), 12 million tons of unburned hydrocarbons, 6 million tons of nitrogen oxides, and smaller quantities of sulphur oxides and particulate matter, including lead compounds.

"Carbon monoxide," he said, "need not reach 'suicide levels' to be harmful. Lesser amounts may reduce your ability to handle complex situations—such as driving. Hydrocarbons in automobile emissions have been shown to produce cancer in animals. Nitrogen oxides irritate the respiratory tract and weaken the body's defenses against respiratory infection . . . Hydrocarbons and nitrogen oxides react in the presence of sunlight to produce photochemical smog, which causes eye irritation, obscures visibility, damages vegetation, and injures human lungs . . . The damaging effects of smog on vegetation have been seen in at least 27 states."

The good doctor estimated that U.S. cars, trucks, and buses would number 103 million by 1971. He was low by 9 million.

Congress passed a Clean Air Act, and the National Academy of Sciences was called upon to conduct a comprehensive study of the technological feasibility of meeting the auto-emission standards prescribed by it. These would be, as compared with 1970, a 90 per cent reduction in five years in emissions of carbon monoxide and hydrocarbons and in six years of oxides of nitrogen.

The Academy's National Research Council announced that the high concentration of lead in the air of central cities constituted a potential health hazard to young children and "certain" groups of workers but currently posed no identifiable threat to the general population. "Due largely," it said, "to the combustion and dispersal of lead additives in gasoline, the air in the

The ultimate in luxury is the perambulating home. This motor home, built on a panel-type truck chassis, sleeps 6 and promises all the comforts of home, even hot showers.

largest American cities has a concentration of lead 20 times greater than air over rural areas . . ."

A group of scientists at Sweden's Wallenberg Laboratory announced that the lead additive in gasoline affected the distribution of chromosomes in lower animals under test and "may" cause human birth defects.

The answers to the motor vehicle's pollution of the air were elusive. Electric cars? In 1959 three independent manufacturers announced plans to produce them. None did. Both GM and Ford did have intensive research programs on electrics under way. Ford actually produced a pollution-free piston engine, but warned that the moment it was fired up its virtues began a progressive deterioration. In California, the Pacific Lighting Company and the Union Oil Company jointly proposed to

More popular are travel trailers, hitched to the family car. The one above can sleep up to 8. Twin axles bear most of the weight.

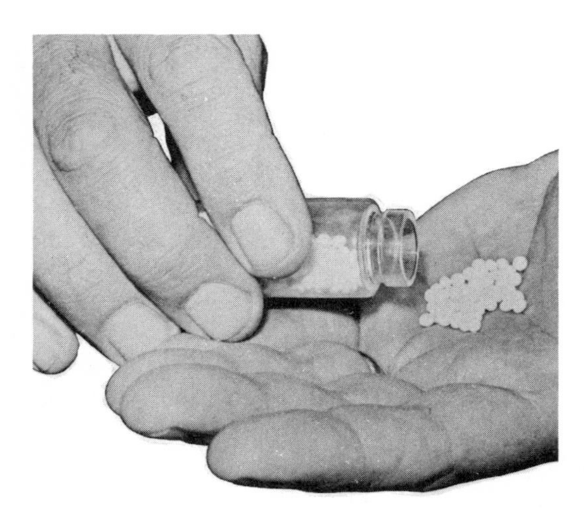

Is this the Pill to curb air pollution from auto exhausts? These pellets are "catalysts" that go into a converter fitted in engine compartment to make noxious gases benign.

This is all there is to the converter. Originating company says it can transform 90 per cent of nitrogen oxides, unburned hydrocarbons, and carbon monoxide. Other cures, such as thermal reactors, vary in principle.

sell compressed natural gas, 90 per cent free of exhaust contaminants, for automotive use.

The steam automobile was disinterred, but its faults had not been mended. A Florida inventor fiddled with a fluid other than water to make the idea work. (By 1972 *Popular Science* was reporting on DuPont's brand-new organic-fluid external-combustion engine.)

More promising was a pistonless rotary engine, the German Wankel, burning jet fuel, with most of its dangerous emissions rendered benign by a catalytic converter. The Wankel had a high mechanical efficiency (low friction losses) but a lower thermal efficiency (degree of fuel conversion to heat) than a piston engine. It balanced that off with simplicity and compactness. Under a Wankel hood was a world of room for components to tame noxious emissions.

David Scott, *Popular Science*'s European editor, wrote the first definitive description of the Wankel for U.S. readers in March 1960. The Germans, said Scott, had "blitzed the auto world." Detroit executives were less enchanted. They accused Eugene Duffield, president of Popular Science Publishing Company, of promoting a joke. But in May 1972, *Popular Science*'s 100th Anniversary issue was able to preview the first U.S.-built Wankel car, a rotary-engine model of Chevrolet's 1974 Vega.

To top the Motor City's woes, a thing called "consumerism" began sweeping the country. For years specialty automotive publications and a particular consumer magazine had been badgering the car manufacturers with criticism of their products. So had *Popular Science.* Safety was a big item in the analyses. A consumer advocate with a missionary zeal, Ralph Nader, wrote a book flaying Detroit.

Discussion of safety was long overdue. Multilane, limited-access roads had spread. The 41,000-mile Interstate Highway System authorized by Congress in 1956 (incorporating cloverleaf intersections that had been pictured in *Popular Science* in 1925 as the wave of the future) was built. Still, deaths from car accidents were in excess of 55,000 a year. So acute had the problem become in 1956 that the Cornell Aeronautical Laboratory in Buffalo, N.Y., designed and built a car that could crash

Smog-free fuels for cars—such as liquefied petroleum gas and natural gas—promise also to clear the air. Instead of gas pumps, LPG stations have storage tanks like this.

Where's the LPG carried? Fuel tank is strapped on its side in a sedan trunk. The conversion kit adds a few valves and hoses to conventional engine. It's easy to install.

Still another idea is dumping the piston engine altogether and using the smog-free gas turbine, the motive power of most airliners. As long ago as 1954 GM produced this Firebird experimental car attired in a racer's sheet metal. Chrysler built 50 turbine cars.

without hurting anyone inside. But, in terms of current design and styling, it was a monstrosity. Who would buy it?

In November 1965 the late U.S. Senator Robert F. Kennedy, writing in *Popular Science,* proposed the creation of a National Highway Safety Center. "Why," he asked, "can't we make cars safer?" The following month the magazine opened its pages to Henry Ford II, Ford's board chairman, to discuss car safety. He pointed out that Ford had been the first automobile company to undertake systematic research on safer design and that the U.S. fatality rate per 100 million motor-vehicle miles had been cut by four-fifths in forty years.

It was, in fact, an abuse of the truth to say that the car industry had been remiss in its obligation to the public. It had done much. It adopted five of the Cornell safety car's design increments, most of which, even the laboratory conceded, were somewhat bizarre. As early as 1950 Nash offered seat belts to protect a car's occupants against the "second collision"—being catapulted into parts of the interior (the windshield especially) capable of inflicting wounds in a crash. Ford followed.

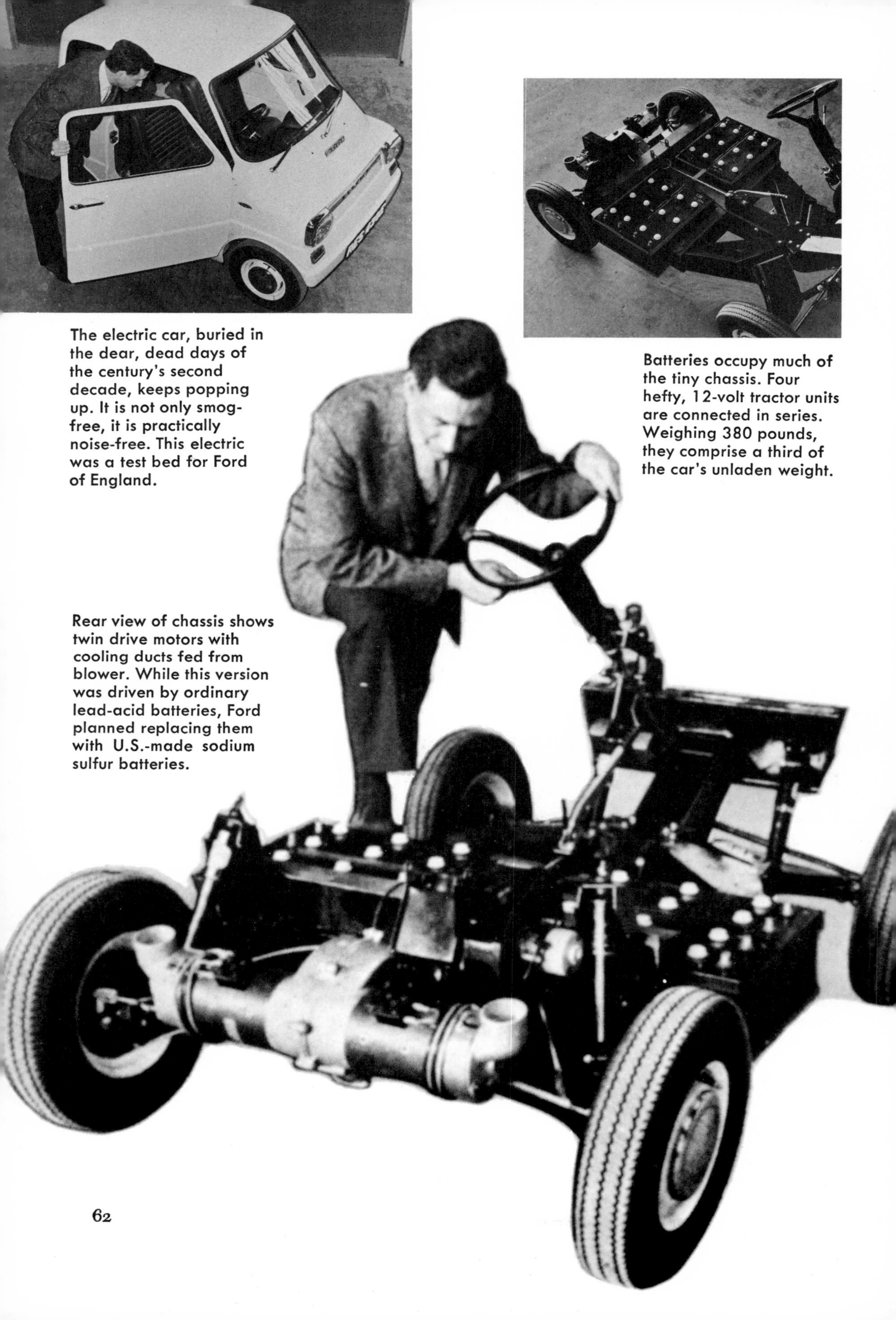

The electric car, buried in the dear, dead days of the century's second decade, keeps popping up. It is not only smog-free, it is practically noise-free. This electric was a test bed for Ford of England.

Batteries occupy much of the tiny chassis. Four hefty, 12-volt tractor units are connected in series. Weighing 380 pounds, they comprise a third of the car's unladen weight.

Rear view of chassis shows twin drive motors with cooling ducts fed from blower. While this version was driven by ordinary lead-acid batteries, Ford planned replacing them with U.S.-made sodium sulfur batteries.

By 1964 seat belts were standard on front seats, by 1966 on the rear ones, and, by government fiat, shoulder harness was mandatory two years later.

Detroit installed energy-absorbing bumpers, safety door latches, seat-back locks, non-glare rear-view mirrors for night driving, side-frame impact bars, padded dashes, recessed control knobs, padded steering-wheel hubs and, in some instances, collapsible steering columns. Selector patterns on automatic transmissions were standardized. Because most stolen cars—600,000 a year—are nabbed by accident-prone teenagers, buzzers were installed to warn a departing car owner if his key is left in the ignition. Steering wheels were made to lock on withdrawal of the key. Tires that sealed their own punctures were improved, almost doubling their life.

Safety legislation multiplied. A National Driver Registration Service required the cross-indexing of data on motorists whose licenses had been revoked for drunken driving or involvement in a fatal accident. States adopted annual-motor-vehicle-inspection laws. Congress passed an omnibus bill on highway safety and established a National Transportation Safety Board. The car industry was given notice that by 1976 it was expected to install in all cars air bags that would inflate on impact to cushion occupants against injury. Periodically factories began calling back thousands of vehicles to mend engineering goofs.

Despite all criticisms leveled at today's descendants of the horseless carriage, the love affair between man and machine has survived.

In his lifetime the average American makes a bigger capital investment in cars than on housing for his family. Four out of five families own one or more. U.S. cars rack up more than a trillion miles of travel a year. Nine out of each 10 miles of passenger travel between U.S. cities is by motor vehicle. Automobiles, plus buses, have cut deep into railroads' passenger business.

Of 1,500 manufacturers that made U.S. automobiles over a period of eighty years, only four, umbrella corporations, remain. Some 3,000 brands of cars have gone into limbo.

Imports have captured 16 per cent of the U.S. market.

If the electric auto refuses to die, so does the steamer. One inventor claimed he sank $10 million into its revival. The car above is —you're right—a VW with a 40 hp Mercury outboard in it converted to steam. Inventor William Smith said he got 250 hp.

Another steamer: with engine and boiler mounted forward. It's a V4—a Chevy V8 sawed in two. The aviation chapter of this book shows a steam airplane built by George and William Besler. The engine of this experimental Chevelle bore the same name.

The horseless carriage, born with a whip socket on the dash, whither is it drifting? Pictured is a 1959 traffic jam. Billions of dollars since have gone into thousands of miles of superhighways to serve more millions of cars. And the jams get worse!

World motor-vehicle production is more than 29 million units annually.

Worldwide, more than 233 million cars are registered annually; in the U.S., 112 million.

Popular Science, in April 1969, reported that the automobile's miles-per-gallon were at an all-time low.

Configurations in motor vehicles are almost endless. Added to the scores of different models in family cars, sports cars, and "sporty" cars with long hoods and short decks are motor scooters (especially in Europe), campers and motor homes (in the hundreds of thousands in the U.S.), snowmobiles, and land-and-water go-anywhere vehicles. Convertibles today are almost extinct.

Speeds that began before the turn of the century at less than 10 miles an hour are an allowable 85 on the more generous turnpikes. If a jet-engined land vehicle, streaking across Utah's Bonneville Salt Flats, can be called an automobile, the world record stands at 622 mph.

In 1969 fatalities ran to 10,000 more than all the deaths among U.S. soldiery in 10 years of the Vietnam war. The National Transportation Safety Board reported that fifteen-to-twenty-four-year-olds were involved in 31 per cent of all fatal accidents on the road.

None of the arguments over safety got at the nub of the problem. That has been all along, and remains, the need for stricter driver licensing and mandatory driver education.

There are portents for the future. As the need for safety devices in cars to protect man from his own foolishness has grown in complexity and expense, it seems likely that the family vehicle will shrink, rather than grow, in size.

The problem of automobile air pollution will be solved; for man's survival, it has to be.

Stricter driver licensing is inevitable.

Anomalous though it seems, the automobile may well be responsible for a growth in common-carrier transport—simply because highway construction can't keep pace with growth in car ownership. And railroads are already piggybacking cars of tourists long distances.

The hour of transition from horse to automobile for personal transport was caught in this nostalgic advertisement showing Ford's first car of 1896 in *Popular Science* in 1932.

Sixty-seven years after Ford's primitive "quadricycle," George De Angelis built this faithful replica. With a magazine staff member—bowler hat, red suspenders and all—at the tiller, it is pictured in front of Ford's first workshop.

through the existing gas pipes in a house.

Any home owner at all mistrustful of the newfangled electric light could play safe by installing combination gas-and-electric light fixtures. A Sears, Roebuck catalog of 1902 offered such a two-in-one brass wall fixture ($2.30) or a gas-and-electric brass chandelier ($3.00 and up). They would have come in handy for modern victims of electric-power blackouts!

In the 1890s the carbon-filament bulb received the finishing touch that made it practical. Squirting a viscous solution of cellulose into a precipitating solution yielded a synthetic fiber for the filament, with the great advantage of perfectly uniform thickness and electrical resistance. Since there were no thin spots to overheat and cause early failure, the squirted-cellulose filament gave bulbs a greatly prolonged life.

These were the turn-of-the-century carbon-filament bulbs that old-timers can remember from their youth—shedding light from a glowing loop that waved on long stalks within the bulb.

Then came the next big advance—tungsten filaments, whose heat-resisting metal could reach the high temperature needed for more light-producing efficiency. The hitch had been how to fashion the notoriously brittle metal into a lamp filament. That difficulty vanished when a way to produce ductile tungsten was developed by Dr. William D. Coolidge in 1910–12.

The resulting lamp bulb contained a tungsten filament that zigzagged between upper and lower supports inside the bulb. Through the bulb's clear glass, the naked filament glowed so vividly that you could close your eyes and still see the afterimage of its cage-like pattern.

Up until now, the vacuum in a bulb had been what kept a filament from burning up. A better way was now discovered—to fill the bulb with an inert gas, such as nitrogen, or the rare gas, argon. Because the gas helped to carry off heat that could damage the filament, it could be run hotter and give a brighter light. The idea was pioneered in 1916 by Dr. Irving Langmuir of the General Electric Company; and a panel of distinguished scientists chose him as the recipient of $10,000 and a gold medal, then the top U.S. yearly science prize, awarded by *Popular Science*.

Pictures of early lamp bulbs remind us of features we've long since forgotten. One was the sharp-pointed tip of an old-time bulb, formed when it was sealed off—a

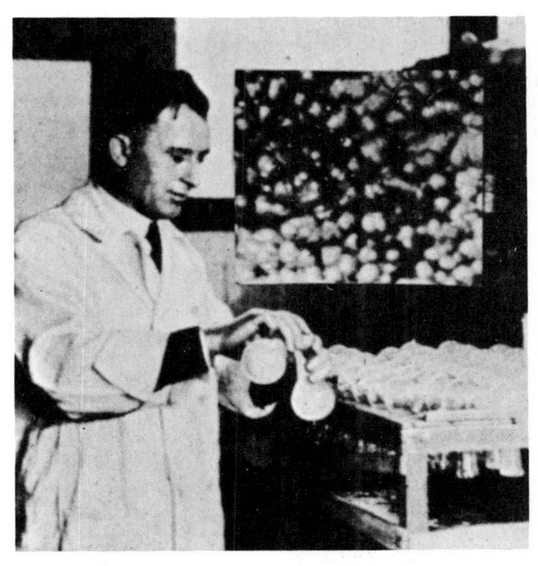

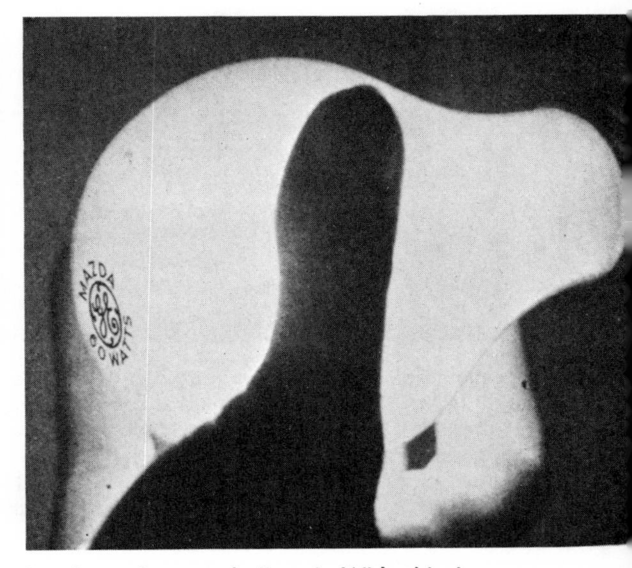

Inside-frosted bulbs that diffused light and banished glare were 1924 invention of Dr. Marvin Pipkin, pictured in his lab. Bulbs on rack were blown with long necks, to facilitate acid treatment that imparted frosting shown in photomicrograph (inset). With this innovation, a bulb's glowing filament vanished from sight (picture at right). Same inventor devised a white silica-powder coating that ultimately replaced inside frosting in 1949.

vulnerable point that led to breakage in handling. By the early twenties the fragile tip had vanished from sight and danger. Newer "tipless" bulbs were exhausted from the opposite end, leaving the tip hidden and protected in the base.

Next to disappear from view was the glaring filament itself—as frosted bulbs replaced those of clear glass. Experimenters first attempted to apply a frosting to the outside of a bulb, but that proved unsatisfactory. Not only did the outside frosting cut down the light, but it collected dust and was almost impossible to clean. The answer was "inside frosting," invented in 1924 by a chemical engineer, Dr. Marvin Pipkin.

Frosting the inside of a bulb was easy enough—etching the glass with strong hydrofluoric acid did the trick—but there was a catch. The etched bulbs were so fragile that they were useless. A lucky accident solved the problem.

To avoid wasting bulbs in his experiments, Dr. Pipkin was using a weak solution of hydrofluoric acid to eat away the etching and leave clear glass for another trial. One day a phone call interrupted him while he was doing this, he told *Popular Science*—and in answering he tipped over a half-cleaned bulb. It fell to the floor. Instead of shattering, as the experimenter fully expected, it rolled under his desk undamaged. And that proved the secret of inside frosting—to follow the strong-acid etching with a treatment of weaker acid, just long enough to round the sharp edges and thus strengthen the etched glass.

NATIONAL BUREAU OF STANDARDS PHOTO

First neon sign was luminous tube exhibited in 1904 at Louisiana Purchase Exposition in St. Louis. It was fashioned by a U. S. Bureau of Standard experimenter, Dr. Perley G. Nutting, and is now a museum piece.

Commercial neon signs, patented by Georges Claude in 1915, suddenly spread over U.S. in the late 1920s. Their scarlet glow soon was supplemented by luminous tubes giving blue, yellow, or green light, obtained with different gases—and, sometimes, with the additional help of tinted glass.

Tiny neon lamps like these found use in home and shop for electrical trouble-shooting and as economical, ultra-low-wattage pilot lights.

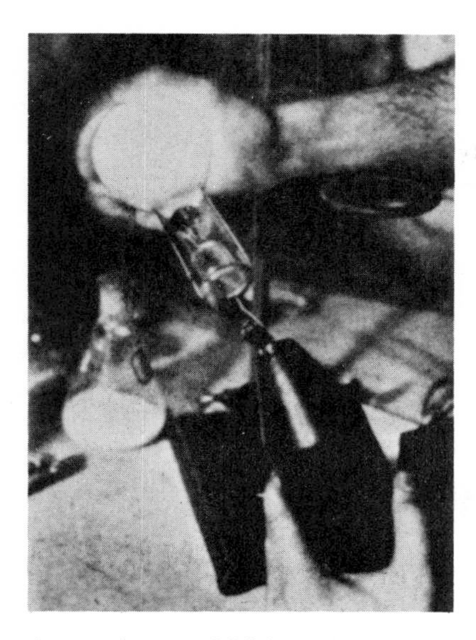

As early as 1934 a Brooklyn researcher was experimenting with this bulb-shaped predecessor of the fluorescent lamp—the next major development in the story of the evolution of electric lighting.

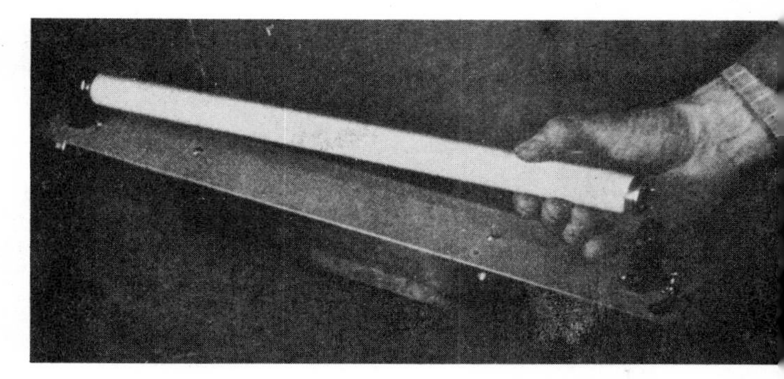

Fluorescent lamps made their public debut in 1938, at previews of expositions in New York and San Francisco. Like the 1939 model pictured above, they had now taken on the tubular form standard ever since. Their comparatively heatless light was nearest approach yet to "cold light" of the firefly.

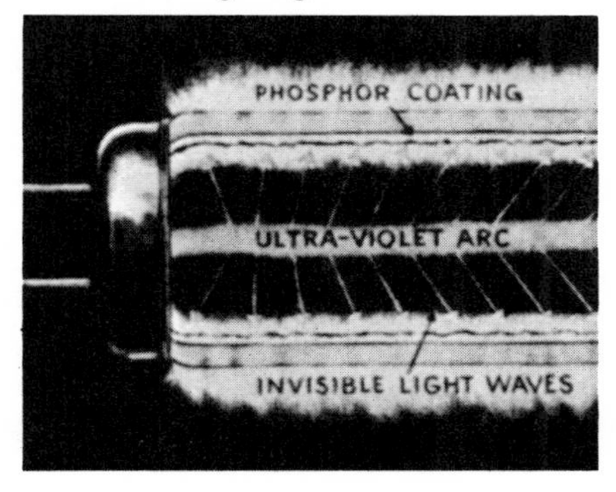

Light of a fluorescent lamp comes from inside coating of materials called phosphors. They glow when exposed to ultraviolet light from electric discharge along center of tube.

Phosphors in a fluorescent lamp are similar to those used in a TV tube, which glow to form a picture when struck by moving electron beam.

A 1939 version of an office ceiling fixture with fluorescent tubes foreshadowed their major use today in business offices. Success story of fluorescent lamps is told by fact that they now produce more than two-thirds of the country's electric light.

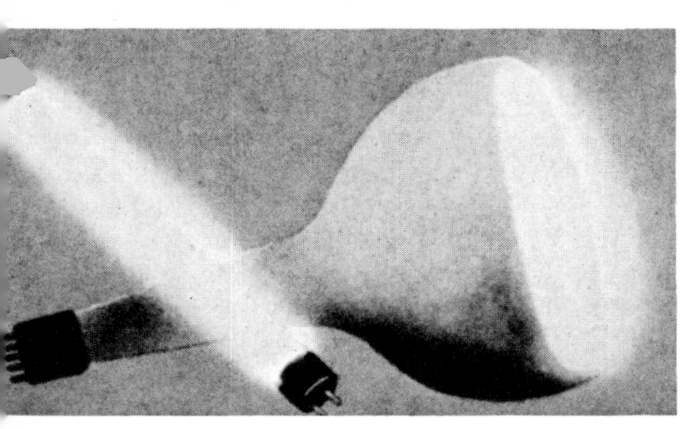

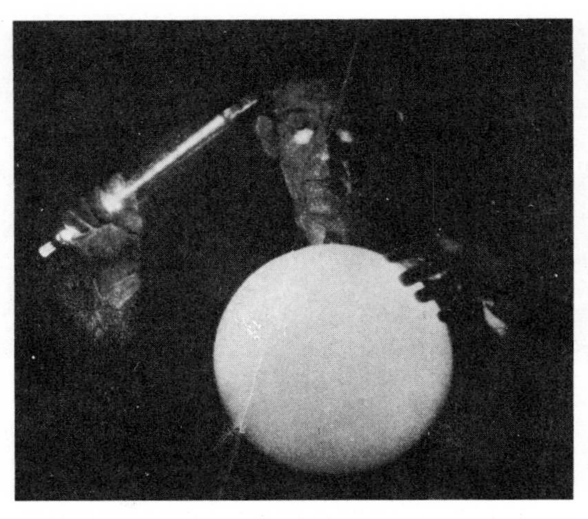

Wireless lamps, radio-powered from sources in walls and floor, light up in magic-like Westinghouse demonstration of 1944. In similar way, Nikola Tesla in 1892 dramatically created "flaming swords" of light.

Today, white silica coating of incandescent bulb at left has superseded inside frosting to diffuse light. New gas-filled incandescent bulb, right, uses desirable krypton gas—formerly considered too rare and costly.

Lucalox sodium-vapor lamp, introduced in 1965 for outdoor lighting by GE, is called the most efficient source of white light yet devised. Yellow hue of former sodium-vapor lamps has been altered to golden-white.
GENERAL ELECTRIC PHOTO

Naturally these advances of incandescent lamps had not displaced arc lighting from uses to which it was especially suited. Arc lamps continued to brighten streets. They found favor for powerful searchlights and for movie projectors in theaters. Their types multiplied—and plain carbon electrodes had minerals added, to produce the luminous vapor of "flame arcs." Automatic carbon-adjusting regulators became simplified and reliable.

Meanwhile, unconventional new kinds of electric light were developing—and it is time to turn back to look at their beginning. Collectively called "electric-discharge lamps," they had no incandescent filament but, instead, produced a glowing light when high-voltage electricity passed through a tube of rarefied gas.

They traced their ancestry back to a nineteenth-century laboratory plaything, the Geissler tube—a glass bauble fancifully twisted into ornate designs, and stimulated by a Ruhmkorff coil to glow in one color or another, depending on the gas that filled it. History records only one example of the use of Geissler tubes for lighting: a rainbow-hued display of them commemorated the Diamond Jubilee of Queen Victoria of England in 1897. By then, their prolific offspring were almost ready to appear.

First practical electric-discharge lamp was Peter Cooper Hewitt's mercury-vapor

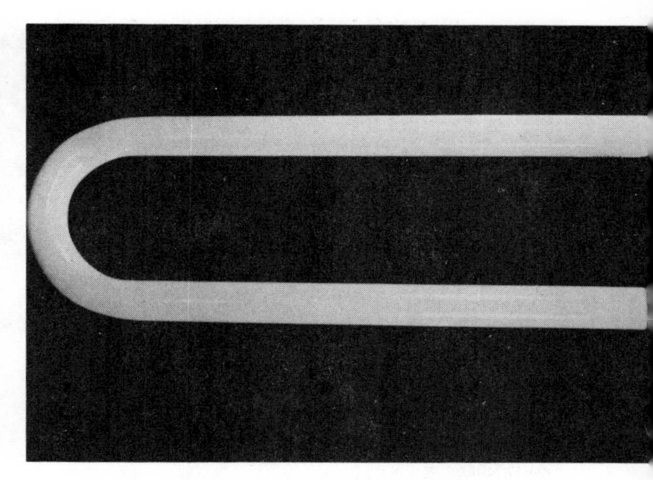

Present-day fluorescent lamps take on many shapes besides straight tubes, to tuck more light into compact space. "Waffle" fixture (upper left) contains tube that turns back and forth within it (lower left). Hairpin-shaped tube (right) and ring-shaped one are other designs. Increased variety of sizes ranges downward to a tiny, pocket-size 4-watter, for patios and walks, that gives as much light as a 100-watt incandescent bulb.

lamp, perfected between 1901 and 1912. Its long tube emitted blue-green light with phenomenal efficiency. Because of the ghastly appearance its color gave people's faces—among other reasons—it would never become a household joy. But factory owners welcomed with open arms its shadowless and economical lighting. When nighttime travels have taken you past an industrial plant aglow with eerie blue-green radiance, you've seen mercury-vapor lamps at work.

The neon lamp, another version, came soon after. As early as 1904 a neon tube was exhibited at the Louisiana Purchase Exposition in St. Louis by a U. S. Bureau of Standards scientist, Dr. Perley G. Nutting. Obviously its brilliant red color unfitted it for home lighting. But a Frenchman, Georges Claude, saw its spectacular possibilities for commercial electric signs—when it was provided with the right kind of transformer to supply and regulate the needful high-voltage current.

In 1910 Claude demonstrated a practical neon sign in Paris. His fundamental patent was granted in 1915, and by 1925 a number of firms licensed to use it had begun operations. Toward the end of the twenties, neon signs suddenly blazed out everywhere. Different gases and tinted tubes gave a variety of other colors, too.

For special purposes, other versions of electric-discharge lamps met with success. One was the sodium-vapor lamp, which found application for highway lighting from 1933 on. Like its relatives, it was disqualified by its color—a brilliant yellow—from household or office use.

Beckoning to inventors all this time was an alluring will-o'-the-wisp—"cold light," the mysterious and elusive light of the firefly. Tantalizing was the dream of an electric lamp that would not waste 93 per cent of its energy—mostly in heat, a fact painfully apparent when you burn your fingers on a hot bulb—but would turn electric current wholly or largely into useful light. Suddenly, in the 1930s, a new kind of lamp appeared that came nearer to "cold light" than any before.

The fluorescent lamp was here.

In the fall of 1935 a tubular 18-inch fluorescent lamp was exhibited to the Illuminating Engineering Society in Cleveland. Its vivid green light gave little hint of the promise of its type. At first it was regarded as a new colored-light source for purely decorative effects. By 1938, when fluorescent lamps were shown at previews of the first New York World's Fair and the Golden Gate International Exposition in San Francisco, they were offered to the public in white and "daylight" hues as well as in five colors—green, blue, pink, gold, and red.

Home above is floodlit by the tiny 250-watt bulb at left—a high-intensity quartz-iodine lamp, most advanced incandescent kind to date. Tungsten "boiled" off filament is caught and returned by iodine, in bulb so hot that quartz must replace glass.

Thirty years later, fluorescent lamps produced more than two-thirds of all the country's electric light.

The secret of this "firefly" kind of light lay in the property of certain materials called phosphors, when bombarded by invisible ultraviolet light, of glowing with visible light—a phenomenon known as fluorescence. A number of natural minerals, and many synthetic ones, possess this property. When the glass tube of a mercury-vapor lamp is coated on the inside with one of these powdered phosphors, its abundant ultraviolet radiation no longer goes to waste, but is turned into useful light by the phosphor coating. You have a fluorescent lamp. Without getting uncomfortably warm to the touch, a 40-watt fluorescent tube of today yields as much light as two 100-watt incandescent bulbs.

Actually, the idea of a fluorescent lamp was proposed in precise detail in *Popular Science* as long ago as 1880. It took so long to come about simply because of the practical difficulty of finding phosphors that fluoresced with sufficient brilliance—a search that met with success in the late 1930s.

True, the new fluorescent lamps brought some complications along with their advantages. They needed certain auxiliaries, special starters and "ballasts," before they could be connected to ordinary lighting circuits. Early fluorescent lamps required, too, a flow of current through their innards for several seconds before they would light up—an inconvenience remedied in 1944 by "instant-start" lamps, and in 1952 by simpler "rapid-start" lamps that light up almost instantly.

One early fluorescent-lamp phosphor, containing beryllium, was found to be a toxic hazard if released from a broken tube. Makers discontinued its use and successfully replaced it with others.

To some critics, fluorescent lamps had other drawbacks, which still persist:

At the end of their days, instead of going

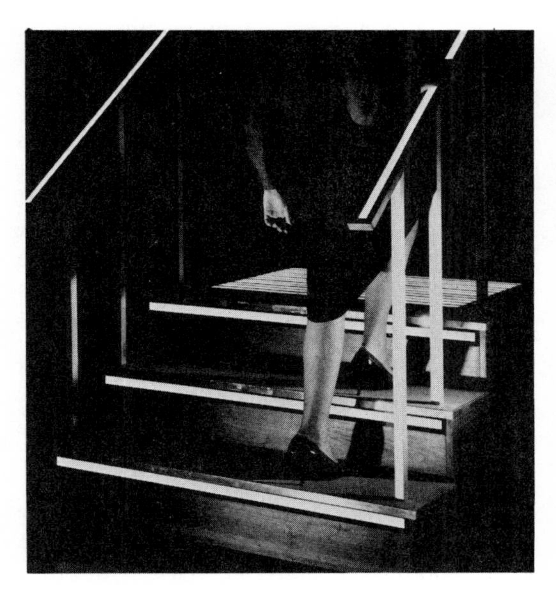

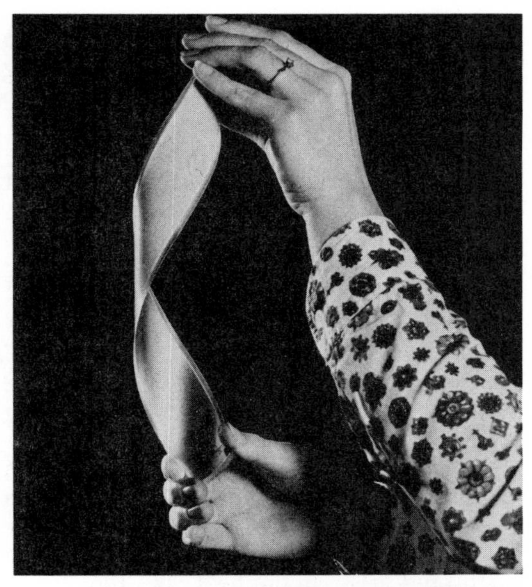

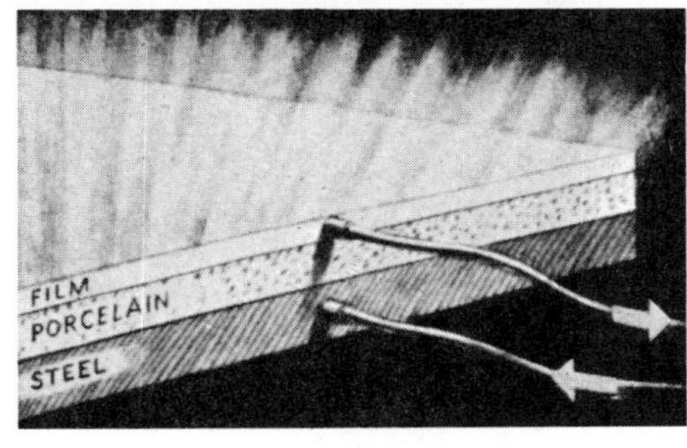

Electroluminescent lamps, a radical new kind dating from 1950, illuminate handrails and stair-tread edges of stairway above. Consisting of light-emitting phosphors sandwiched between conductive layers, they may be either stiff panels or flexible tape (upper right) that stays lit even if tied in knot. Diagram shows make-up of rigid kind: phosphor-containing porcelain backed with steel and faced with transparent film.

out with the finality of an incandescent bulb, fluorescent tubes had an eye-plaguing way of flickering on and off—with the attention-compelling effect of a flashing electric sign—until someone got around to replacing them.

Fluorescents were vocal, too. Their special electric gadgetry—commonly built into their fixtures—emitted a characteristic hum, which could be objectionably audible in an otherwise quiet room. Makers developed hushed-up models that they recommended for such locations; apparently, those responsible for some installations never got the message.

But these complaints were outweighed by fluorescent lamps' decisive advantages, their rapid acceptance plainly showed.

Those of us who work in office buildings have seen fluorescent lighting virtually take over for business quarters. While fluorescent lamps have invaded homes more slowly, they serve today to illuminate many a household kitchen, bathroom, and home workshop.

"Lamps of tomorrow" that magically lit without wires were exhibited in 1944 to a *Popular Science* reporter and cameraman by Samuel G. Hibben, Westinghouse researcher. They were powered by ultrashort radio waves from high-frequency sources that could be hidden in walls and floors, to free portable lamps of trailing cords. More than fifty years earlier, Nikola Tesla had dramatically demonstrated the same possibility, when he turned long vacuum tubes into luminous "swords of fire" by waving them between tinfoil sheets connected to a high-frequency generator. Both Tesla and Hibben proved far ahead of their time. The wireless lamps of "tomorrow" were seemingly for day-after-tomorrow.

But the story of electric light comes up to the present with a rich variety of other advances—refinements in all existing forms of electric lamps, the advent of one brand-new kind, the promise of another:

• Today's incandescent bulbs have something even better than inside frosting to diffuse their light—a white silica coating for their inner surface, developed in 1949 by the same inventor, Marvin Pipkin.

• Gas-filled incandescent bulbs are reported to burn more brightly and last longer when filled with the rare gas krypton—once considered too scarce and expensive, but now employed by a leading maker.

• Incandescent floodlights apply a new technological advance, the "iodine cycle," to operate at intense brilliance. Iodine molecules, added to inert gas surrounding the tungsten filament, intercept "boiled-off" tungsten molecules and return them to the filament before they blacken the lamp wall. Typically a lamp of this new kind is a tiny tube, made of quartz to withstand the high heat of the lamp. Moviemaking and outdoor floodlighting are among its uses.

• Fluorescent-lamp refinements crowd more light into compact space with hairpin-shaped tubes and others that double back and forth within "waffle" fixtures.

• A descendant of the sodium-vapor lamp claims the distinction of being the most efficient source of white light yet developed. A 1,000-watt lamp of this new Lucalox kind was announced in 1971 to have the fabulous light output of 130,000 lumens, as much as it would take about 800 100-watt household bulbs to provide.

• About 1950, an entirely new kind of electric lighting appeared, in the form of electroluminescent lamps—popularly described as "flat bulbs" or "ribbons of light."

Actually an electroluminescent lamp is no bulb at all, but a flat plate or a flexible ribbon—made up of two layers of electrically conductive material, with light-emitting phosphors sandwiched between. Apply an alternating current to the conductive layers (of which the outer one is trans-

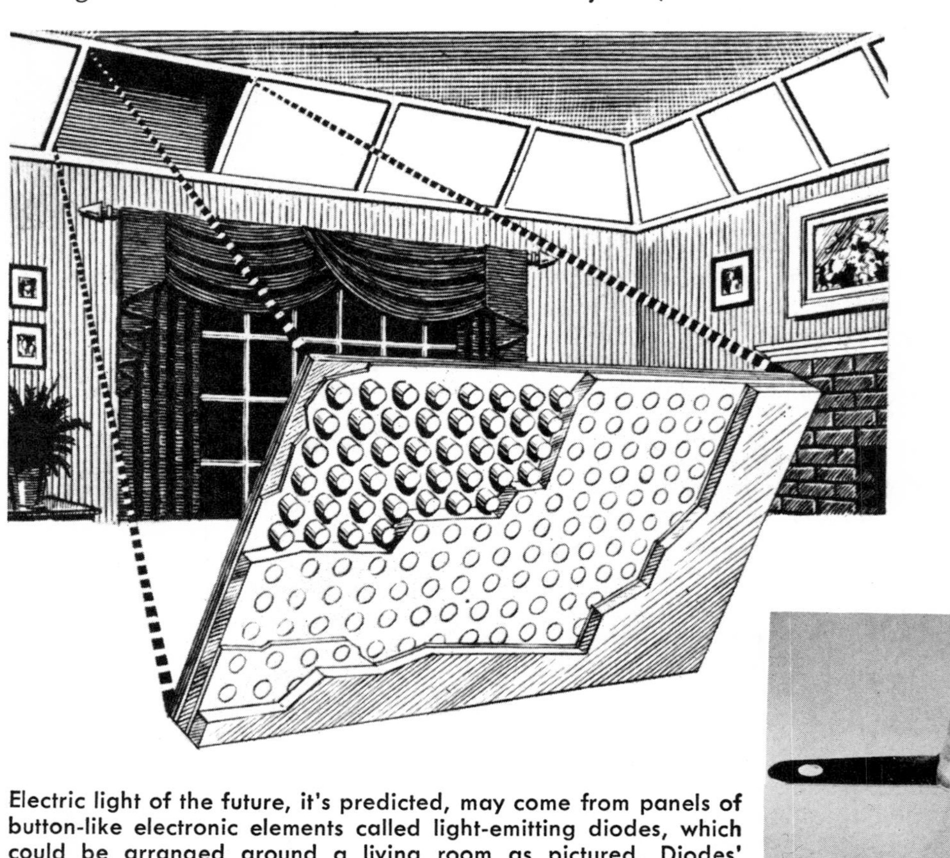

Electric light of the future, it's predicted, may come from panels of button-like electronic elements called light-emitting diodes, which could be arranged around a living room as pictured. Diodes' various hues—red, blue, yellow, green—could be mixed for over-all effect of white light. Red diode at right, for illuminating a photo darkroom, is example of one now available.

parent), and light from the phosphors shines forth—white, green, yellow, or blue, according to the kind of phosphor used. An electroluminescent lamp thus has the novel ability to turn electrical energy directly into light.

In the form of flexible tape, this new source of illumination literally offers "light by the yard." Using the luminous tape to outline the handrail and stair treads of a dark stairway has been one early application. Small solid panels serve as night lights, and arrays of larger ones have been experimentally used to light whole rooms. Fantastic decorative effects with light become readily feasible.

• And finally, it looks as if the future will bring still another rival to the electric lamps of today: electronic light, from light-emitting diodes. Introduced elsewhere in this book, in the chapter on electronics, these tiny luminous buttons have already become available for such special purposes as lighting a photographic darkroom. Panels of the buttons, suited to bathing a living room in light as pictured in an accompanying illustration from *Popular Science*, are in the future.

Coming, too, perhaps, for the future may be homes independent of the electric company and its central lighting station. Radical and promising new home generating plants already are well along in experimental development.

These experimental home outfits employ a fuel cell, an ultramodern kind of generator that converts a fuel—without burning it—directly into electricity. In one of the most advanced versions, no larger than a small domestic furnace, the fuel is natural gas. Thus it opens the possibility that electricity's one-time rival, gas, may have the last word after all—no longer for lamps, but for a domestic source of the electric current that lights them.

Along with electric light came an inno-

Will tomorrow's homes generate their own current? Ultramodern device to do it is this experimental fuel cell, as compact as a small home furnace, developed by Pratt & Whitney Aircraft Corp. Into it goes natural gas—and out comes electricity. Demonstration model is cut away for peek at its innards, where gas mixes with air and, in presence of a catalyst, reacts to produce current. Reportedly it could be readied to be marketed by 1975.

Marvels *of the* Electric Home

How New Inventions Utilize Electric Light, Heat, and Power in Every Room of the House

ELECTRICITY is changing our homes so swiftly and yet so smoothly that the younger generation today can scarcely conceive of homes as they were thirty years ago.

One marvelous device after another has been imagined, invented and perfected. One irksome household task after another has been lightened or eliminated. The electric washer and ironer are producing a new race of laundry-girls—young, trim, girls who "wash by machine only." The vacuum cleaner has brought order out of the annual chaos of spring housecleaning. The electric refrigerator promises to make the iceman a picturesque memory. Electricity cools us, warms us, and cooks our food.

Every nation has its ideal home. In the electrical home, it seems, America has found its own ideal. Partly responsible for this is our lack of groove-bound traditions; partly, the activity of our scientists, electrical men and inventors. At least the ideal we are offering the world is something entirely new—that of a home beautiful and comfortable, but, above all, easy to run.

It is not often given to inventors and scientists to see the direct fruits of their labors so strikingly benefiting their own generation. Even more inspiring is the thought that there are countless electrical inventions yet to be made

In the picture, our artist has suggested a few of the ways that electrical invention is changing traditional methods of doing household tasks. Even the children play with electric toys, while the housewife vacuum-cleans to radio music. Every room has benefited by electricity's service of light, heat and power

vation that changed our way of living no less profoundly—electric appliances for the home.

Before homes were wired for electricity, "domestic motors" to lighten household chores had been ponderous steam or gas engines or were quaintly turned by running tap water. Portable heaters were as primitive as a Sears, Roebuck offering—a wire-handled slab of soapstone, to be preheated on a stove top or in an oven.

With electricity on tap, a householder could enjoy the convenience of a host of appliances that rapidly appeared—electric irons, electric fans (as early as 1885), washing machines, toasters, percolators, vacuum cleaners, and many, many more.

Up to the 1920s, he plugged an appliance into the nearest light fixture. He had to unscrew a lamp bulb first—until a two-

Electric clock, keyed to pulses of alternating-current power, was innovation announced in 1928 item.

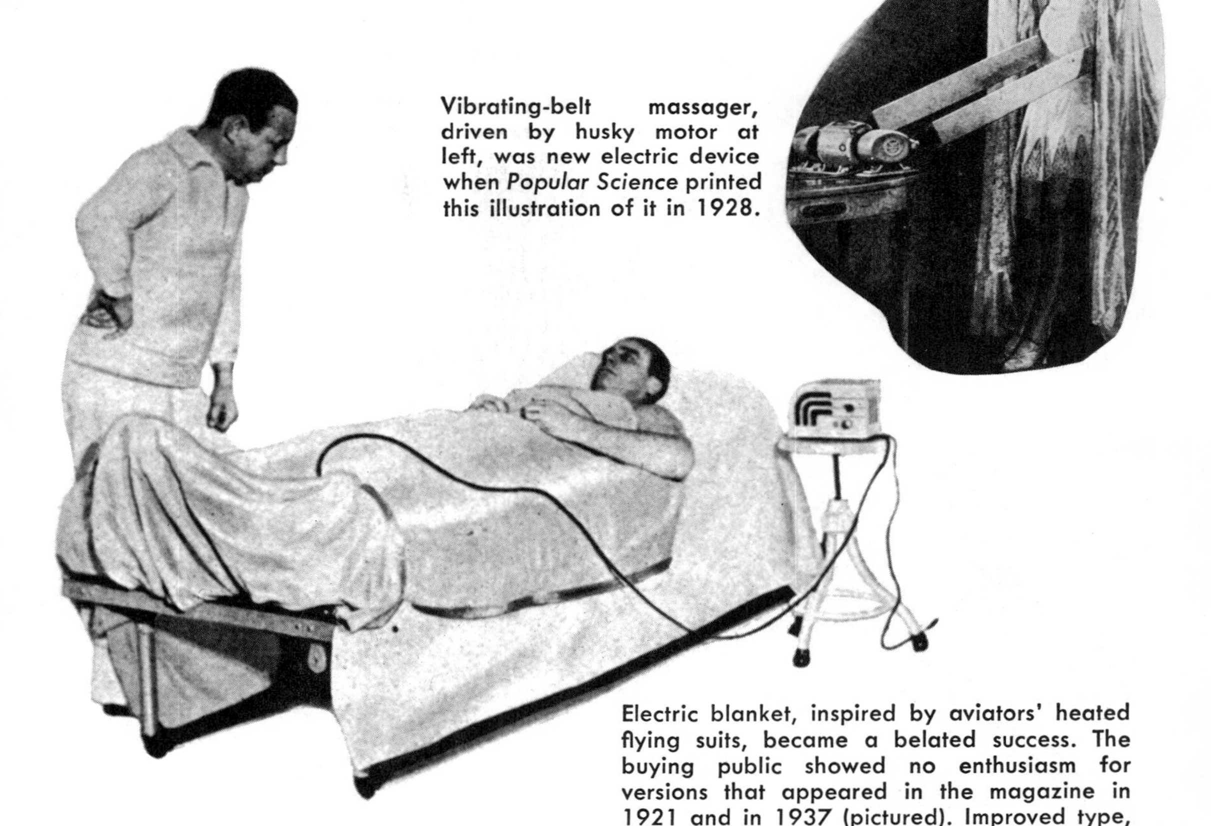

Vibrating-belt massager, driven by husky motor at left, was new electric device when *Popular Science* printed this illustration of it in 1928.

Electric blanket, inspired by aviators' heated flying suits, became a belated success. The buying public showed no enthusiasm for versions that appeared in the magazine in 1921 and in 1937 (pictured). Improved type, offered in 1945, caught on.

Electrical goodies for the 1926 home were illustrated by the contemporary *Popular Science* page reproduced opposite. Nine succeeding examples show how its text's promise of "countless electrical inventions yet to be made" has been fulfilled.

Postwar electric refrigerators of 1940s added a frozen-food locker at top, kept near zero by separate cooling system and shown in use in this picture of a 1945 Frigidaire.

Deep-freezers for homes arrived, too, in post-World War II years. This 1947 Sears, Roebuck model and other stored 800 million pounds of frozen food then bought yearly.

Controversial electric garbage disposers of 1950s, to shred waste and flush it down kitchen-sink drain, met with ban by some cities—but others favored their use.

Dish washer by 1958 was sophisticated machine with automatic flush-wash-rinse-dry cycle set by timer. Buyer could choose between type loaded at front (photo) or at top.

way attachment made a double socket out of a single one and spared him the trouble. "Every wired home needs three or four," said a 1920 *Popular Science* advertiser.

Wall outlets came soon after—but, at first, there was a catch. "Unfortunately, electric wall sockets are not uniform," chided the magazine in 1921. "Some are slotted to receive parallel blades; others call for perpendicular blades; a few are made for T-shaped blades; and some call for blades in line with each other." Finally makers settled on the kind in use today.

By the mid-1920s the "electric home" could boast the 21 aids depicted on a preceding page from *Popular Science*. The concluding pictures offer just a few examples of the many that have followed.

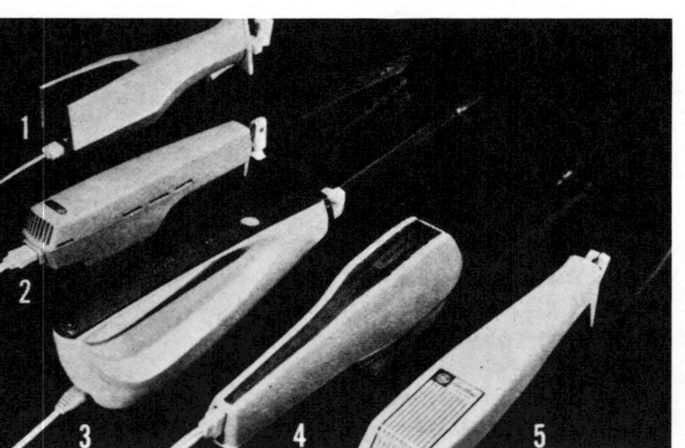

Electric carving knife, said to turn rankest amateur into an expert, first appeared in early 1963. Here is array of models put on sale by various makers soon after.

Room-type air conditioners of 1950s have by now been so beefed up that 1971 model pictured can cool a whole house.

CHAPTER 5

The Phonograph: From Tinfoil to Tape Deck

THE first words spoken into the first practical device for recording and reproducing sound, in 1877, were these immortal lines:

"Mary had a little lamb,
Its fleece was white as snow,
And everywhere that Mary went,
The lamb was sure to go."

The verse (written, in case you have forgotten, by Sarah Josepha Hale) has been inextricably entwined with the origins of the phonograph. Its unweighty character suggests the almost nonchalant way in which Thomas Alva Edison first regarded his tinfoil phonograph, which was to become his favorite invention and was to startle and astound the world.

The centuries saw many efforts to mimic human speech via "mechanical men" and some apparently had tolerable success. But no one was able to record and play back such sounds.

In 1857 Leon Scott developed his Phonautograph, probably the first practical means of recording sound—but with no way of playing it back. He used a cone-shaped ear-trumpet to collect sounds with a diaphragm stretched across the small end. The diaphragm, in turn, moved a bristle stylus, which traced the complex patterns of sound waves on a revolving cylinder coated with lampblack. But there the record was frozen—it could not be transformed back into sound.

(Curiously, the word "phonograph" was coined in 1863 in a patent by F. B. Wenby of Worcester, Massachusetts, for "The Electro Magnetic Phonograph," a complicated device that bore little resemblance to the "real" phonograph, and, as a matter of fact, was never built.)

As with so many other developments of the late nineteenth century, the phonograph grew out of parallel researches into the telephone and telegraph. In 1877 Edison was working on a telephonic repeater device, whose purpose would be to receive voice signals and then reinforce them so they could be transmitted to the next station. (Bell had invented the telephone one year before.)

Then one day in July, in a note at the bottom of a sheet of sketches describing his work, Edison penned excitedly: "Just tried experiment with a diaphragm having an embossing point and held against paraffin paper moving rapidly. The speaking vibrations are indented nicely and there is no doubt I shall be able to store up and reproduce automatically at any future time the human voice perfectly."

By August, Edison's thoughts had apparently ripened. Now he was after a recording and reproducing device for its own sake—not as part of an electrical telephone or telegraph system. On August 12, 1877, he handed John Kreuzi, one of his faithful assistants at his Menlo Park, New Jersey, laboratory, a sketch of the first phonograph, inscribed, "Kreuzi . . . Make this . . . Edison." (Drawing is reproduced on p. 84.)

If the sketch was crude, the inspiration was not. This earliest form of the phonograph remains one of the most elegantly simple of great inventions.

In 1877 Edison invented the tinfoil phonograph—first practical device for recording and reproducing sound. The original model was built by his assistant Kreuzi following sketch at bottom of page. Soon after, *Popular Science* reported event (see cover, left) with illustrations (below) of hand-cranked cylinder machine.

[PRICE 50 CTS.

No. LXXII.] APRIL, 1878.

THE
POPULAR SCIENCE
MONTHLY.

CONDUCTED BY E. L. AND W. J. YOUMANS.

CONTENTS.

		PAGE
I.	Evolution of Ceremonial Government. III. Mutilations. By HERBERT SPENCER	641
II.	The Eucalyptus in the Future. By Prof. SAMUEL LOCKWOOD. (Illustrated.)	662
III.	Introduction and Succession of Vertebrate Life in America. By Prof. O. C. MARSH	672
IV.	The Wicked Weasel	697
V.	The Dissipation of Energy. By GEORGE ILES	701
VI.	Illustrations of the Logic of Science. IV. The Probability of Induction. By C. S. PEIRCE	705
VII.	On Edison's Talking-Machine. By Prof. ALFRED M. MAYER. (Illustrated.)	719
VIII.	The Marpingen Miracles	724
IX.	The Source of Muscular Power	72
X.	Living Corals. By W. E. DAMON	
XI.	Poisons of the Intelligence.—Chloroform. By CHARLES RICHET	
XII.	Sketch of Professor Secchi. (With Portrait.)	
XIII.	CORRESP	

...ctrical Science—Mr. Wallace and Cl...
...onograph...
...rs of Chauncey Wri...
...of Muscular Pow...
...'s What was He?...
...th American Ethn...
...honograph—Garden...
...Plants—New Process of...
...Causes of the Chinese Famin...
...tning-Rods attract?—Cookin...

NOTES...

NEW YORK:
D. APPLETON AND COMPA
549 & 551 BROADWAY.
1878.

COPYRIGHT BY D. APPLETON & CO., 1878.

In this laboratory at Menlo Park, Edison, at right, operates his phonograph before a bedazzled group, one of the first public exhibits—which were to create a sensation.

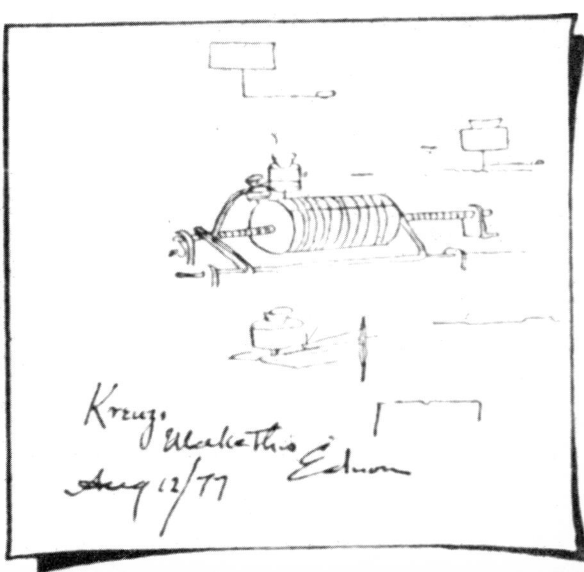

It consisted of a grooved cylinder mounted on a feed screw and covered with tinfoil, which was to be turned rapidly with a hand crank. The spoken voice actuated a diaphragm attached to a stylus, which indented an undulating trace of varying depth in the soft metal foil (a *vertical,* or "hill-and-dale," method of recording rather than a lateral one), as the drum revolved and inched past the stylus. For playback, the groove made in the tinfoil was simply allowed to wiggle a stylus attached to a diaphragm—the reverse of the recording procedure.

When Kreuzi brought Edison a finished model, the inventor spoke into it the deathless doggerel and then played the cylinder back. He was amazed to hear his own words repeated clearly, not in the hoarse murmur he had wistfully expected. Years later, Edison described his reaction this way: "I was never so taken aback in my life—I was always afraid of things that worked the first time."

Edison's tinfoil phonograph was first shown to the press in December 1877, and public demonstrations soon followed. The reaction was instantaneous and delirious. *Popular Science,* in an editorial in its issue of April 1878, called the new invention the "acoustical marvel of the century . . . as simple as a grindstone."

In the same issue appeared both a complete technical explanation of the apparatus (from which some of the illustrations shown here are drawn) and also "Anticipations Concerning the Phonograph" by Dr. William F. Channing, a Providence, Rhode Island, savant who had been helpful in developing Bell's telephone. Here are some of his surprising prophecies:

"We shall have galleries where phonotype sheets [his name for records] will be preserved as photographs and books now are. The utterances of great speakers and singers will there be kept for a thousand years. . . . Certainly, within a dozen years, some of the great singers will be induced to sing into the ear of the phonograph, and the electrotyped cylinders thus obtained will be put into the hand organs of the streets, and we shall hear the actual voice of Christine Nilsson, of Miss Cary, or even of Miss Jenny Lind and Alboni, ground out at every corner!

Edison's portrait hallmarked Blue Amberol records. The cylinder slipped onto a revolving drum and reeled off 4 minutes of sound.

As the country became phonograph crazy, cylinder models like this 1915 Amberola flourished.

First disc player, Berliner's 1887 Gramophone (above), was followed by these 2 Victrolas of 1900 (used in trademark) and 1919.

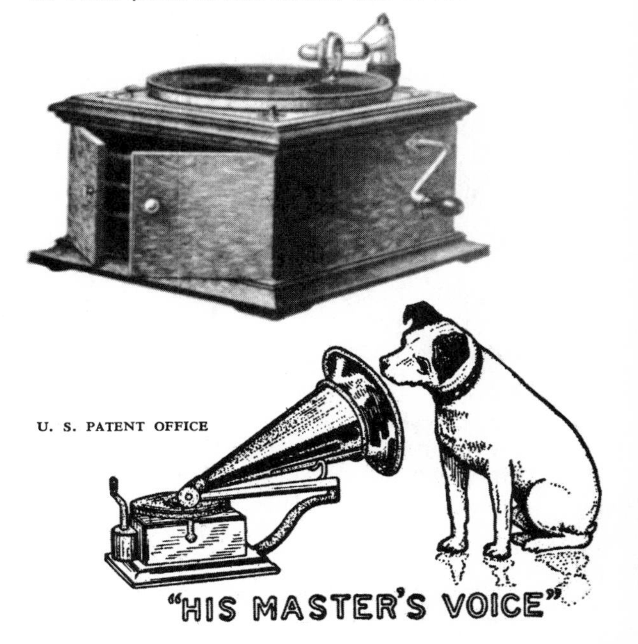

has be-
une, $8,-
opolitan
City, and
s given
sity for
arnegie's
tish uni-
sum for
tly been
at he is
nce for

agazine,
the Car-
r. Car-
he trus-
ary were
draw up
nd they
e, which
commit-
obert H.
ngineer-
y; Pro-
the Col-
sity of
Gray, of
ate, and
the Ar-
outlines
he whole
ughness,
hool and

THE TELEGRAPHONE.

Mr. Poulsen, of Copenhagen, has given the name telegraphone to an instrument in which he has most ingeniously combined the telephone and the phonograph. Its general construction will be understood from the illustrations, originally published in the London 'Electrician.' The details of the two instruments differ, a short wire being used in the one and a long steel ribbon in the other, but the general principle is the same. The steel wire or ribbon passes before the poles of an electro-magnet in a telephone circuit, and is thus magnetized in a manner varying with the current in the telephone circuit produced by the voice of the speaker. When the steel wire is then passed over the poles of an electro-magnet, the same undulations will be set up in the current passing through its coils, and the sounds will be reproduced in the receiver. The reproduction is as definite as in a good telephone and much superior in quality to that of any form of phonograph. The record can be used as often as desired, and is said to last indefinitely, but it can be wiped out by passing the wire over an electro-magnet. The wire can be passed over any number of re-

Ancestor of wire and tape recorders was the Telegraphone, an 1899 invention. Text excerpt and adjacent pictures are from 1901 issue of *Popular Science*.

Wire-recorder Telegraphone put sounds on a steel wire by passing it before poles of electromagnet in phone circuit. Electromagnet and receiver played back magnetized wire.

Tape-recorder version of the Telegraphone used a steel ribbon in just same way that other machine employed wire. Recordings of both of them could be magnetically erased. The originator was a Danish inventor, Valdemar Poulsen of Copenhagen.

Ancestor of record changers, circa 1918, played cylinders one after another without changing or rewinding. Item said it "does everything but provide smokes and drinks."

Acoustic recording of bygone days (here 1920) was strictly low-fi—artists bellowed into the horn. Behind horn: a diaphragm lever, jeweled point, and recording disc.

From the Laboratory
of
Thomas A. Edison
Orange N.J. Aug 14 1918

In reply to your question, let me say
was the first person to speak into the
graph. The first words spoken by
the original model and that were
were "Mary had a little lamb"
other three lines of that verse

Yours sincerely
Thos A Edison

Edison in 1921 dictates into his Ediphone office machine. His 1918 letter to telephone pioneer J. J. Carty, reproduced at left, confirms that the first words he spoke into his tinfoil phonograph were "Mary had a little lamb. . . ."

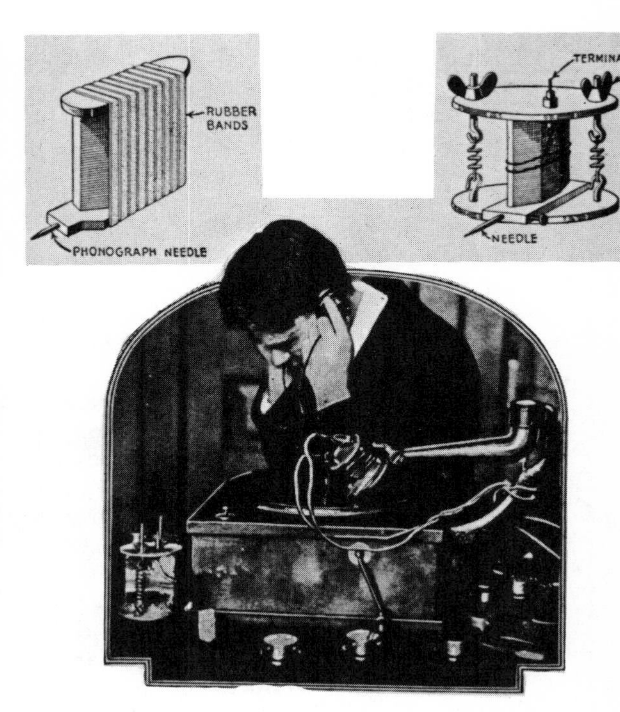

This 1921 audiophile is listening to a crystal reproducer, the forerunner of the modern electrical phono pickup. Crystals generated electricity. Two versions of device appear.

Caruso immortalized
A vast heritage of arts and literature has been bequeathed to the world by the passing centuries, but it remained for the Victrola to perform a similar service for music.
It has bridged the oblivion into which both singer and musician passed. The voice of Jenny Lind is forever stilled, but that of Caruso will live through all the ages. The greatest artists of the present generation have recorded their art for the Victrola, and so established the enduring evidence of their greatness.
There are Victrolas from $25 to $1500. New Victor Records on sale at all dealers on the 1st of each month.
Victrola
REG. U.S. PAT. OFF.
"HIS MASTER'S VOICE"
REG. U.S. PAT. OFF.
Victor Talking Machine Co.
Camden, New Jersey

Victor
"HIS MASTER'S VOICE"
Any time is dancing time wherever there is a
Victrola
Whenever you feel like dancing, when a few friends stop in, when soldier and sailor boys are home on furlough, the Victrola is always ready with the music.
Music so superb as to take the place of an orchestra, and yet so accessible that you can have an impromptu dance at any time.
In camp and on shipboard the Victrola enables our boys in the service to have their little dances, too.
Everywhere the Victrola and Victor Dance Records are a constant invitation to dance—a source of keen wholesome pleasure.
Hear the newest Victor Dance Records today at any Victor dealer's. He will gladly play any music you wish to hear and demonstrate the various styles of the Victor and Victrola—$12 to 950.
Victor Talking Machine Co., Camden, N. J.
Berliner Gramophone Co., Montreal, Canadian Distributors

Ads for Victor records and Victrolas extol the treasures and pleasures of the phonograph, from great artists eternally recreated to the joy of jazz summoned at will.

"In public exhibitions, also, we shall have reproductions of the sounds of Nature, and of noises familiar and unfamiliar. Nothing will be easier than to catch the sounds of the waves on the beach, the roar of Niagara, the discords of the streets, the noises of animals, the puffing and rush of the railroad-train, the rolling of thunder, or even the tumult of battle."

For a generation used to the sight of individuals with "transistors" perpetually fixed to ear, it is hard to imagine a world in which there was absolutely no "canned" sound and hard therefore to envision the sensation that the phonograph produced. In fact, the invention was a moneymaker from its earliest days, in striking contrast to the telephone, which had terrible financial problems for years.

To capitalize on the enormous popular interest that had been aroused by the first demonstrations of the phonograph, a group organized the Edison Speaking Phonograph Company on April 24, 1878. The company, had five stockholders, including Alexander Graham Bell's father-in-law and others intimately involved with the telephone. (Both the telephone and phonograph companies shared the same New York offices.) It paid Edison $10,000 for his tinfoil phonograph patent and guaranteed him 20 per cent of the profits.

In its first year the company arranged public demonstrations around New York, for which an admission was paid. Later, entrepreneurs split the nation into territories and leased demonstration rights. For the most part, the phonographs used in these demonstrations were on lease, not sold. Machines leased to the public were mostly used in business for dictation.

Here is part of an early advertisement for the Edison Speaking Phonograph Company:

"The first recorded attempt to make a Talking Machine was 2,600 years ago; though diligent efforts have been made ever since, it remained for Prof. THOMAS ALVA EDISON, of Menlo Park, N.J., to finally solve the problem, and place within the reach of everyone a machine that not only talks, but will record sounds of all kinds, and REPRODUCE THEM INSTANTLY, with FIDELITY AND DISTINCTNESS.

"The adaptation of this wonderful invention to the practical uses of commerce not having, as yet, been completed, in all its mechanical details, this company is now prepared to offer to the public only that design or form of apparatus which has been found best adapted to its exhibition as a novelty.

"The "PARLOR SPEAKING PHONOGRAPH" is intended for use in the parlor or drawing room, and will hold 150 to 200 words. The cylinder is so arranged that the foil can be taken off and replaced at any future time, thereby reproducing the same sounds that have been imprinted upon it. It speaks loud enough to be heard in any ordinary room. [The original announcement of the phonograph said it could be heard up to a distance of 175 feet.] We have a limited number now ready which we will sell for $10 cash, packed for shipment, with all needed appliances ready for use."

For some years thereafter, Edison did little to improve on his invention, which, elegant and sensational as it was, did leave much to be desired. For one thing, the reproduction was relatively poor, and the capabilities, limited to speaking into the machine and hearing yourself back, were bound to pall upon the user sooner or later. Edison realized these limitations, but was immersed in a demanding and—to him—far more important project: the development of electric lighting. The result was that he allowed other inventors to gain a march on him.

Bell was able to take up the challenge through a quirk of history. Some years earlier, Napoleon III of France had established a prize for scientific accomplishment in honor of the Italian electrical genius Alessandro Volta. In 1880 Bell won the award—the sumptuous amount of $20,000. Because the telephone was yet to turn a profit for him, this was just what Bell needed to go on with his researches. He opened a laboratory in Washington, D.C., and summoned his cousin Chichester and Professor Charles Sumner Tainter to form the Volta Laboratory Associates.

The triumvirate developed the Graphophone, an instrument that looked almost exactly like Edison's phonograph and operated in exactly the same way, with one important exception. It used cardboard cylinders coated with wax, and a stylus that *incised* a groove of varying depth into the wax. This was an advantage over *in-*

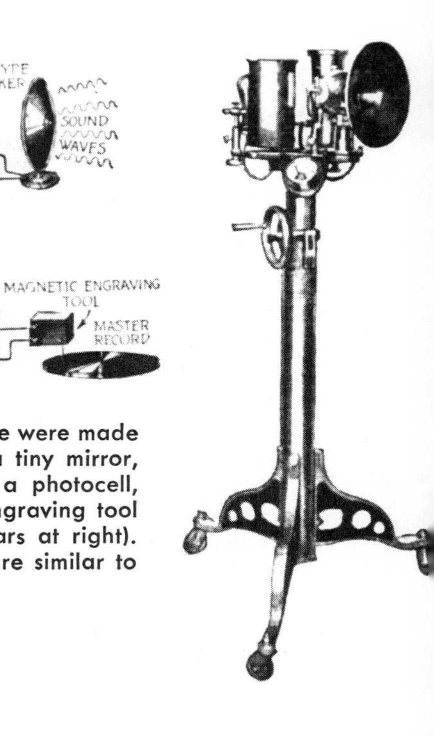

Recordings for the Panatrope were made when sound waves struck a tiny mirror, whose vibrations actuated a photocell, eventually controlling an engraving tool (recording apparatus appears at right). Playback techniques (top) are similar to those used today.

The shortcomings of acoustic recordings were swept aside by this "remarkable new electric phonograph" of 1926. Dubbed the Panatrope, it was a joint product of General Electric, Westinghouse, and Brunswick-Balke-Collender. The since-abandoned "light-ray" recording method is shown above, right.

A long-playing record in 1927? Yes—a 40-minute 12-inch disk invented by Thomas Edison, whose son Charles (at left below) exhibits it with its special reproducer.

The early thirties heard the resin-coated paper "Hit-of-the-Week" record bought at newsstands (left). Superior vinyl records (below at right) made the scene in 1945.

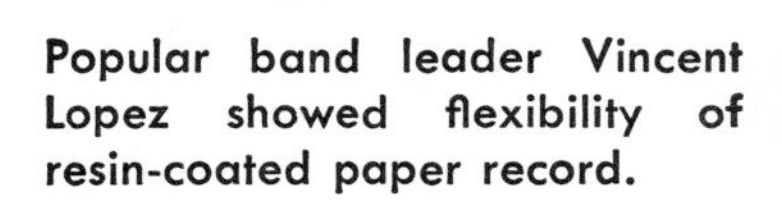

Popular band leader Vincent Lopez showed flexibility of resin-coated paper record.

denting tinfoil; the resulting reproduction was far clearer and the wax cylinder was easier to handle. This and other events spurred Edison, beginning in 1886, to improve his phonograph, and he did so. In 1888 he marketed a much more sophisticated design using an all-wax cylinder.

The Graphophone and the Edison phonograph continued to compete into the 1900s, under a system of intertwined corporate agreements and associations so complex that it resembled an octopus fight. The important thing is that the phonograph craze was launched.

In England, Edison had appointed as his agent a Colonel Gouraud, who began selling the machines in London in 1888. In true publicist's spirit, he persuaded celebrities of the time including P. T. Barnum, William Gladstone, and Oscar Wilde to speak into his phonograph and record messages for Edison on the cylinders. Here are some samples:

William Ewart Gladstone: "The request that you have done me the honor to make, to receive a record of my voice, is one that I cheerfully comply with, so far as it lies in my power, though I lament to say that the voice I transmit to you is only the relic of an organ, the employment of which has been overstrained. I offer to you as much as I possess and so much as old age has left me, with the utmost satisfaction, as being at least a testimony to the instruction and delight I have received from your marvelous invention."

Florence Nightingale: "When I am no longer even a memory, just a name, I hope my voice brings to history the great work of my life. God bless my dear old comrades of Balaklava, and bring them safe to shore."

And from Sir Arthur Sullivan, the most noted English composer of the day, in the earliest of these recordings to survive, made in Colonel Gouraud's home, "Little Menlo," comes this telling appraisal: "For myself I can only say that I am astonished and somewhat terrified at the results of this evening's experiments. Astonished at the wonderful power you have developed, and terrified at the thought that so much hideous and bad music may be put on record forever."

And, indeed, the reaction to the "acoustic marvel of the age" was not unmixed. Dogs, especially, seemed unappreciative, some going wild, others merely diving into the horn to get at the mysterious, disembodied voice. Simple, country folk thought that recordings were unnatural and therefore probably inspired by the devil. Many were offended by the fact that "spicy" songs were being brought out of the music hall and into the home and were bringing blushes to the cheeks of the young ladies who typically staffed record stalls.

In one case at trial in 1904, a man was accused of stealing a horse and selling it to buy a phonograph. The defense argued that the man was not in full possession of his faculties and therefore could not be held accountable. The judge agreed, stating that anyone fool enough to buy a phonograph was obviously of "weak intellect."

And one man refused to believe that the device was other than defective because when he heard his voice played back it sounded "affected and obnoxious."

In 1896 Edison placed on the market his first spring-motor phonographs (earlier models had been turned by hand or driven by battery-powered electric motors). By this time, there were thousands of music cylinders available and sales of the new machines were excellent.

There were, however, many problems that threatened to still the phonograph in its infancy. Probably the most important was the inability to mass-produce the recordings on cylinders. To make multiple copies of his performance, a singer, say, would bellow into the horns of an array of recording phonographs ranged on shelves before him. Obviously, the number of "originals" that could be made this way was very small.

As the next step, each original could be copied, but only by playing it into another machine, a one-to-one, time-consuming operation and one that also resulted in a decline in quality with each transfer. Moreover, the master cylinders did not last very long. A singer, instrumental soloist, or band with a popular number destined to sell in the thousands would have to record masters day in and day out for eight hours a day in order to fill the demand.

Edison was working on a method of making duplicates by a molding process, but apparently the molding of cylinders was too formidable a technical problem to become practical in commercial quantities until 1901.

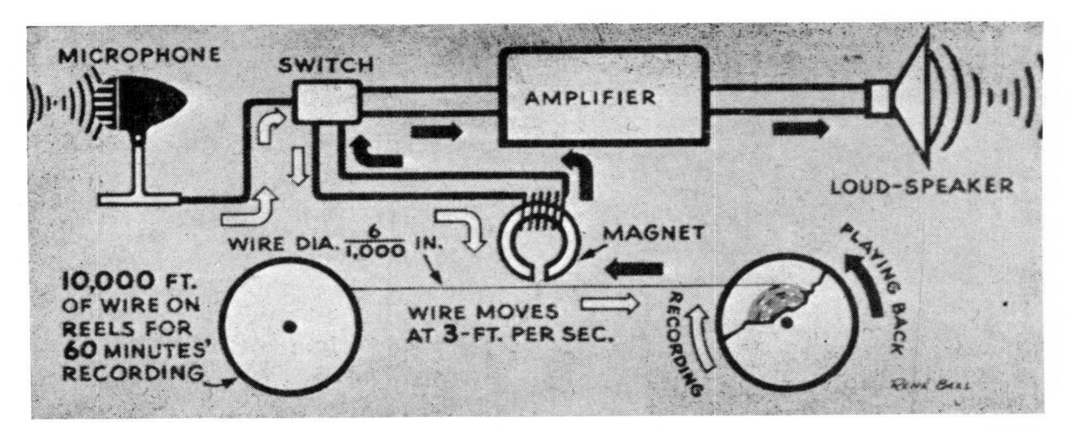

Wire recorders made it big in World War II. This 1946 how-they-work diagram indicates how little they departed from the principles of 1899 Telegraphone. Sound is picked up by a mike, fed to head that magnetizes wire from reel. Magnetism on wire varies with sound fluctuations. For playback, process is reversed; sound emerges from speaker.

In the meantime, a formidable contender had made its appearance on the phonographic-record scene—the disc.

Discs had been conceived of by several of the pioneer inventors of the time, including Edison and the Bell-Tainter group. But it was Emile Berliner, a German-American retail-store clerk, who was the first to produce such records commercially and the first to turn out stamped or molded recordings commercially.

First wire recorder to reach the home market after World War II was this St. George unit. Below is a 1947 tape machine—the Brush. Tape supplanted wire in home recorders.

Berliner began tinkering with discs in 1887 and developed a model of a machine, the Gramophone, to reproduce sound from the discs the next year. It was a hand-operated machine and, instead of the hill-and-dale or vertical recording method used in both Edison's phonograph and the Graphophone, it used a groove made with a horizontal, sideways motion of the cutting stylus.

By 1949 there were 3 speeds to play records at—78 rpm, the old standby; 33⅓ rpm, Columbia's long-playing entry; and RCA's 45s. Below: same music 3 ways.

One advantage of this system was a sound level far greater than that obtainable from cylinders of any kind or size. This let the listener hear through a funnel rather than tubes placed in the ears. In 1889 a firm of German toy manufacturers produced a model that played discs of 5-inch diameter at a speed of 150 revolutions per minute.

A contemporary account of 1891 describes Berliner making a recording in his Washington, D.C., workshop. Dressed in a "monkish frock," he was found speaking

Binaural records of the early fifties had a pair of tracks, one inside the other. Both tracks were played simultaneously by a double-headed tone arm. This version of stereophonic recording never really caught on.

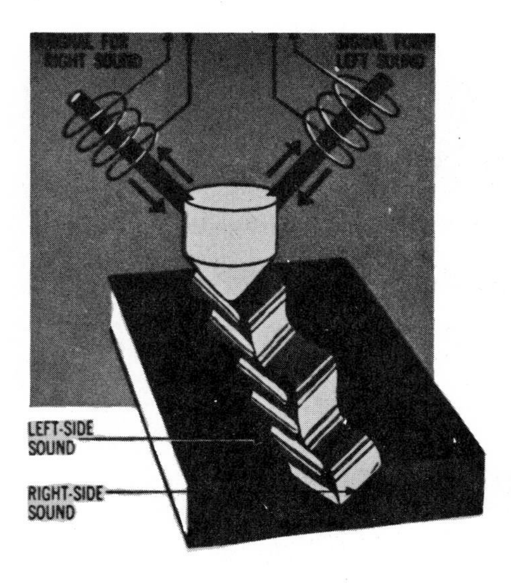

Modern stereo record packs 2 sound tracks into 1 groove by using each side of groove separately. This 1958 sketch shows the left sidewall with high frequencies (ripples close together) and the right sidewall with lower ones (ripples farther apart). Ripples in left wall move stylus along line at 45 degrees to record's surface, to right. Ripples in right wall wiggle stylus along line at right angles to first line. Result of combined pushes on stylus is complex stylus motion that is "decoded" by devices in pickup that convert motion to electricity. They separate out original left and right tracks for separate reproduction.

into a diaphragm. Apparently with the example of Edison's epic first phonographic words firmly in mind, he chanted, in guttural, accented English, "Tvinkle, tvinkle, liddle shtar, How I vunder vot you are."

Berliner originally used zinc discs, coated with wax, for his masters. After the recording process cut a wavy "groove" into the wax coating, the disc was dipped into a bath of chromic acid, which etched the exposed parts of the zinc. Duplicates were then pressed from vulcanite—hard rubber.

By 1898 both cylinders and the new mass-produced discs were freely available. The Gramophone had been equipped with a spring motor, and the prices of the various competing machines had come down enough—and the quality of recordings had risen enough—to take the device out of the category of a mere toy or a business dictating machine.

Then Berliner took the step that sounded the death knell for cylinders, although they continued to be issued for decades. The zinc-etch process had proved less satisfactory in musical quality than Edison's new wax cylinders, and the hard rubber used in making duplicates developed gases in the hot molding, resulting in flat spots. Berliner found a substitute material for the hard rubber in a button material composed mostly of shellac, lampblack, barytes, and cotton flock as a binder.

This material proved vastly superior to the vulcanite and was used to make standard 78s into the 1950s. He combined this with a new, all-wax master disc that far excelled the zinc masters in quality. The result was a high-quality recording medium capable of mass reproduction.

In 1902 Edison, Berliner, and Bell and Tainter, each of whom held the most important phonographic patents, pooled their patents to allow continued progress for the fledgling invention and keep its wings from being clipped.

The advent of Berliner's Gramophone discs (eventually the word became generic and lower-case) assured the continued success of phonography, to which poets sang their paeans of praise.

And a writer in *Popular Science* in 1919 notes: "Years ago . . . the parlor organ was the principal source of music in the home . . . Today, however, the phonograph holds sway. . . ."

These early years of the phonograph were filled with the activity of ardent gramophonists who remarkably resemble today's breed of audiophile in the incessant search for technical perfection. Enthusiasts argued over the merits of one or another machine, the dimensions of acoustic horns, the superiority of needle materials. Then, as now, many a wife complained bitterly of a husband who could think of nothing but "his new toy." And, believe it or not, Edison and others carried out live versus recorded comparisons in which the audience couldn't tell the difference!

Some of the more daring technical experiments were in vain. One Professor MacKendrick, for example, devised a filter to eliminate record scratch by passing the phonograph's sound through 40 feet of tubing filled with peas. Unfortunately, the filter removed not only the scratch but every last decibel of the recorded sound as well.

There was a real technical revolution in the wings, however. Until 1925 the whole recording and playback system was a purely acoustical one in which mechanical vibrations governed. This kind of system is extremely limited in frequency response and volume. Deep bass is lost on the low end, and sibilants and overtones on the high end; and really high volume is difficult or impossible. This handicap was finally offset with the advent of electrical recording and reproduction.

Electric recording grew out of the techniques of radio broadcasting, which began in 1919 (and incidentally threw record sales into a nose dive in the early 1920s). It depended on converting mechanical energy to electrical (via the microphone) and vice versa, using the vacuum-tube amplifier as a means of controlling and amplifying the vibrations of sound.

In heyday of the 78s, the Capehart was the reigning monarch of record changers. With a marvelously intricate series of motions, it could play the first side of successive disks to end of a stack, then turn them all over, or play each side in turn.

Although there were several competing methods of electrical recording and playback originally, including the Brunswick Panatrope—which used a photoelectric cell in making the recording—the most important development, and the one that was to survive, was the Western Electric system pioneered by J. P. Maxfield and H. C. Harrison of the Bell Telephone Laboratories. The new records had a dynamic and frequency range that was staggering. One of the most impressive of the first electrical recordings was "Adeste Fideles" sung

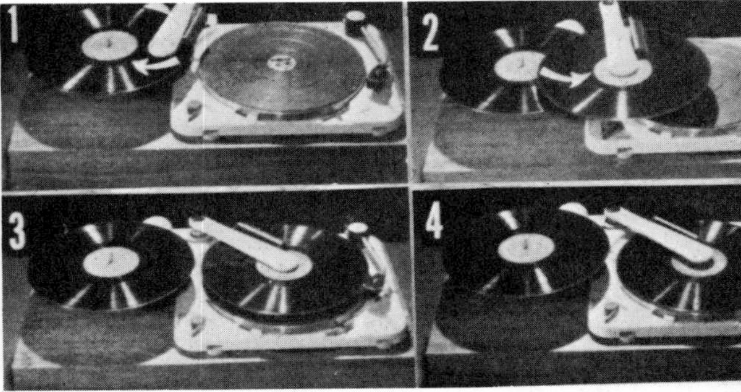

Popular Science **editor describing this contemporary successor to the Capehart—the Thorens TD-224—wrote that his friends burst into admiring laughter at sight of its incredible antics. At far right, changer goes through some of its paces in sequence: arm moves to the stack, deftly removes a record, swings it over, and deposits it on turntable.**

New battle of formats—this time in tape. In 1969 *Popular Science* discussed these competing systems (reading clockwise from left): reel-to-reel, 8-track cartridge, 4-track cartridge, Playtape, and cassette.

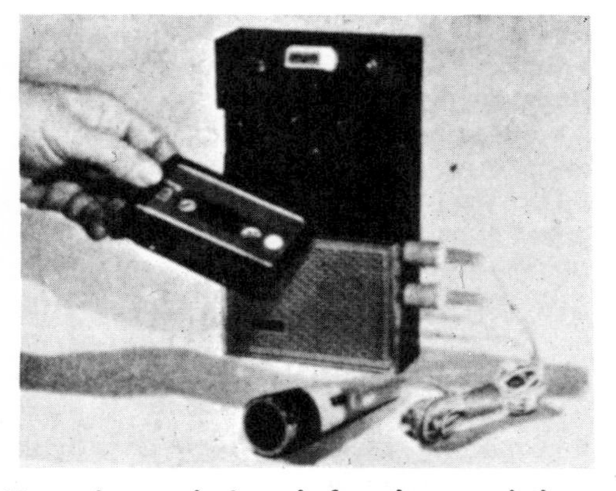

First player designed for the revolutionary tape cassette was introduced in 1964.

by some 800 voices at the Metropolitan Opera House.

Ironically, *Popular Science* celebrated the new development by concentrating on the machine—the Orthophonic Victrola—that Victor had brought out to play the new records. Ironically, because the Orthophonic was actually the last gasp of *acoustic playback* phonographs, and the writer—as well as Victor's Orthophonic—ignored the electrical origin of the discs themselves.

The success of electric recording was phenomenal. Acoustic recordings simply couldn't compete, and within two years the acoustic era had ended. By 1927 the phonograph industry, only fifty years old, was a giant, stamping out more than a million records a year.

The next decades saw lean and fat years for Mr. Edison's famous invention (he was forced to discontinue production of Edison recordings and music players in 1929, on the eve of the Great Depression). The decades were studded with occasional technical improvements that yet did not substantially raise the quality of sound reproduction.

The novelty of the early thirties, as noted by *Popular Science,* was a new light, unbreakable record, made of resin-coated paper. The records had a bad habit of warping and curling with time.

Edison had developed a 12-inch disc that played for 40 minutes in 1927; a *Popular Science* description that year revealed that its groove was a mile and a quarter long and as thin as a human hair. In 1930 RCA Victor introduced a 12-inch long-playing disc that played 30 minutes. Neither made a permanent mark on the industry.

In 1939 a critic complained of the rel-

Inside a tape cassette: tape runs from reel to reel over rollers and guides and past a small pressure pad in front edge, which presses tape against player's magnetic heads.

Inside an 8-track stereo cartridge: a single endless loop of tape feeds from the reel's inner edge, past the player's magnetic head, and back to the outside of the reel.

atively few improvements in sound quality from the phonograph and said that record materials were much the same as they had been twenty-five years before—and as noisy.

The first real breaks came after World War II. After eleven years of research, observed *Popular Science* in November 1945, RCA Victor began making records of transparent red vinyl resin plastic (Vinylite) which were unbreakable and had far less surface noise than the shellac pressings.

At about the same time, the London Decca Company began making FFRR records (full-frequency-range recordings) which had an extended response.

The biggest bombshell was that dropped by Peter Goldmark of Columbia in 1948—the LP (long-playing) record. Designed to be played at 33⅓ rpm instead of the standard 78 rpm, they were pressed in black Vinylite and played about 20 minutes per side. Columbia's LP avoided some of the problems of earlier long-playing discs by being the first lateral-cut record designed for a permanent stylus only and seemed to offer higher-caliber reproduction as well as convenience.

Many record companies swung to the LP system, but RCA Victor did not—unfortunately, in the eyes of many industry observers. Instead, Victor brought out its own integrated record and player system —7-inch, 45-rpm discs with virtually the same microgroove as Columbia's, to be played on a special changer. The "war of the speeds" that ensued was short-lived; Victor succumbed, making its own 33⅓-rpm LPs in 1950, as does every company today, and the 45-rpm record was used primarily for short, popular selections.

One development that contributed greatly to the high fidelity of the new records was, oddly, the tape recorder, because in 1948 all major American record companies began to use tape as the first step in the recording process. Work by German experts during World War II had perfected the tape recorder to the point where it was free from many of the limits of disc recording. (Wire recorders, which had had a flurry immediately before, during, and after the war, were of limited fidelity.)

Curiously, the magnetic wire and tape recorder had been invented exactly fifty years before, in 1899, by the Danish scientist Valdemar Poulsen, who called his invention the Telegraphone. In a 1901 story, *Popular Science* described both of Poulsen's models, one using a short wire; the other, a long steel tape; both employing the same principles of magnetic recording as are used today. But the absence of a practical method of electrical amplification kept the device from being used for almost half a century.

Tape was also the simplest medium for a practical system of stereo—two-channel —recording and reproduction. Tape players with this capability appeared in 1949. But stereo, too, had been anticipated by the disc pioneers long before; there were stereo discs in the early thirties that used a combination of lateral and vertical cutting to produce separate sound tracks in a single groove, quite similar to the system now in use. *Popular Science* noted the background in a 1953 article on stereo broadcasting, which also cited Emory Cook's strange binaural records; these had two separate tracks cut in the disc and had to be played with a cantankerous double-headed arm.

Stereo tape machines for the home ar-

Cassette changers? Why not? This ingenious snap-on "circulator" converts a Norelco player into a changer by flipping a stack of cassettes to give you 12 hours of music.

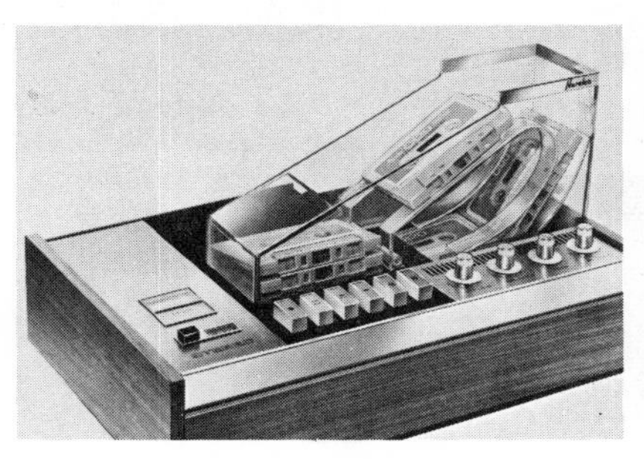

The incredible video disc, made of plastic foil, is good for 1,000 plays. Played on a special machine, it feeds a 15-minute color, stereo sound program directly to your TV set.

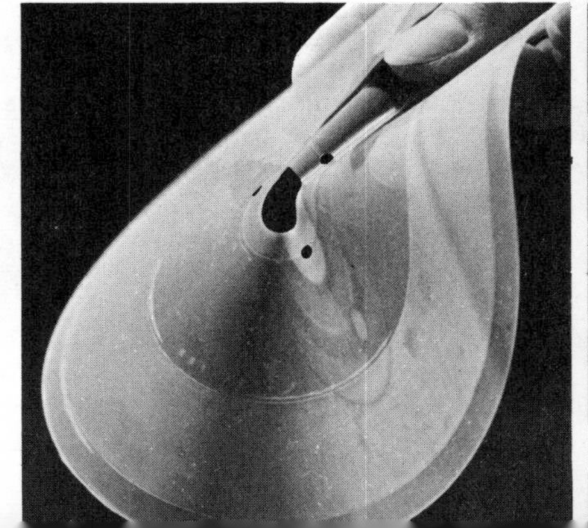

rived in 1955, and after strenuous and confusing efforts, the commercial stereo disc was made available in 1958. It has been so successful that most major record manufacturers no longer release any monaural material. Four-channel sound, from both tape and discs, arrived in 1970.

Almost from the introduction of home tape machines, there have been gloomy predictions of doom for the disc. And indeed, the new wave of excellent tape-cassette machines, which have attained true high fidelity while bringing an unprecedented convenience and simplicity to playback, bids to be powerful competition to disc phonography.

Now we are on the verge of a new revolution in audio-visual home entertainment—the video cassette. These compact packages of program material will be played over special devices hooked up to the antenna terminals of a TV, to give a color program with stereo sound. For a while, magnetic tape was the only medium considered feasible for such systems. But in 1971 *Popular Science* covered a remarkable new development—the *video disc.* Marketed by Teldec, the new disc is virtually indestructible, and can be stamped out at high rates, thus preserving the low manufacturing cost typical of records—one of their great advantages over tape. Made of plastic foil as thin as newsprint, the disc rides on a cushion of air as its special player twirls it at 1,800 rpm. Its grooves are packed in, 120–140 per mm, ten times the density of conventional records, and they are hill-and-dale. Its makers expect it to be comparable in price to a standard long-playing record.

In its article on this amazing turn in the evolution of the phonograph, *Popular Science* started with the sentence: "The disc is not dead."

That is as good a way as any to conclude this chapter.

Where will it all end? Four-channel sound systems, disparaged when they first moved onto the home entertainment stage, now promise to steal the show. Here an 8-track 4-channel cartridge plays over an RCA system. Front speakers give left-and-right stereo channels; rear speakers reproduce sound quality of original recording location.

CHAPTER 6

Conquest of the Air

DID the story of aviation begin with Mme. Montgolfier's petticoat?

So went this tale passed along by *Popular Science* to its readers in 1885: One spring day in the year 1783 Mme. Jacques Montgolfier, wife of a French paper manufacturer, washed a petticoat and hung it over a chafing dish to dry. The hot air, swelling out the folds of the garment, lifted it up. Astonished, she called her husband to witness the sight. It didn't take Jacques long to catch its significance.

Then and there was born the idea for the first human flight in history. Jacques and a brother, Joseph, proceeded to launch some hot-air balloons. On June 5, 1783, the Montgolfiers publicly exhibited their device, unmanned, for the first time. On October 15, another Frenchman, Jean François Pilâtre de Rozier, made mankind's first ascent, in a tethered balloon.

Conquest of the air had long fascinated the human mind. In Greek mythology, Daedalus and Icarus flew on waxen wings. Before the time of Christ, the Chinese flew miniature helicopters. Roger Bacon, in the thirteenth century, proposed filling a large copper globe with "ethereal air or liquid fire" for flotation. In the sixteenth century, the Florentine Leonardo da Vinci diagramed a device powered by human muscle that would fly by "compressing" the air.

Then, as recorded by *Popular Science* in its February 1885 review of a hundred years of aviation, came Mme. Montgolfier's petticoat. Aeronautical research by generations of scientists followed. Ballooning became a craze.

Inevitably, experimenters—some firmly grounded in science, some as wild as March hares—began crocheting on the fabric of man-made flight. They fell into two classes: those who wanted to add power, and thus a means of direction, to lighter-than-air (aerostatic) craft, and the proponents of heavier-than-air (aerodynamic) craft.

A Frenchman tried to propel a balloon by oars. A Professor G. Wellner enunciated a new principle in a "sail-wheel flying machine," a cross between a propeller and a kite. Then Jerome B. Blanchard—of Colorado—patented a flying machine sustained by "aeroplanes" (wings) and powered by a pilot pedaling furiously to turn a propeller. Beecher Moore of Buffalo claimed to have built a working model of a heavier-than-air machine propelled by an engine fed a mixture of saltpeter, sulfur, and charcoal. And, ah yes, somewhere in the contraption he had a tank of liquid air. He was a strong advocate of rocket power.

William S. Henson of England fiddled fruitlessly with a small steam engine for heavier-than-air flight. One fellow actually had published an idea for directing the flight of a balloon with four trained eagles.

The inventor of the machine gun, Hiram S. Maxim, built an enormous flying machine, to carry three men, at a cost of $100,000. It had wings, two propellers, and a 363-hp steam engine. It ran on a track; another, above, aimed to keep it from taking off. Straining to rise, it wrenched free and lurched upward, to the crew's consternation. Unprepared to fly, they cut its power and it crashed, a wreck.

France's Gaston Tissandier used batteries to drive an elongated balloon with a propeller at 7-mph speed. A brace of French army officers sharp enough to lay hands on 100,000 government francs got 15 mph out of the Tissandier idea. John Stringfel-

SMITHSONIAN INSTITUTION

First aerial voyage in history was made over Paris in a hot-air balloon by Jean François Pilâtre de Rozier and a passenger in 1783.

Flying machine with flapping wings, powered by a dozen cartridges ignited by candles, was described before a science society in 1870. Contraption never got off ground. Illustration appeared in 1894 *Popular Science*.

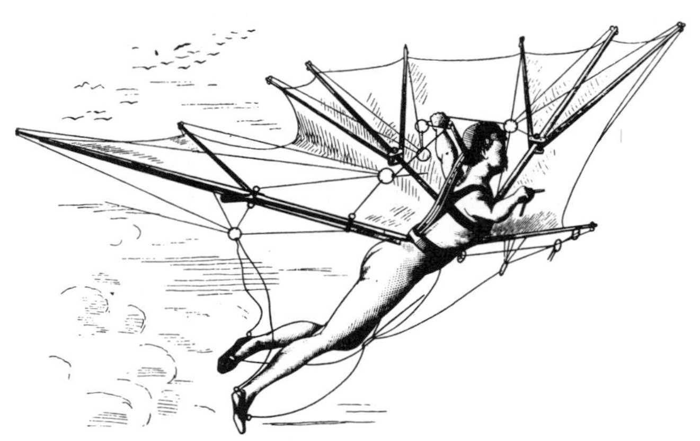

Drawing of man flying by muscle power, on file at the U. S. Patent Office. It was 1885 dream.

Gaston and Albert Tissandier were a bit more practical. In 1883 they drove a balloon against a brisk wind with a propeller turned by an electric motor and storage battery.

Hold your hat and pedal like mad! Jerome Blanchard patented this flying machine in 1894. Drawing illustrated article on aerial "progress."

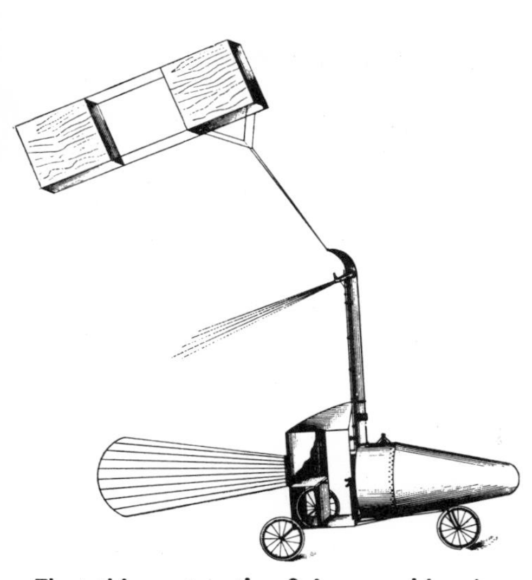

That thing atop the flying machine is a kite, or wing. That thing at apex of pipe on the fuselage is a stream of propulsive gases. Tank in front is for fuel. Out back is rudder. A man named Beecher Moore designed it.

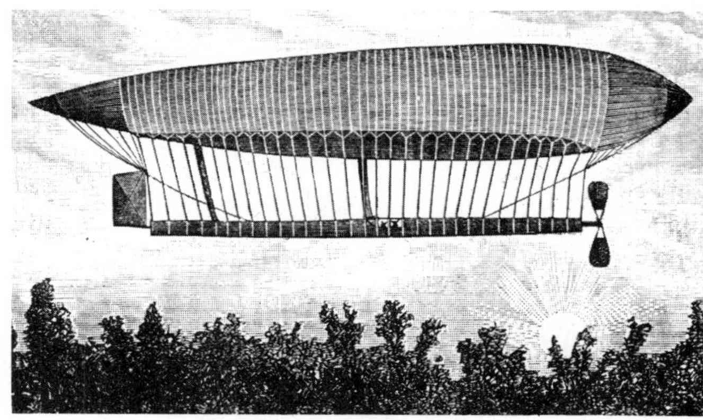

This airship of Renard and Krebs used the Tissandiers' drive.

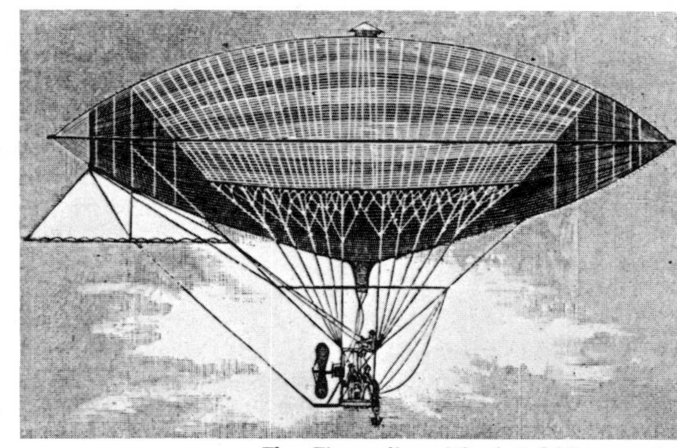

The Tissandiers lifted 2,800 pounds, half that of Renard and Krebs, but were first to give balloons steering.

A hundred years ago, this model ornithopter, propelled by twisted rubber band, was flown in England. Inset shows wings starting downbeat. Tail was peacock feathers.

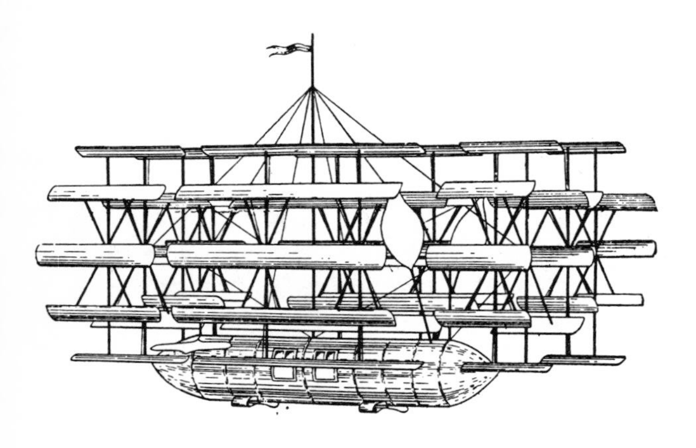

One of last of nineteenth century's wild ideas for flight: Wellner's sail-wheel with giant paddles like a side-wheeler steamer to "compress" the air. It didn't.

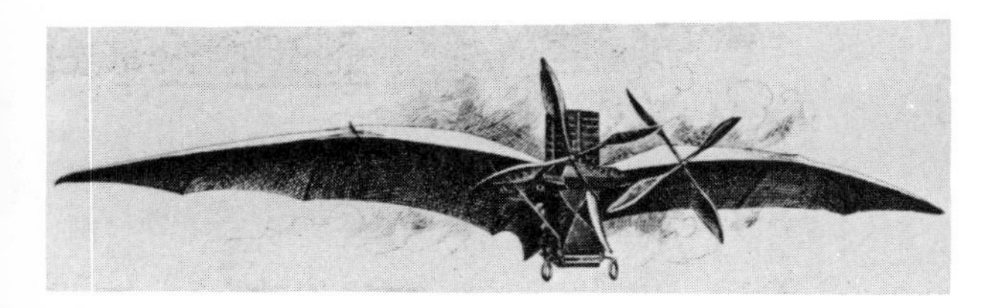

Dream of a French engineer, Ader, the "Avion" had enormous batlike wings and 2 propellers to pull it along. Taxied on 4 wheels, it rose, but crashed. It was pictured in 1901 *Popular Science*.

Otto Lilienthal was the first man to establish that a wing must have a curved surface to create lift. After hundreds of glider flights, a fatal crash from 50-foot height ended his pioneering research.

low of England flew an unmanned, steam-powered model. The first real success in maneuvering a gas bag came in 1852 when Henri Giffard of France produced a 3-horsepower steam engine weighing 110 pounds.

Sir George Cayley, the "father of British aeronautics," contributed much to the theory of flight, including that by helicopter, though he produced no hardware. Less may be said for expounders on the future of flight.

Joseph Le Conte, University of California professor of geology and natural history, in 1888 wrote: "A flying machine . . . —i.e., self-raising, self-propelling—is impossible in spite of the testimony of the birds." In six years he recanted: "It is no longer justifiable to say . . . that a flying machine is physically impossible . . . But the engineering difficulties are enormous and possibly insurmountable."

As late as November 1903, on the eve of the first flight by the Wright brothers, fathers of the powered airplane, one self-appointed clairvoyant stated, "Aeroplanes will probably be used for military purposes and for adventure, but not for . . . transportation and commerce." Even Octave Chanute, American aviation pioneer and authority on the principles of gliding, wrote of heavier-than-air flight, "The machines will eventually be fast . . . used in sport, but . . . not to be thought of as commercial carriers."

Three generations later, airlines had all but annexed railroad passenger business.

If the detractors had in mind the risk to life and limb, they had a point. It was dangerous; the air was terribly unforgiving of mistakes. Balloonists fell to their deaths with depressing regularity. "More than one aëronaut," remarked *Popular Science* in 1885, "has lost his life by being carried out to sea." In Germany a Dr. Wolfert and his engineer died when the engine ignited their balloon.

Two companions of Tissandier on a flight to a height beyond human endurance expired before he could pull the balloon's ripcord. John Wise and two friends drowned crossing Lake Michigan.

In an ascent at Philadelphia, amateur James Wilcox was luckier. For $50, he agreed to soar aloft under 47 small hydrogen balloons in a contraption dreamed up by two estimable gentlemen of the Philosophical Society. It was perfectly safe—the calculations showed he could not rise above 97 feet. But he did—and panicked. He began puncturing the balloons and plummeted into a fence, winding up with only a sprained wrist.

All this frenetic experimentation began paying off. The internal-combustion engine, infinitely lighter than the steam engine, had been invented by Nikolas Otto in the mid-1870s. This not only released balloonists from the wind's whims; it gave a leg up to heavier-than-air enthusiasts. A Brazilian, Alberto Santos-Dumont, went to Paris to buy an automobile and stayed to astound the French with his exploits in petroleum-powered, cigar-shaped balloons in 1901.

But the most notably successful lighter-than-air experimenter was a German count, Ferdinand von Zeppelin, who was exposed to ballooning while serving with the Union army in the Civil War. Originator of the "rigid" airship—great metal forming rings and longitudinal stiffeners enclosed a series of separate gas bags, with everything inside a fabric skin—he launched the first in 1900. Underpowered for its 22,000 pounds, it was driven by two 16-horsepower engines. In all, 130 zeppelins were built. Some were used even before World War I as airliners, and after the war came passenger, mail, and freight transport in highly sophisticated airships.

Meantime, the lunatic fringe of experimenters in heavier-than-air flight had been succeeded by scientists. Sensible research had discarded ornithopters—machines with flapping wings, aping the birds. Among the researchers was one of the great tragic figures of aviation, Samuel Pierpont Langley, secretary of the prestigious Smithsonian Institution. He launched steam-powered model airplanes as early as 1894. In 1903 he had a gasoline-propelled machine ready to carry a man in flight. It was a twin monoplane, with sets of wings fore and aft.

A paper by Langley published posthumously by *Popular Science* described the ribs of the wings: "Comparative measurements were made between these large cross ribs, 11 feet long, and a large quill from the wing of a harpy eagle, which is probably one of the greatest wonders that nature has produced in the way of strength

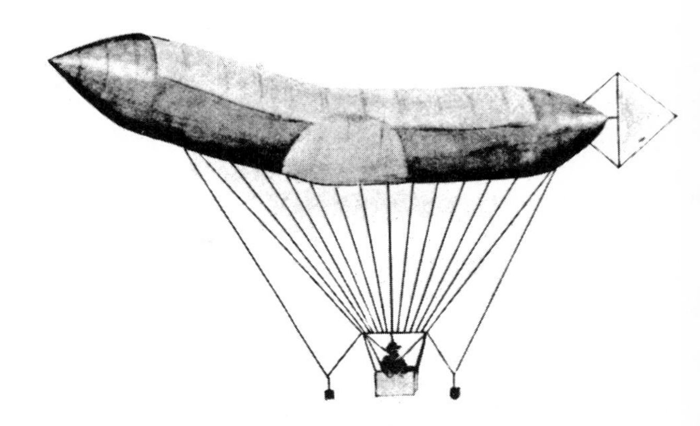

Alberto Santos-Dumont often used air-cooled "petroleum" engines, adapted from tricycle cars, in experiments. This illustration appeared in *Popular Science* in 1901.

Era of giant, rigid-framework lighter-than-air ships was ushered in by Ferdinand von Zeppelin. This ship, his second, was partly financed by a Württemberg state lottery.

This is Langley's "Aerodrome" about to crash in the Potomac after being launched from a houseboat in 1903. It was tripped up by a guy post. Repaired, it crashed again. Langley died embittered by his failures.

for weight. These measurements showed that the . . . ribs were equally as strong, weight for weight, as the quill of the eagle . . ."

Langley's engine, with 5 cylinders arranged radially around a crankshaft to drive 2 propellers, was a work of pure genius. Built by Charles H. Manly, a Langley assistant, it remained unmatched—and uncopied—for more than two decades.

On October 17, 1903, the Langley "Aerodrome" was catapulted from a houseboat on the Potomac with Manly at the controls. It struck part of the catapult and plunged into the water. Repaired, it was relaunched on December 8 and crashed again under almost identical circumstances. That did it. Langley was through. He had used $50,000 of government money. Criticism was shrill. He died in 1906—friends said of a broken heart.

As a footnote to history, Glenn Curtiss, pioneer airman, flew the Langley machine, "reconstructed and modified," in 1914.

Langley set the stage for Orville and Wilbur Wright, who never got through high school. *Popular Science* in March 1890 contained these lines: "The supposition that the vehicle must be lighter than the air . . . is a mistaken one. The skill which has produced . . . the modern bicycle will not find the task of designing such a structure too difficult."

The Wrights ran a bicycle shop in Dayton, Ohio.

To gain their needed knowledge of aerodynamics, the Wrights had conducted an extended series of experiments with gliders. They were elated and discouraged by turns. Their biggest problem was lateral balance. Then one day a man came into the bicycle shop for an inner tube. Wilbur held a cardboard box by the ends while the customer looked at the contents. Suddenly he noticed how he was twisting—warping—the box in his hands.

The Wrights had their answer—they could warp the wingtips of their glider for lateral control. This led to the design of their first really successful biplane glider. The next year they built a biplane hefty enough to tote an engine that would turn twin, contrarotating pusher propellers. Like Langley's, this engine, with 4 cylinders in line, was homemade. It was the product of their mechanic, Charlie Taylor. Water-cooled, it weighed 170 pounds, turned 1,200 revolutions a minute, and produced 16 horsepower.

Nine days after Manly was fished from the Potomac for the second time, the Wright airplane, with Orville aboard, flew off the sands of Kitty Hawk, North Carolina, their seaside laboratory. The flight lasted 12 seconds and covered 120 feet. Only five persons besides the Wrights witnessed the flight. A life guard from the nearby Kill Devil station snapped a picture of the event, the most famous photograph in aviation history, reproduced in these pages. Other flights followed on that same day. In the fourth and last Wilbur covered 852 feet.

Orville telegraphed his father in Dayton: SUCCESS FOUR FLIGHTS THURSDAY MORNING ALL AGAINST 21-MILE WIND STARTED FROM LEVEL WITH ENGINE POWER ALONE AVERAGE SPEED THROUGH AIR 31 MILES LONGEST 57 SECONDS INFORM PRESS HOME FOR CHRISTMAS. OREVELLE WRIGHT. A telegraph operator's errors clipped two seconds from the time of that flight and misspelled Orville's name.

The newspapers were informed. They yawned. What news was there in a couple of nuts in a flying machine? An article on "aerial navigation" by Octave Chanute devoted four paragraphs to the Wrights' feat in *Popular Science*, and the magazine let it go at that.

The U.S. and British governments turned down the Wrights' offer of their invention. But the French ordered 30 machines when Wilbur flew 56 miles in 91½ minutes, collecting a prize of 20,000 francs. The French Academy of Sciences awarded the Wrights a gold medal. They sold the French their patents for $100,000.

By 1909 the U. S. Government capitulated. A flight by Orville at Fort Myer, Virginia, exceeded specifications of the War Department—speed of 40 mph and endurance of 125 miles—and won a $25,000 contract.

True to one of its own, the Smithsonian displayed the Langley Aerodrome placarded as the first heavier-than-air machine "capable" of carrying a man. Wilbur, meantime, had died in 1912. Orville, indignant, dispatched the Kitty Hawk airplane to a museum in England. It was not until 1948, after Orville's death, that the

The most important moment in the history of human flight (top photo) occurred on December 17, 1903, when the Wright brothers flew a powered airplane for the first time. Their patent papers, issued in 1906, bore their main contribution to the world's knowledge of flight —a rudder, an elevator for up-down control, and warpable wingtips for lateral stability. The entire machine was homemade, including a 4-cylinder, 12-horsepower engine and opposite-rotating propellers.

Orville Wright

Wilbur Wright

Another historic "first" was Louis Blériot's flight from France to England in 1909 in a home-built monoplane.

Henri Farman, another noted French flier, frankly built planes on Wright designs.

Glenn Curtiss built U. S. Navy's first airplane (above). Below: his June Bug. "Wings are not arched this way nowadays," said 1918 *Popular Science*. "It's inefficient."

Smithsonian reworded its placard and the Wrights' airplane was returned to the U.S.

The Wrights opened the floodgates to powered flight. Glenn Hammond Curtiss won international fame with aircraft of his own design and became the "father of U.S. naval aviation." For the Wrights' wing-warping, he substituted hinged ailerons.

Abroad, Santos-Dumont, among others, won new laurels in "aeroplanes." Louis Blériot, in 1909, was the first to fly the English Channel without lifting gas. There was such a surfeit of flying feats that a second cross-Channel flight a year later by Jacques de Lesseps was scarcely noted.

In 1914 airplanes were still wood-and-rag contraptions, suitable for little more than scaring the wits out of spectators who flocked to see their pilots—"birdmen"—perform. The U. S. Army hadn't a dozen usable machines. But the flying enthusiasts were dreaming big. Rodman Wanamaker, a well-heeled New Yorker, had commissioned Curtiss to build a twin-engine flying boat to cross the Atlantic. It was to attempt it in July. Then, on June 28, the Archduke Francis Ferdinand of Austria and his wife were murdered at Sarajevo, and just over a month later all Europe was at war.

As in all wars, technology mushroomed. And aircraft had a tremendous potential for destruction. Planes could observe the enemy, carry bombs, and be mounts for those romantic warriors, the aviators. Early on, opposing airmen, scarves flung to the wind, waved cheerily to each other as they passed on reconnaissance. But one day a pilot emptied a revolver at his enemy counterpart, and the air war was on. Revolvers gave way to shotguns, then machine guns.

Most of the airplanes were "tractors"—engine and propeller in front. This posed the problem of shooting at the enemy without shattering your own prop. The first try at a cure brought metal deflector sleeves for the blades. The second was more sensible. Anthony Fokker, a Dutch aircraft maker, devised a synchronous gun that fired when a blade was not in the way.

News dispatches from the front rang with the names of air heroes—Bishop, Immelmann, Guynemer, Richthofen and—when the U.S. got into the war—Luke, Lufbery, and Rickenbacker. Baron Manfred von Richthofen, the most famous German ace, scored 80 kills before getting his comeuppance at the hands of a Canadian pilot who never before had fired a gun in anger.

The flying machines kept apace of the demands of the generals. No less than 82 designs were socked into hardware in four years of fighting. Not one that originated in the U.S., birthplace of the airplane, saw combat. American pilots had to fly French and British machines. Most famous French fighter plane was the Spad that Captain Eddie Rickenbacker (leading American ace with 26 kills) flew in combat. Best of the British fighters was the Sopwith Camel. Pilots coaxed as much as 130 mph out of these planes, a speed described by *Popular Science* as "tremendous." For the times, it was.

As the war progressed, the scout-observation plane and the bomber appeared. The bomber was especially needed by the Germans. At war's start, Count Zeppelin began grinding out gas bags to raid England. But of 51 zeppelin raids, only a dozen penetrated the defenses of London. The hydrogen-filled targets were sitting ducks for British fighter pilots firing incendiary bullets.

In 1914 the only multi-engine planes were Russians. They were big, lumbering, 4-engine machines designed by Igor Sikorsky. Within a year Sikorsky was producing 100-mph bombers armed with 6 machine guns and 1,000 pounds of bombs. Then the Germans brought out the Gotha, twin-engined, carrying 850 pounds of bombs. The British saw some and upped the ante—Handley Page bombers carrying 3 tons. The French had their Farmans that exceeded 80 mph and the Italians, their incredibly slow triplane Capronis.

The scout-observation planes were a mixed bag, 2-seaters making 100 mph and armed with a fixed gun in front and a flexibly mounted one in a rear cockpit. They were fairly easy prey for fighters, because slower. The French and British navies used them fitted with floats. A hulled U. S. Navy plane came too late for combat.

On engines, designers outdid themselves. At first, the commonest on both sides was a monstrosity called the rotary—not to be confused with Manly's radial. It was a radial, too, but the entire engine revolved around a fixed crankshaft. It had one speed

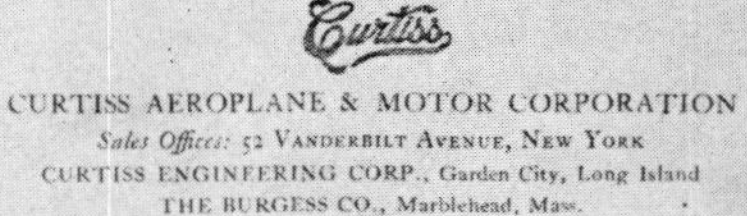

As early as 1919, Curtiss was selling "aeroplanes" for civilian use. Full-page ad appeared that year in *Popular Science*. The first licensed U.S. pilot as well as a designer, Curtiss left his imprint on aviation.

German Taube pursuit (or fighter) plane of World War I had birdlike design.

Greatest German ace, Manfred von Richthofen, was shot down at controls of Fokker triplane, a design that the army had rejected.

—full out. Torque forces, in consequence, were sublime beyond description.

Any man who got a rotary-powered airplane aloft and returned whole was automatically a veteran. The rotary coughed up great volumes of its lubricant, castor oil, and this 2-way emetic, blowing back into the cockpit, incapacitated many a pilot trying to draw a bead on an enemy.

Something had to be done, and it was. France produced a V8 Hispano-Suiza; the U.S., a 12-cylinder Liberty; the Italians, a Fiat; the Germans, an Argus and a Mercedes; the British, a Rolls-Royce. These engines ran from 160 to an astonishing 400 horsepower.

Again the pundits spouted. "Very soon," wrote Eustace L. Adams, journalist, in 1916, "aeroplanes will be carrying our mails to inaccessible spots. Shortly after will come the carrying of passengers on a schedule as regular as that of our Twentieth Century Limited." (If he had only known! The Twentieth Century Limited expired in 1967, a victim of transport planes eight times as fast.)

And a cheeky chap named Carl Dienstbach, billed as Aeronautical Editor, laid down in *Popular Science* specifications for a transatlantic airliner—a flying boat with wings spanning 100 feet. Passengers wouldn't walk upright in it because the interior would be a bit cramped, but they could lie comfortably on couches. Electrically warmed food would be "prepared by a small cook, probably a Japanese."

On May 15, 1918, before the war's end, the U. S. Army inaugurated scheduled airmail service between Washington and New York via Philadelphia. Results were less than spectacular. Thirty minutes after President Wilson bade Lieutenant George Boyle godspeed at the national capital he was downed by fog at Waldorf, Maryland. The load of mail in his Curtiss JN-4 ("Jenny") training plane had to be trucked to Philadelphia. The next day, Lieutenant Stephen Bonsal encountered fog and cracked up his plane at Bridgeton, New Jersey. Again the mail had to be trucked. It really wasn't all that bad. It just wasn't as good as it should be, and the Postmaster General, with even less knowledge of instrument flying than the pilots, wrung his hands and demanded, "Why

Eddie Rickenbacker, leading American ace, with his French Spad fighter. Planes were getting sturdier. Only six years before, military planes—in the U.S.—were averaging 1 crash for every 4 flights.

U. S. AIR FORCE PHOTO

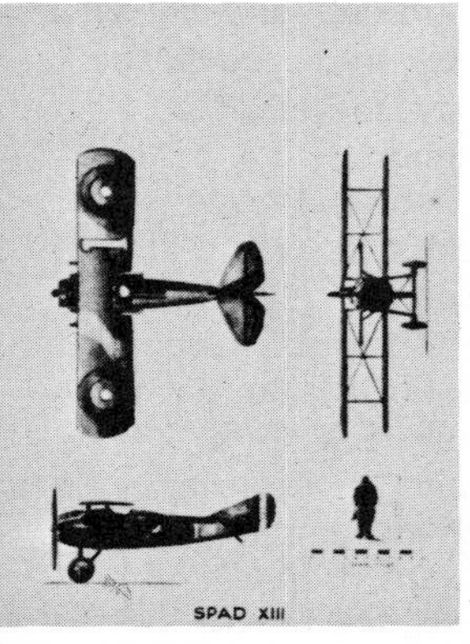

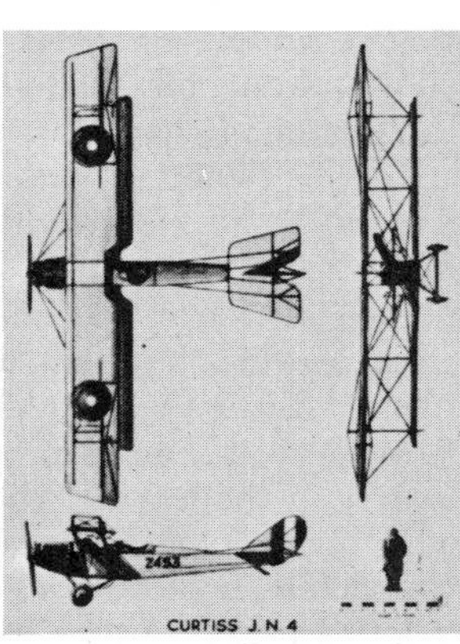

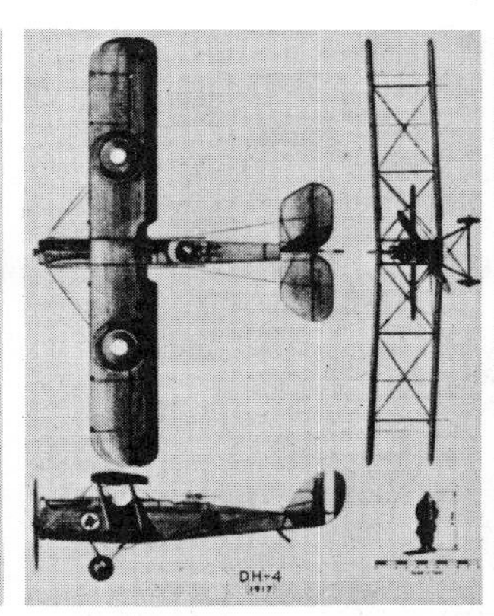

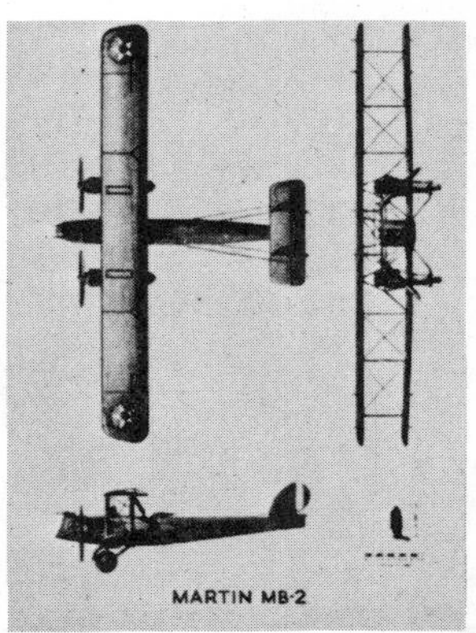

Derring-do spirit of air combat by individual warriors is caught in a *Popular Science* painting of Rickenbacker attacking a German Fokker D-7 fighter in World War I. Panel above shows Allied planes of the period.

New York postmaster hands pouch to Lt. Torrey Webb at start of airmail service.

British 1920 airliner had 7 wicker seats. U.S. passenger service came in late 1920s.

don't they fly by compass?" But most of the planes got through.

Three years later the Army began transcontinental airmail service—with British-designed airplanes—and by dint of grim determination kept clockwork schedules that were the envy of the world.

The British began scheduled passenger service, London to Paris, at war's end with twin-engine Handley Page biplanes seating 14. In the U.S., magazines published details of a 7-passenger airliner with upholstered wicker chairs capable of 100 mph. But it was another decade before airlines began service on this continent, and then they fought off passenger trade because hauling mail was more profitable.

In the twenties, a Spaniard, Juan de la Cierva, came along with a craft he called an autogiro that could land at a walk. "The name," said *Popular Science* in 1923, "is derived from the self-stabilization accomplished by a four-bladed horizontal screw turned by the wind produced in flight." It wasn't a helicopter, because the blades auto-rotated, and it had an engine (not connected to the blades) and propeller out front. In helicopters, power was applied directly to the "parasol." Gentry of that persuasion were hard at work, too, but not having much luck.

The lighter-than-air people were nothing if not persistent. Back they came at war's end. In 1919 a British dirigible, the R-34, made the round trip across the North Atlantic with 31 persons. In 1924, the zeppelin ZR-3 flew from Friedrichshafen to Lakehurst, New Jersey—part-payment of war reparations. Renamed the *Los Angeles,* she joined the U. S. Navy's air arm. As a joint venture of the U.S. and Norway, an Italian-motored gas bag, the *Norge,* flew over the North Pole in May 1926.

But airplanes were getting more robust and the flyboys were itchy for challenge. The first American transcontinental flight, in 1911, took 59 days. In 1923, a couple of army lieutenants, O. G. Kelly and J. A. Macready, flew from New York to San Diego in 26 hours, 50 minutes; and non-stop, too. Three U.S. flying boats of the NC (Navy-Curtiss) category took off for England in 1919 by way of Newfoundland, the Azores, and Portugal. Destroyers stationed each 60 miles beamed guiding searchlights

First transatlantic airplane flight was credited to U.S.N. NC-4, New York to England via Nova Scotia, Newfoundland, the Azores, and Lisbon. 1919 flight took 19 days. Above: flying boat landing at Lisbon.

British dirigible R-34 flew North Atlantic round trip at end of World War I. Longer than 3 city blocks, it had 1,000 horsepower. Top photo: arrival at Roosevelt Field, N.Y., after 108-hour trip.

First non-stop transatlantic airplane flight, in 1919, was made by Capt. John Alcock and Lt. A. W. Brown in a converted Vickers-Vimy bomber powered by 2 engines of 350 hp each. Takeoff from Newfoundland ended in a "controlled crash" in an Irish bog 1,960 miles later.

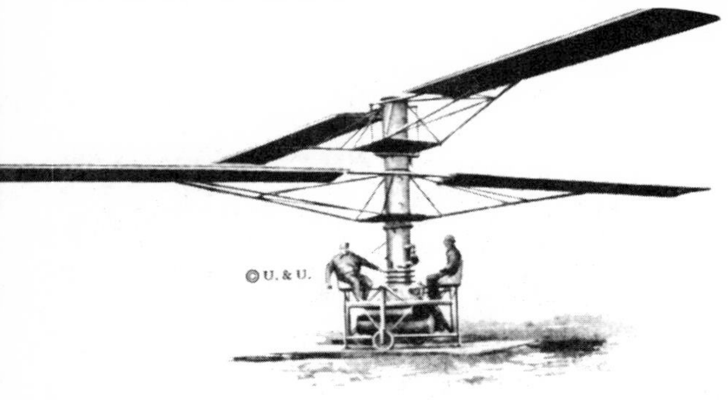

Early U.S. helicopter, the Perry, was a monstrosity with contrarotating props.

Similar machine of an Austrian, Von Karman, set record of sorts, by ascending 150 feet—tethered by cable for safety—with 2 men aboard the contraption.

into the night sky. One plane made it—in 19 days.

In 1919, Captain John Alcock and Lieutenant A. W. Brown of the Royal Air Force flew from Newfoundland to Ireland—1,960 miles—in 16 hours, 12 minutes. For this first non-stop Atlantic crossing by airplane, they pocketed a London *Daily Mail* prize of $50,000. Four U. S. Army airplanes started around the world in 1924. Two made it—in 175 days. The U. S. Navy's Commander Richard E. Byrd with Floyd Bennett as pilot flew over the North Pole round trip from Spitzbergen in a Fokker trimotor in 1926.

All this notwithstanding, U.S. aviation languished. Flying could work wonders as long as the government was picking up the tab. Making a commercial success of it was something else again. Airlines were not in being. Manufacturers of personal airplanes, like Curtiss, bravely advertised their wares, but the public was having none of their ex-

1925 *Popular Science* drawing contrasted U. S. Navy German-built *Los Angeles* with GZ-1, planned at Akron, O., and to become model for twin ships *Akron* and *Macon*. Artist also showed RS-1, our first semirigid ship, and tiny non-rigid Pony blimp.

pensive, limited-utility—and above all, dangerous—machines.

Flying was for the adventurer or the foolhardy. Apart from the military services, themselves on short rations now that the war was over, and the Post Office, all that kept flying alive for years after 1918 was a breed of men known as "barnstormers." Wartime planes were in surplus. The barnstormers grabbed them and ranged the land in "air circuses." Any handy cow pasture became a flying field. Daredevil stunt men walked the wings and swung from landing gears, performing at county fairs. They even changed planes in flight. Pilots carried passengers at $5.00 a head when they could get it, at $2.00 when they couldn't. It was a happy life if, for many an aviator, a short one. In 1923 alone, barnstorming killed 85 persons—pilots and passengers. Mainly, the airplanes used were Jennies.

By 1927 the world of flight was nearing a climax. As early as 1919 Raymond Orteig, a hotel owner, had posted $25,000 for anyone who would fly an airplane non-stop between New York and Paris. No airplane built had been capable of it. Late in 1926 a Frenchman, René Fonck, piled up a Sikorsky biplane taking off from Long Island for Paris, killing two of his crew. Two Americans lost their lives testing their *American Legion* preparatory to a transatlantic attempt. Charles Nungesser and François Coli, Frenchmen, flew westward from Paris, to be lost at sea.

The crashes did not deter the commanders of three other aircraft bent on gold and glory. One was Byrd in a trimotor Fokker monoplane, the *America.* He had a crew of three. The second was Clarence Chamberlin, pilot of a single-engine Bellanca, the *Columbia,* owned by Charles A. Levine, who wanted to be a passenger.

Then there was a third man: Charles Augustus Lindbergh, twenty-five years old.

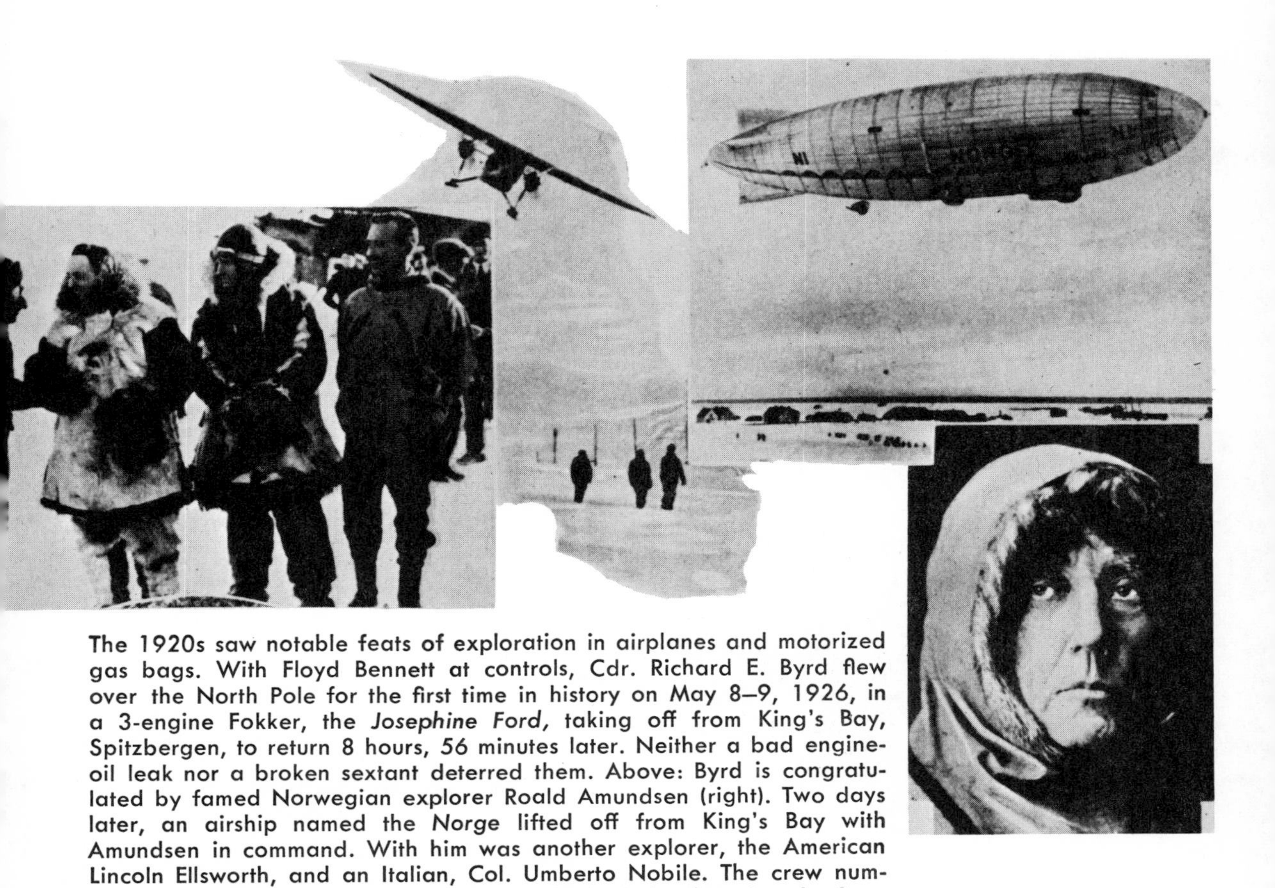

The 1920s saw notable feats of exploration in airplanes and motorized gas bags. With Floyd Bennett at controls, Cdr. Richard E. Byrd flew over the North Pole for the first time in history on May 8–9, 1926, in a 3-engine Fokker, the *Josephine Ford,* taking off from King's Bay, Spitzbergen, to return 8 hours, 56 minutes later. Neither a bad engine-oil leak nor a broken sextant deterred them. Above: Byrd is congratulated by famed Norwegian explorer Roald Amundsen (right). Two days later, an airship named the *Norge* lifted off from King's Bay with Amundsen in command. With him was another explorer, the American Lincoln Ellsworth, and an Italian, Col. Umberto Nobile. The crew numbered 16 and a dog. The *Norge,* too, made the Pole, dropping the flags of Norway, the U.S., and Italy, but was damaged at Teller, Alaska.

Lindbergh had started out as a stunt man; a wing-walker, and parachutist. But he yearned to fly, and learned. He was a loner. In May 1927 he arrived at Roosevelt Field, Long Island, the takeoff point, without fanfare. He had made it from San Diego with one stop for fuel. His mount was a Ryan monoplane named *Spirit of St. Louis* in deference to his backers. He proposed to fly the ocean solo.

Nobody paid him much heed. Lindbergh's chances of taking off ahead of his formidable opposition, much less of getting to Paris, were considered scant. But early on the morning of May 20, he calmly climbed aboard his plane, made a harrowingly low ascent over obstacles at the far end of the sod field, and disappeared over a misty Atlantic. Thirty-three hours, 29 minutes, and 3,605 miles later he landed in darkness at Le Bourget field outside Paris. The Orteig prize was his.

He was mobbed in France by celebrants gone wild. America sent a cruiser, the U.S.S. *Memphis,* to bear him home in style befitting a national hero. Ticker tape buried him in a New York parade on his return. Stock in a railroad called the Seaboard Air Line skyrocketed because investors mistook it for a passenger airline.

Chamberlin and Levine did fly the Atlantic, all the way to Eisleben, Germany, a world's distance record of 3,911 miles, and Byrd and his crew did fly to France, landing in the surf at Ver-sur-Mer when fog closed in on Paris.

But it was Lindbergh who singlehandedly gave the U.S. aviation industry the shot of adrenalin it needed. Flying schools sprang up. Investment money was coaxed into the airlines that spread, for a time, across the land without proper safety supervision. Many, faced with crashed aircraft, went broke.

Engine manufacturers had been quietly at work improving reliability and increasing power and were ready for a new era. It was one of these latter-day engines, engineered on exactly the same principle as that employed by Charles Manly to power Langley's Aerodrome, that got Lindbergh to Paris.

Even the gas-bag fraternity seemed to get the fever, though lighter-than-air had, in fact, a dismal record. In 1922 an Italian semirigid airship *Roma* exploded at Langley Field, Virginia, killing 34. The U.S.N. *Shenandoah* was wrecked in 1925 in a

Lindbergh's *Spirit of St. Louis* (below and at right) was built with flier standing over factory's workmen day after day for 2 months. At one point he insisted that engine lubricating lines be broken each 18 inches and spliced with rubber hose to reduce danger of fracture from vibration. Of the *Spirit*'s gross weight of 5,150 pounds, 3,000 was fuel. Lindbergh was only 3 miles off course when he crossed coast of Ireland, thanks to an earth inductor compass. He drew route on library globe, now a museum piece.

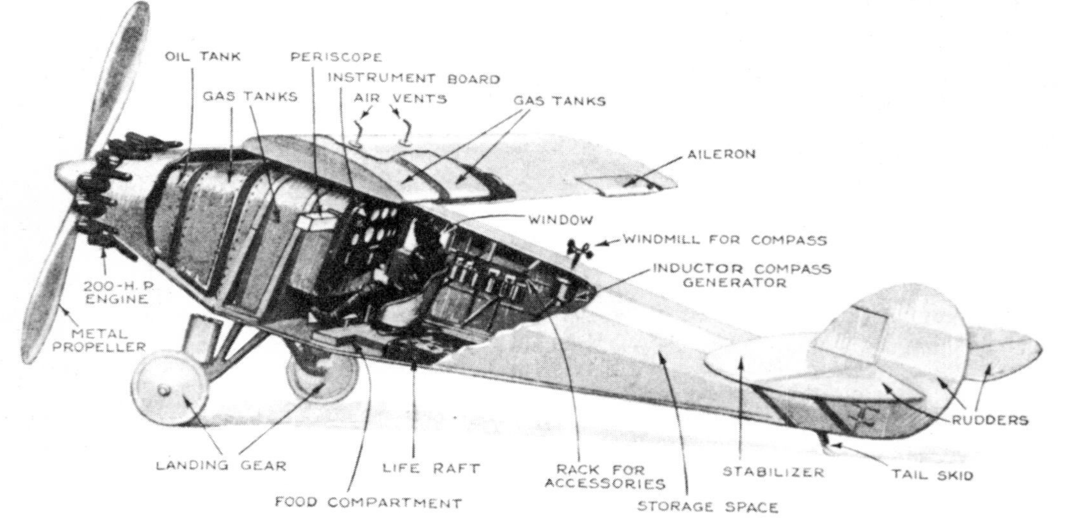

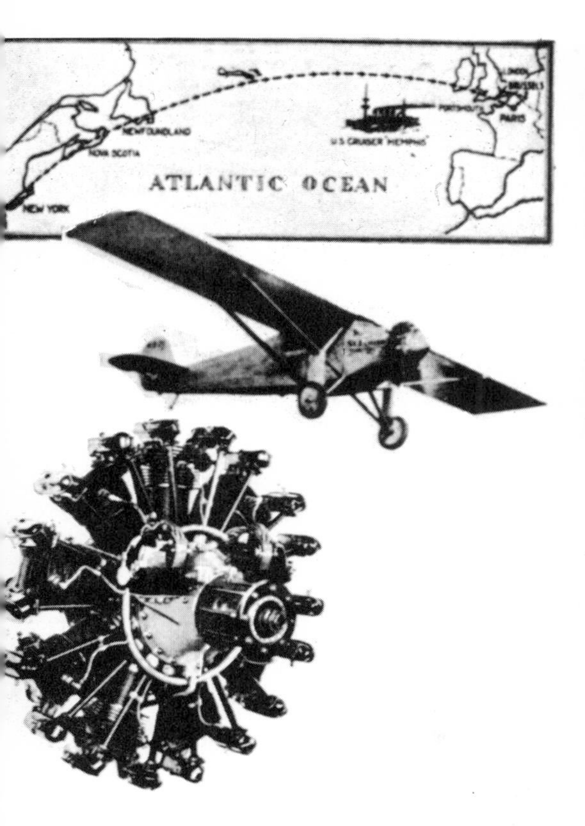

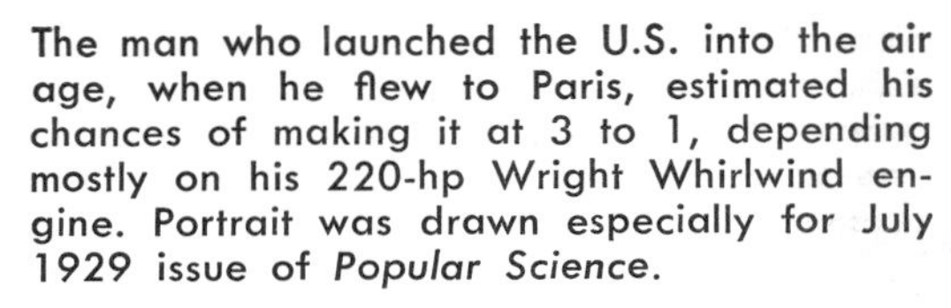

The man who launched the U.S. into the air age, when he flew to Paris, estimated his chances of making it at 3 to 1, depending mostly on his 220-hp Wright Whirlwind engine. Portrait was drawn especially for July 1929 issue of *Popular Science*.

SMITHSONIAN INSTITUTION PHOTO

CONSOLIDATED AIR PASSENGER TIME-TABLES

All service daily except where otherwise indicated

TRANSCONTINENTAL

Westward (read down)		(National Air Transport. Emergency passenger service*)		Eastward (read up)
12.15(ET)	Lv	**Hadley Field, N.J.**†	Ar	4.45
4.20	Ar	**Cleveland**	Lv	12.15
4.35	Lv	**Cleveland**	Ar	(ET)12.00
5.20(CT)	Lv	Toledo	Lv	9.50
7.00	Ar	**Chicago**	Lv	8.00
		(Boeing Air Transport)		
7.50	Lv	**Chicago**	Ar	5.45
9.40	Lv	Iowa City, Iowa		1.30
		Des Moines, Iowa	Ar	12.30
12.20	Ar	**Omaha, Neb.**	Lv	
12.35	Lv	**Omaha, Neb.**	Ar	(CT)12.15
2.50	Ar	**North Platte, Neb.**		
3.00	Lv	**North Platte, Neb.**		7.30
4.30(MT)	Ar	**Cheyenne, Wyo.**		(MT)7.15
4.45	Lv	**Cheyenne, Wyo.**		
7.05	Lv	Rock Springs, Wyo.		2.05
10.00	Ar	**Salt Lake City, U.**	Lv	1.45
10.20	Lv	**Salt Lake City, U.**	Ar	11.00
11.15(PT)	Lv	Elko, Nev.	Lv	9.00
1.30	Ar	**Reno, Nev.**	Lv	8.45
1.45	Lv	**Reno, Nev.**	Ar	7.45
2.45	Lv	Sacramento, Cal.	Lv	(PT)7.00
4.30	Ar	**Oakland, Cal.**	Lv	

*One passenger transported daily, in open mail plane, subject to weather conditions and loading. Additional passenger service to be provided soon.

†Terminal subject to change.

NEW YORK-BOSTON

(Colonial Air Transport. Summer passenger service, daily except Sunday)

Eastern Standard Time

To Boston (read down)				To New York (read up)
5.00	Lv	**Hadley Field, N. J.**	Ar	9.15
6.30	Ar	**Hartford, Conn.**	Ar	*7.35
8.05	Ar	**Boston**	Lv	6.15

*No passengers carried from Hartford to New York until Newark, N. J., airport available.

TERMINALS

HADLEY FIELD, N. J. Transcontinental passengers from New York via 10.15 AM Pennsylvania R. R. train, changing at Newark for Stelton (flag station near New Brunswick, N. J.); thence by N. A. T. auto to field.

OAKLAND, CALIF. Transcontinental passengers from San Francisco via 6.15 AM Southern Pacific ferryboat to Alameda, thence by Boeing auto to field.

CHICAGO (Municipal airport at Cicero). Passenger furnishes own transportation to or from loop district; allow 1½ hours by street car, "L," or motor.

CONNECTIONS

AT CLEVELAND: Detroit via Stout Air Service. (Except Sunday.) Albany and Buffalo via Colonial Western Airline. (Except Sunday.)

AT CHICAGO (Municipal Airport): Kansas City via Nat'l Air Transport. Minneapolis via Northwest Air- (Except Sunday and day and M St. Louis vi Sunda Cincinnati Memphis v

AT CHEYENN Pueblo, Colo

AT SALT LAKE Los Angeles, Express.

N

AM time is show figures; **PM time in bold fa**

(ET) means Eastern Time; (CT), Central Time; (MT), Mountain Time; (PT), Pacific Time. All time shown is Standard Time, one hour earlier than Daylight Saving.

These tables subject to change without notice.

SEATTLE-SAN FRANCISCO-LOS ANGELES†

(Pacific Air Transport, daily except Monday; West Coast Air Transport, daily except Sunday and as noted)

Pacific Standard Time

Southbound (read down)							Northbound (read up)		
PAT	WCAT	PAT	WCAT				WCAT	PAT	WCAT
....	8.30	2.00	4.00	Lv	**Seattle**	Ar	8.15	*2.00	3.45
....	8.45	s	4.15	Lv	Tacoma	Ar	7.50	*1.30	3.20
....	9.05	....	4.35	Lv	Chehalis	Ar	7.30	....	3.00
....	9.45	4.00	5.15	Ar	**Portland**	Lv	7.00	11.30	2.30
7.00	a10.00	....	....	Lv	**Portland**	Ar	....	....	b 2.15
	a12.30	....	....	Ar	**Medford**	Lv	....	....	b11.45
	a12.45	....	....	Lv	**Medford**	Ar	....	9.00	b11.30
					Corning	Ar	....	....	b 9.30
								5.00	b 8.00

Primitive beginnings of U.S. air transport in 1928 put the ticket agent right beside the plane—here, a Travel Air cabin plane on a run from Kansas City to Chicago, or 410 miles.

Coast-to-coast flight took 19 hours 15 minutes, with 13 intermediate stops! But never a stackup. Only 1 passenger a day could be accommodated "in open mail plane, subject to weather conditions and loading." Cost of your ticket: $400.

Ticket for a trip on the National Air Transport mail plane, New York to Chicago, cost the lucky rider $200. Upper portion of ticket is the passenger's receipt and identification. Note the air map across the top.

Around *the* World *by* Zeppelin

Walking the *Graf Zeppelin* into its home port hangar at Friedrichshafen, Germany, at the end of its epochal voyage around the world, completed in 20 days and 4 hours. The average speed was 67 miles an hour; actual flying time 13½ days.

The great dirigible floating above Seville, Spain, in the course of a tour over southern Europe a few months before the record-breaking voyage around the world.

Over the Pacific—one of the officers using a sextant to determine the position of the huge airship. A violent electrical storm threatened the ship on this leg of the flight.

Anchoring the *Graf's* nose to the mooring mast at Los Angeles, Calif., at the end of the first nonstop flight across the Pacific, completed in 78 hours, 58 minutes. On the take-off next morning the ship was slightly damaged.

Looking down from the Zeppelin's cabin upon Wolodga, Siberia, during the long flight from Friedrichshafen to Tokio, Japan.

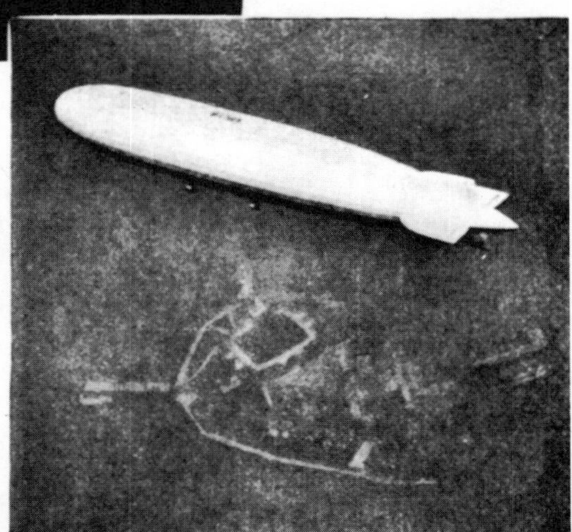

Back again above New York City after circling the globe. This aerial view shows the dirigible saluting the Statue of Liberty to which it had said good-bye just 21 days, 7 hours, and 33 minutes before. The *Graf* broke the previous round-the-world record by 2 days, 7 hours, and 48 minutes. Magellan's world voyage, completed in 1522, required 3 years and 29 days. After this photo was taken, the dirigible sailed back to Germany.

Crowds in Tokio welcoming the *Graf Zeppelin* to its hangar after the nonstop flight of 6,800 miles from Germany. The ship was in the air 101 hours and 58 minutes. Damages during the take-off the next day delayed the *Graf* 35 hours.

Historic voyage of the *Graf Zeppelin* in 1929 was saluted by this picture page in *Popular Science*.

storm over Ohio, killing 14. The Italian semirigid *Italia* flew over the North Pole in 1928 but was wrecked on the return voyage. A British dirigible rammed into a French hillside en route to India in 1930, killing almost all aboard. Yet in 1929 the German *Grap Zeppelin* had girdled the globe in 21 days, gaining world plaudits.

It was a short-lived renaissance for the gas bag. The Goodyear-Zeppelin Corporation constructed two titanic airships for the Navy—the *Akron* and the *Macon*—so sophisticated that scouting planes carried in their bellies were launched and recovered in flight. These dirigibles held 6½ million cubic feet of gas, almost twice that of the *Graf*, and it was non-combustible helium. Within a span of three years, the *Akron* was wrecked in a storm off the New Jersey coast, carrying 72 to their deaths, and the *Macon* broke up off Point Sur, California. Only two lives were lost in the *Macon's* structural failure, but that ended the romance between our Navy and lighter-than-air. Non-rigid "blimps" did continue to fly for a time.

The *Hindenburg*, with a capacity of 7 million cubic feet, was successor to the *Graf*. The gas cells were filled with hydrogen since non-combustible helium was not available to them. Airships received their *coup de grâce* at the start of the *Hindenburg's* second season of transatlantic flight in 1937. In landing at Lakehurst, New Jersey, she took fire and burned, and 36 died. A. A. Hoehling, who investigated the tragedy years later, reported in his book (*Who Destroyed the Hindenburg?* Little, Brown & Co.) that evidence pointed to sabotage by a crew member allied with Communists and anti-Nazis.

In 1935, a free balloon, the *Explorer II*, soared to 72,395 feet with Captains Albert W. Stevens and Orville A. Anderson of the U. S. Army aboard in a sealed gondola —a height never before reached by man.

Meantime, the elder Henry Ford caught the aviation fever. He began to turn out the country's first reliable all-metal transport, the Ford Tri-Motor, or "Tin Goose," that more than 40 years later was still toting cargo in South America and passengers on a short-run airline in the U.S. Presently the Fords were replaced by more capacious craft. First came the Boeing 247-D low-wing; then, the all-time champion workhorse of the airways, the 24-passenger Douglas DC-3.

That resident prognosticator of *Popular Science*, Dienstbach, was pretty well on target when in 1919 he forecast transoceanic travel by flying boat. On the heels of bigger planes for domestic travel did indeed come flying boats.

The concept of the flying boat was based on a fallacy—that a hulled craft could survive wind and wave if forced down at sea. Several flying boats broke up in choppy waters before airlines became disenchanted with the idea, but the thin skin necessitated by weight considerations was not their chief drawback. They were fat, huge, and just too slow. Tidying their silhouettes was like trying to streamline a brick.

The first ambitious attempt at a flying boat after the Lindbergh flight was the German Dornier design with 12 engines—6 driving tractor propellers and six more, pusher propellers—all mounted atop a 158-foot-long wing. Called the DO-X, it had a three-deck hull and, fully loaded, weighed 52 tons. It had pulled off a spectacular stunt before facing up to the Atlantic. It carried 169 people aloft at Lake Constance. The largest dirigibles hadn't carried half that many.

On November 4, 1930, the DO-X set off for the U.S. Mercurial weather led to playing it safe; it took a southern route. Misfortune plagued the airplane. Fortunately, for the initial legs of this flight, it carried no passengers. Between Southampton and Bordeaux it was forced down at sea. It taxied to port. At Lisbon fire destroyed half its wing. Repairs took more than two months. During tests at the Canary Islands, the hull caved in. It was June before the plane was ready to challenge the South Atlantic.

It made the flight to Brazil without a hitch. There it was refitted. Finally, on August 27, 1931, the DO-X landed with 72 passengers and crewmen in New York harbor to the cheers of 50,000 spectators. Months later it flew home at its lumbering 100 mph. That was it for the DO-X. It rocked at its mooring, unloved, until it was blown to bits during an Allied bombing raid in World War II.

Then the French had a go at the world's biggest airplane. Weighing in at 37 tons,

First of operational U.S. "airliners" was the Ford Tri-Motor above, the "Tin Goose." Seating 12 passengers, she flew 100 mph. Some Geese still fly passengers and cargo.

Most famous of transports was the Douglas DC-3 below, "workhorse" of airlines. New in mid-1930s, it carried 21 passengers at unprecedented speed of 3 miles a minute.

SMITHSONIAN INSTITUTION PHOTO

In forty years the Cub grew up. Single-engine jobs cruised at 160 mph; twin-engine, 9-passenger models at 250 for 1,250 miles non-stop.

First of "air flivvers" was the Taylor (later Piper) Cub, capable of 70 mph on 40 hp. It brought rental planes within reach.

Steam-powered plane, wreathed in vapor from its motive power, carried a man in 1933 for first time. Engine was built by Besler brothers.

this bird—the *Lieutenant de Vaisseau Paris*—had 12 de luxe, double-bed cabins, each with bath, and seats for 42 second-class transatlantic passengers, plus a bar and kitchen. The *Lieutenant,* too, vanished from sight after only three survey flights.

Igor Sikorsky, who produced the mammoth Russian bomber for World War I, fled to the U.S. after the Revolution and produced a series of well-behaved flying boats, notably the S-42 that surveyed air routes across the Pacific and Atlantic during the thirties. But they couldn't carry enough payload to be profitable.

The winged boats that captured the public imagination on the North Atlantic run were the Boeing 314s. Wide-hulled, with wings so thick that a mechanic could crawl inside to service the 4 engines in flight, they were slow. With moderate fuel consumption, they cruised their 42 tons at about 150 mph compared with 180 for the DC-3. But the seats could be turned into berths for a dozen or so passengers, and for dinner the steward presented hot dishes on linen with a silver service. It was in a Boeing 314 that President Roosevelt flew to a wartime conference with Churchill in North Africa in 1943.

Glenn Martin, a U.S. aviation pioneer, produced the Mars. It was colossal. On a San Francisco–San Diego flight one Mars transported 301 passengers and 7 crewmen. But the biggest flying boat ever built was Howard Hughes's plywood Hercules. This monster grossed 150 tons, had a wingspan of 320 feet, and a hull 220 feet long. Hughes's press agents said its 8 engines could haul 750 people (standing, of course). The Hercules flew once, about a mile at an altitude of 70 feet, before it went into limbo.

For all their devil-may-care operation, the barnstormers of the twenties spawned an interest in "private" flying. And it grew. The small-aircraft makers learned to fabricate machines that you or I could afford to fly. The first of these was from the Aeronautical Corporation of America—the Aeronca. Known disparagingly as the "bathtub," it in fact resembled one. Then a man named William Thomas Piper came along and with C. G. Taylor began grinding out a machine called the Cub. The Cub was cheap, and if you bent the throttle it could go 70 mph, almost as fast as a Ford V8 of those years.

Piper turned out Cubs like cupcakes. He was no fool. He offered flying lessons to his workers for pennies against a going commercial flying-school price of $12 to $15 an hour. If his workers on the production line slighted their chores, their own necks were at stake.

Others entered the field. But the Cub dominated its market, and four out of every five combat pilots in World War II got their initial training in machines made by the man who became known as "the Henry Ford of aviation." He lived to see privately and corporately owned airplanes, grown sleek and fast, carry as many people annually as all the domestic airlines combined.

There were variations on heavier-than-air flight. Hybrid "flying automobiles" appeared and disappeared because they were neither good airplanes nor good cars. Both the U.S. and Germany produced diesel-engined airplanes. Eighty-five years after John Stringfellow propelled a model by steam, two brothers, George and William Besler, built and flew the world's first full-scale steam airplane. Compared with the raucous machines driven by internal-combustion engines, it was eerily silent.

Then there were the rotating-wing disciples, paced by Cierva. For a time, the Spaniard's flying machines were manufactured in the U.S. under license. An airline used them to transfer airmail between the Philadelphia airport and the post office roof. But they were inefficient, and their fore-aft balance was critical. Give their proponents credit. Long years later, in the 1970s, gyrocopters, assembled in the backyard from kits, were no novelty.

The true helicopter—dubbed an "agitated palm tree"—held more promise. The Germans engineered a "chopper" that flew inside the Deutschland Halle, Berlin, in 1938. But the design of this machine, the Focke-Achgelis, was ungainly and its rotor mechanism primitive. It remained for Igor Sikorsky, the immigrant airplane builder, to produce a chopper that the military services and business interests could live with. His first, the VS-300, flew in 1940. More than thirty years later the spawn of this machine was being used by the thousands throughout the world.

By the 1940s aviation technology was on the doorstep of the second biggest single

Dirigible *Hindenburg* burned at Lakehurst, N.J., in 1937, ending era of giant gas bags, helping usher in one of giant airplanes.

First of rangy transoceanic flying boats in regular service was 41-ton Boeing 314.

Igor Sikorsky designed first commercial helicopter. Successor, below, is a commonplace.

invention since the Montgolfiers. The art and science of flight had been waiting for it ever since the Wrights spread their wings at Kitty Hawk. It was a new method of propulsion based on Newton's Third Law of Motion: for every action there is an opposite and equal reaction.

The principle employed was simple—demonstrated by the writhing nozzle of a garden hose when the water is turned on or the antics of a toy balloon when its contents are released. The invention was propulsion by jet reaction. Hero of Alexandria used it to spin a ball. A man named John Barber obtained a patent on a jet engine as early as 1791.

The Germans were experimenting with reaction airplane engines in 1939 and, in World War II powered fighter planes with rockets and gas-generating turbines. Notable among the latter was the Messerschmitt 262. But it was Frank Whittle, a Briton, who brought practical jet propulsion, specifically the gas turbine, to fruition in 1941.

Though the development was under way, jet engines had no impact on the air war, and it was eleven years abroad and eighteen in the U.S. before airlines began capitalizing on their speed potential. The military were ahead of them by a dozen years. This was due in part to a lack of heroic-size engines required by big transports, which was due in part to a lack of demand for them by the airlines, which was due in part to a chronic airline headache—the cost of aircraft that grew ever larger and more expensive.

The prototype of U.S. 4-engine landplane transports was the Boeing Stratoliner, produced before this country was catapulted into World War II. It had successively bigger ones on its heels—the Douglas DC-4, the Lockheed Constellation, the DC-6, the Boeing Stratocruiser, and finally—last of the great passenger planes propeller-driven by reciprocating engines—the DC-7. Speeds kept shooting up. Whereas the DC-3 took 20 hours to go from New York to Los Angeles, with 5 intermediate stops, the DC-7 made it in 8½ hours nonstop.

Meantime, the jet engine was undergoing astonishing refinements in military planes. Whittle's gas turbine swallowed air at the front, routed it through a radial-flow compressor, and spewed it into combustion chambers where, fed fuel, it expanded and roared out a tailpipe. When the air reached the perimeter of the compressor, it had to turn a 90-degree corner to reach the combustion cans. America's first jet, the Bell P-59, or Airacomet, was driven by such an engine as early as 1942. U.S. designers improved on it. For the radial, they substituted an axial-flow compressor with several stages. This not only sped up the air for greater tailpipe velocities—and therefore higher aircraft speeds—but also shrank the frontal area of the engine, reducing "air drag."

High speeds, by now, were no novelty. As early as 1933, an Italian, Francesco Agello, flew a scanty-winged seaplane over a measured course at 423 mph. The rocket-powered Bell X-1 on October 14, 1947, smashed the myth that man could not pierce the sound barrier, a speed of 760 mph at sea level. With the versatile gas turbine to toy with, designers of military aircraft had a field day. They built huge, 6-engine, 10-mile-a-minute bombers. They built supersonic bombers.

Before they were through they had reached a speed that boggled the mind prior to space flight—almost 4,000 mph, or better than a mile a second. They achieved this, true, not with gas turbines but rockets. The airplane was the X-15. Carried aloft under the belly of a jet bomber, it dropped like a stone for a few seconds after release. Then the pilot fired his rockets and streaked for an altitude never before reached by man—67 miles. In an almost airless environment, the pilot had to control his "attitude"—pitch, yaw, and roll around three axes—with bursts of hydrogen peroxide steam (working, again, on Newton's Third Law).

Nor were the designers through. Now they tackled fighter aircraft with wings that were more or less at a right angle to the fuselage at takeoff and landing for maximum lift and were swung into a sharp sweepback at high speed that gave the plane the attributes of an arrow. The swing-wing design culminated in an all-purpose, supersonic military airplane, the F-111, built by General Dynamics, that had a number of unfortunate crashes and was the subject of bitter congressional dispute.

Two decades before the first successful jet planes flew in the 1940s, they were anticipated in this remarkable concept of F. Mélot, exhibited at a Paris aeronautical show. It was left to Wing Cdr. Frank Whittle (below) to design and build the first gas-turbine (as opposed to rocket) engine for aircraft. First flight of a plane so powered occurred in 1941. So secret was the work that Whittle's own family learned of his involvement only upon announcement of American and British jets in January 1944. He was knighted for his research.

Whittle's engine (below), which instantly added 200 mph to plane speeds, was far simpler than a piston engine. Air entered impeller (A) where it was directed into and compressed by diffusion chamber (B). At chamber's perimeter, a scroll (C) routed air to combustion chamber (D). Here, injected fuel (E) was burned to vastly expand air-fuel mixture. Seeking an outlet, this raced through turbine (F), spinning it. A solid shaft connected turbine to compressor. Free of turbine, gases shot rearward out of tailpipe (G) to thrust plane forward, satisfying Newton's Third Law of Motion.

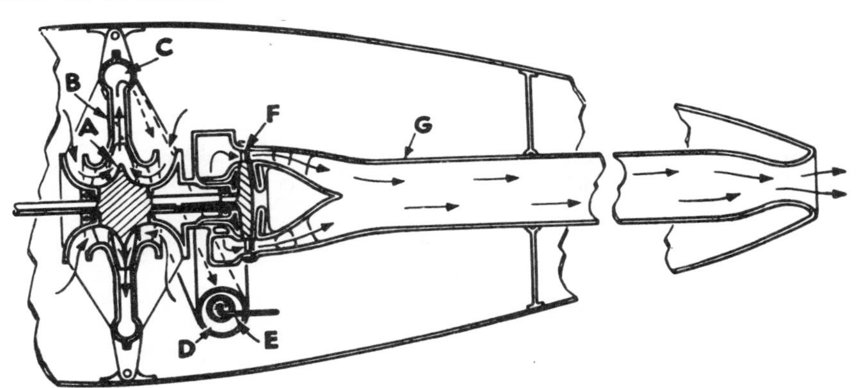

Meantime, the airlines were eyeing the inevitable advent of jets warily. First, many of them bought not true jets, but turboprops—aircraft with gas turbines driving propellers. The gas turbines were almost endlessly adaptable. They drove ocean-going boats, railroad trains, trucks, and (experimentally) even cars.

Their partisans said they would burn practically anything, from stove wood to Chanel No. 5. While that was abusing the truth, they certainly were not finicky about their fuel, and they whirled props as readily as they drove aircraft by jet reaction. Lockheed here and Vickers in England provided the first of the free world's turboprop transports. They were fast. Limited only by the tip speeds of their propellers, some flew at more than 400 mph.

Then, at the end of the fifties, came the pure-jet airliners. These transports were unlike anything the world had ever seen. They flew at 500—some even 600—mph. They were almost vibrationless. Gone was the drumming of the piston engine. Whereas previous aircraft had had a critical "center of gravity" in loading passengers and cargo, the jets minded much less about weight distribution, and the *amount* of weight they could absorb was tremendous. Jets added four miles to operational altitudes. Now airliners flew above 40,000 feet, free of most weather.

The jets *were* noisy. They thundered. But they were reliable, they were economical, and they were fast.

Soon their motive power was refined for even higher speeds and lower seat-mile costs. "Bypass" gas turbines, routing air around the combustion chambers for discharge at the tailpipe, appeared. Then fans were added up front, and airplanes were right back where they began, with propellers. The fans added to the volume of gases at the tailpipe, and hence to the power, and in bypassing the combustion stage performed a fringe chore of cooling the turbine power blades.

First of the free world's jet transports was the British De Havilland Comet 1, in 1952. It was a hard-luck airplane, beautiful to behold in flight but plagued by structural weaknesses that caused two Comets to explode from the internal air pressure necessary to high-altitude flight.

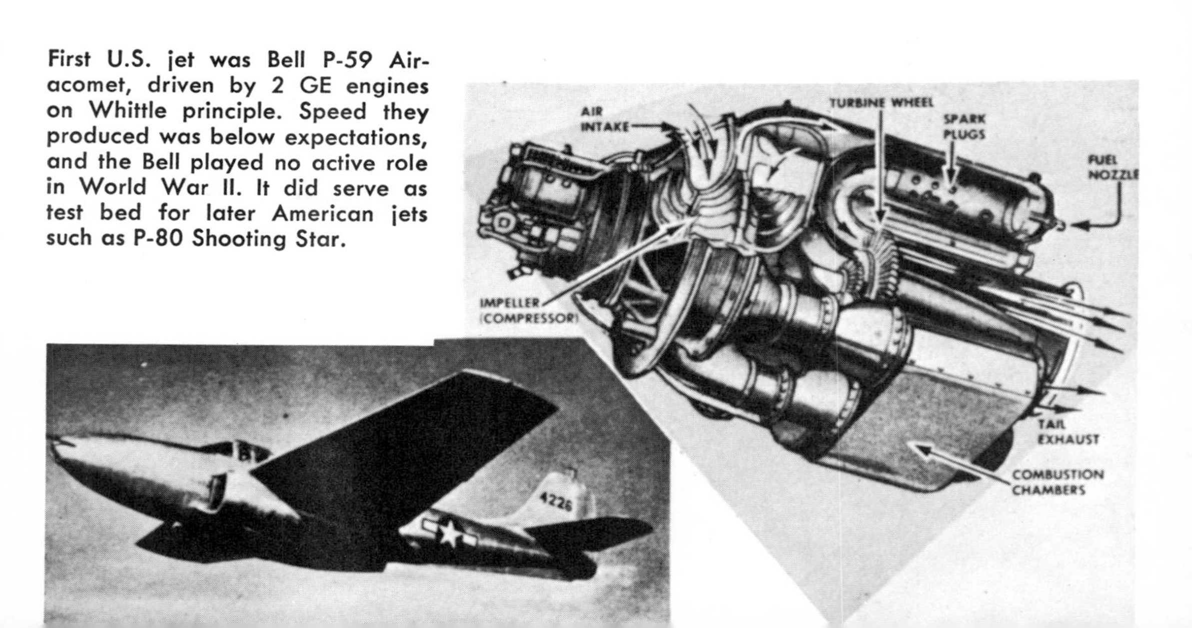

First U.S. jet was Bell P-59 Airacomet, driven by 2 GE engines on Whittle principle. Speed they produced was below expectations, and the Bell played no active role in World War II. It did serve as test bed for later American jets such as P-80 Shooting Star.

The Boeing 707, seating 179 passengers and cruising at 550 mph, was the first U.S. commercial jet, followed by Douglas and Convair models. Engines, suspended on wing pods, now had axial instead of radial flow of air to boost their efficiency.

Like all jets, it had sweptback wings to reduce air resistance. The 4 gas turbines were submerged in the wings for the same reason. It could tote 100 aboard at 500 mph. The Comet's faults were mended in time, but not before U.S. transports—faster, more capacious—flew in to dominate the world's airways.

Boeing led the way with its 707 for 179 passengers. Douglas followed with its DC-8, of about the same capacity. With London only 6½ hours from New York, well-heeled citizenry were flying the Atlantic just for weekend grouse shooting in Scotland. "Stretched" versions of the big jets boosted seating to 250, and at cut-rate fares even low-income adventurers could pony up enough money to vacation abroad.

The British beat the U.S. into the air with a jet transport, the De Havilland Comet 1 above, with 4 engines submerged in the wings. Ill-starred from start, the plane suffered 2 in-flight explosive decompressions, necessitating engineering changes.

The gas turbine soon proved its versatility. Ninety per cent of power was used to turn props, and only 10 per cent for jet propulsion, in the Canadian turboprop passenger airliner below, a Vickers Viscount.

Tail-mounted engine pods—next engineering wrinkle—eliminated a potential hazard in pedestal wing-mountings: dragging a pod on the ground when landing. With engines that far aft, as on this British Vickers VC-10, high tailplane was needed.

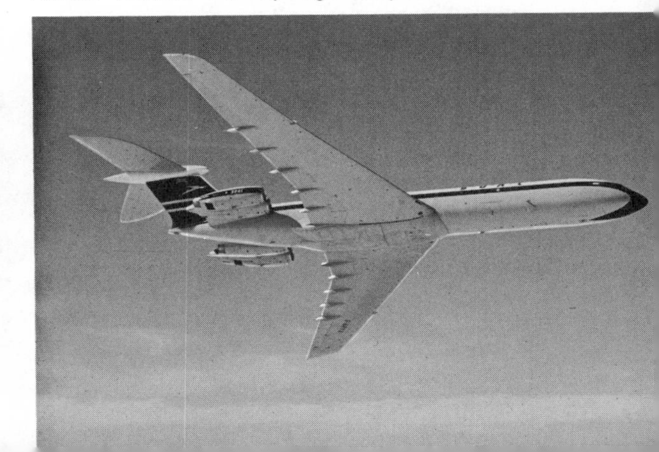

Assaults were made on the "sound barrier" that "could tear a plane to pieces." Bell X-1, fueled with liquid oxygen, alcohol, and water, passed speed of sound unscathed.
NASA PHOTO

Next development was inevitable—pivoted wings that could be swept forward for take-off or landing and swept back for high-speed flight. Bell X-5 was first of breed.

U.S. jet transports differed from Britain's first ones in more than speed. For safety, the engines were mounted on pedestals under the wings. If an engine failed structurally and at 10,000 revolutions a minute began chewing itself up, with attendant fire—and now and then that happened—it could jettison itself by burning off its mount. With fuel lines stoppered and inert gases smothering fire in the wing, a transport could proceed to the nearest airport. But some engines were put in pods on the tail, and the variety of silhouettes proliferated.

The Russians produced the Tupolev with wing-submerged engines, seating 100. Britain produced a Vickers series with 4 on the tail and the passengers, up forward away from their roar, were treated to sylvan quiet. France brought out a short-range Caravelle with a brace of tail-mounted jets. U.S. Convair 880s and 990s began competing with Boeing and Douglas for airline favor.

The long-range jets, steadily replacing propeller planes, were only openers for a galaxy of others. Airlines eliminated more and more stops by the big craft because, with the gas turbine's prodigious appetite for fuel at low altitude, landing and taking off again was costly. Scores of communities found themselves without any service at all. Their plight was compounded as railroads off-loaded unprofitable passenger trains. Then airlines began buying smaller jets to serve stricken communities.

Boeing produced a little transport, its 737, as well as a somewhat larger 727. Then somebody got the idea for "air buses," holding a lot of people, with no fancy meals at the airlines' expense. Instead, the stewardesses would sell sandwiches and coffee. Douglas and Lockheed designed planes for that trade.

Nor was that all. Boeing introduced a jet to end all subsonic jets—the 747 seating 360—with a hoity-toity cocktail lounge on an upper deck. The airplane was such a whopper that one airline put its kitchens—several of them—on a *lower* deck. Another airline tore out some seats in its tourist section and built a lounge and stand-up bar. The customers, it argued, liked to drink, too.

The power required to fly these behemoths ran off the graph paper—180,000 pounds of thrust in 4 engines. (One pound of thrust equals 1 horsepower at a speed of 375 mph. That much power would drive 900 pretty zippy automobiles.)

Nor was *that* all. Lockheed came out with the C-5A, a military jet transport with two full-length decks, the lower one longer than the Wrights' first flight. The upper deck carried 100 troops; the lower, 141 tons of cargo. Lockheed proposed a commercial version seating 500 passengers.

The airlines began wishing that all this mania for bigger airplanes and more speed, requiring billions of dollars in capital expenditure, would stop. But international pressures for commercial air superiority were too great to throttle. And, as if it didn't have enough problems already, something else began bugging the airlines.

There were not enough airports, or airports big enough, at major cities. The skies were crowded. Airlines moaned of private and corporate planes using major airports. So airplane manufacturers began designing short takeoff and landing (STOL) planes for use on short, airport-perimeter landing strips, not only to free the 2-mile-long runways for the big jets but to get into short fields that would serve additional hundreds of towns.

STOLs came in several varieties, but they all had characteristics in common—huge wings so that each square foot carried less weight, leading-edge "slats" to permit control at low speeds, and trailing-edge flaps, extensible wing segments as big as barn doors, all designed to permit takeoffs and landings at speeds no greater than the legal highway speed of a car (as against 125 mph and up for the mammoth jets).

For the military, the manufacturers went further. They experimented with vertical takeoff and landing (VTOL) airplanes. These came in even greater variety. One, a Ling-Temco-Vought, tilted its wings through a 90-degree arc so that propellers attached thereto pointed skyward at takeoff and landing. Some used jets for ascent and descent. French, German, British, and American aircraft factories all got on this kick, and all their aircraft were afflicted with the same ailment—a lack of stability in hovering and low-speed flight. Several crashed.

The manufacturers had another flying

It was only a step to first operational supersonic fighter, North American's F-100.

Next step was prodigious—to the rocket-propelled X-15 (above and below) that flew 4,500 mph at 67-mile altitude. Launched from a B-52 jet-bomber "mother," it had liquid oxygen for interior cooling to keep air-friction heat from frying the pilot.

Witness the "convertiplane." This plane's wings pivot to point propellers upward for landing or takeoff; forward, for flight.

The rocket belt for personal short hops is powered by nozzles jetting steam. Hand controls regulate fuel and flight direction.

device, really far-out. A woman motorist driving just east of Niagara Falls, New York, one day heard an unfamiliar sound, looked skyward, and, shocked out of her wits, promptly drove into the ditch. Emerging from a billowing cloud of steam, a goggled and helmeted man was rising into the air as jets of vapor spewed from tanks on his back. This was man-flight by "rocket belt."

There were variations—an individual jet platform, and a jet chair. The object of all this, on the sober statement of the Army's Transportation Command, was to give the individual soldier mobility. He could skim through the air at 35 mph for hundreds of feet, jump rivers (or skip over to the Post Exchange for a short beer).

An idea for hurtling airline passengers across the face of the globe at supersonic speeds began germinating as the 1960 decade began ripening. This was the final straw for harried airline executives. The president of one transoceanic carrier said privately that he wished the supersonic transport (instantly jargoned into "SST") would go away. Publicly he made noises of approval.

He had to. The British and French were collaborating on one design, the Concorde, and the Russians and the Americans had their own versions. The Kremlin's SST—much like the Franco-British one in that it was planned to fly at about twice the speed of sound, or 1,400 mph—was the TU-144. The American SST designers saw their competition and sweetened the pot. The U.S. product would fly at almost 1,800 mph.

On paper the SST looked like a dandy idea. The U.S. transport could leave New York at noon, local time, and land in Los Angeles a little after 11 A.M., local time. It could span the Atlantic in 2 hours 40 minutes. Then the laws of physics began asserting themselves, and it was a different story. An ugly aircraft ground-borne—it had a droop-snoot, raised for streamlining at high speed and lowered for pilot visibility for takeoffs and landings—the SST would create sonic booms that could be heard 20 to 30 miles on each side of the flight path.

It would abuse the ears and nerves. It could even damage buildings. The verdict was that its supersonic speeds would have to be confined to over-ocean legs. At the heights it proposed to fly, 12 to 14 miles, it

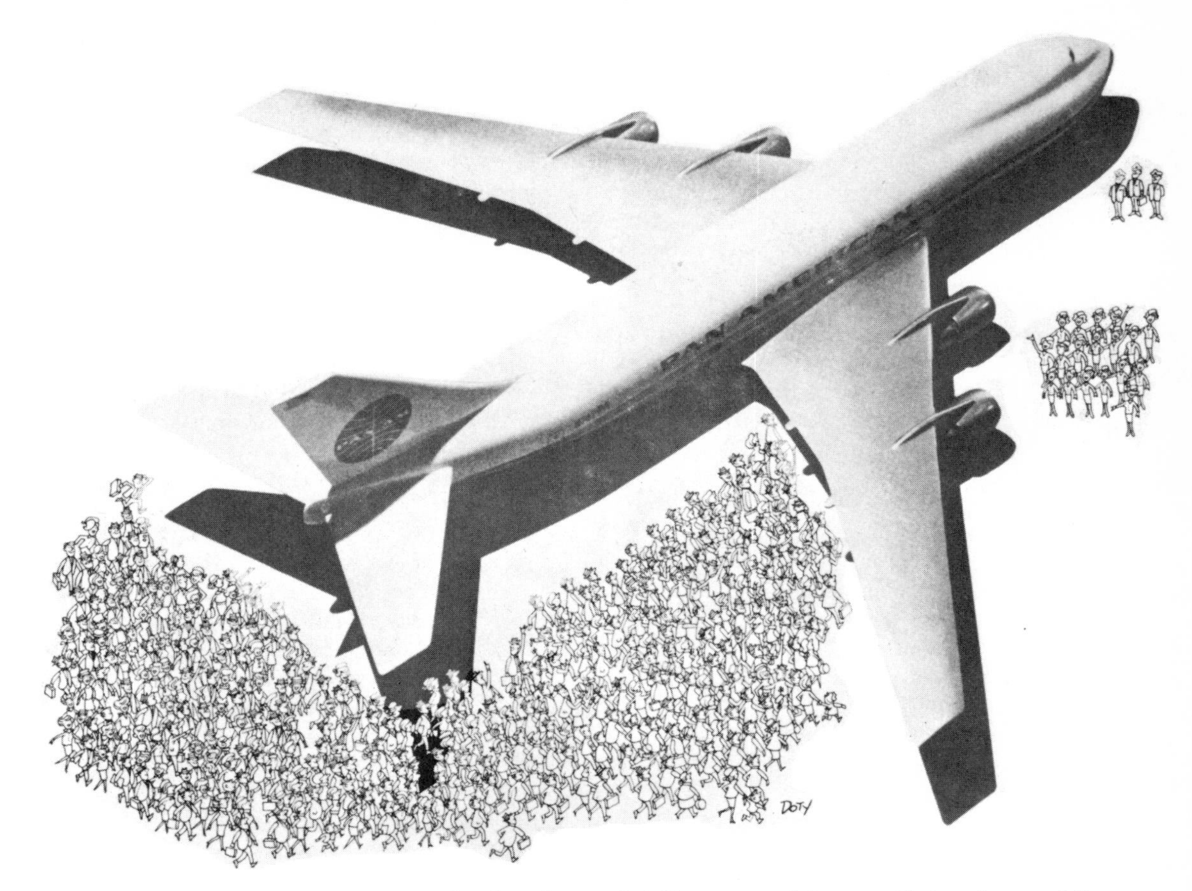

At the close of 189 years of human flight, the world's biggest airliner was a 360-seat, 600-mph behemoth—Boeing's 747. The landing gear alone weighed 11 tons.

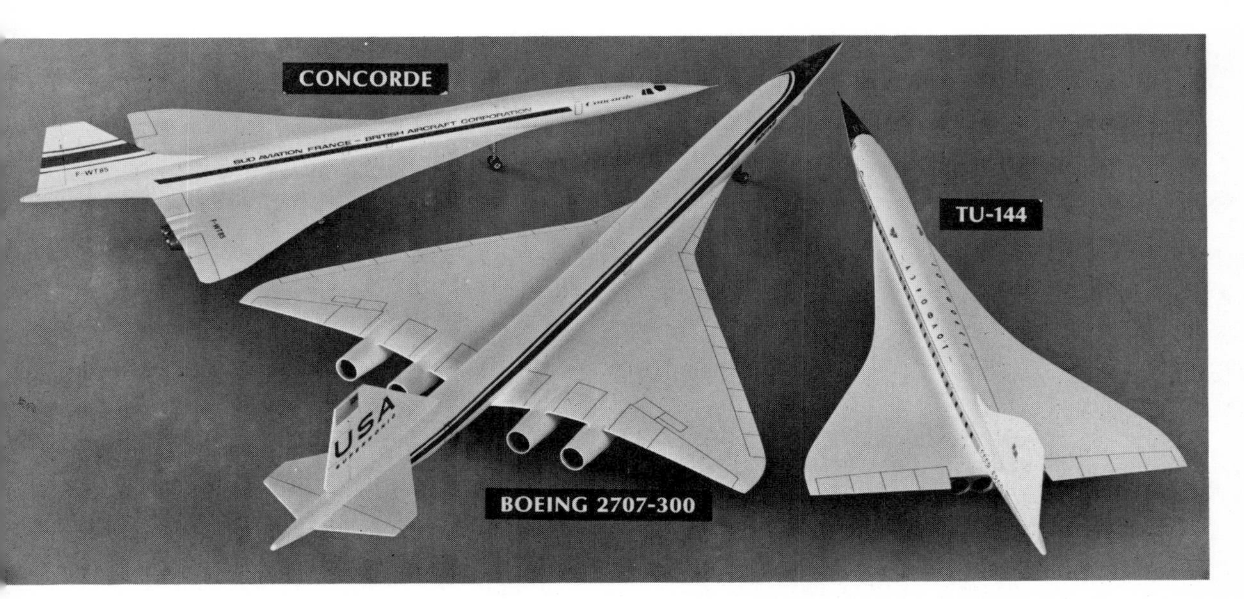

Where to from here? The supersonic airliner left many ecologists cold. Boeing's 1,800-mph entry is dormant, if not dead. The Franco-British Concorde and Russia's TU-144, both designed to fly at 1,400 mph, are beset by developmental ills.

might subject passengers and crews to doses of cosmic radiation. Polluting the air, it could destroy some of the ozone shield for earthlings and encourage skin cancer. It would cost fabulous sums to build and buy. Finally, who wanted it? The "jet set" that would streak to Paris for dinner at the Tour d'Argent and be back in New York by mid-evening, local time?

The U. S. Congress, increasingly concerned over the ecology of the planet, early in 1971 began looking at the American SST with a jaundiced eye. Its decibel output at airports promised to be monstrous. A hue and cry already was in progress over "noise pollution" around jetports, and a serious proposal was advanced that on the seaboards they be built offshore as "wetports."

In March the Senate killed a stupendous appropriation necessary to continue design and construction of the SST. The French and British were having birthing troubles. The tight-lipped Russians said merely that their SST would be in service soon.

As the 189th year of human flight was celebrated in 1972, in the twentieth century—on the 100th anniversary of the birth of *Popular Science*—interest had been revived in an idea proposed in the sixteenth century: flight by human muscle in a craft heavier than air. In England, the U.S., and Japan designers were trying to overcome man's unfavorable power-to-weight ratio as compared with that of the birds.

Three hundred years before, a renowned Italian scientist, Giovanni Borelli, scoffed at flight by human muscle. More than half a pigeon's bulk, he noted, is in the muscles it employs in flight. He wrote: "It is impossible that men should be able to fly craftily by their own strength."

Yet flight by human muscle actually took place in the Space Age. Two man-powered planes, with propellers spun by bike pedals, were built and flown in England in the 1960s. One set a record of 993 yards. Now, with man-powered planes of refined new designs, experimenters in Britain, Japan, and the United States are seeking to better that mark.

A dream that had been born, perhaps, when man first walked upright on his planet—every human being as at home in the air as in the sea, and free as a bird—appeared to be close to realization.

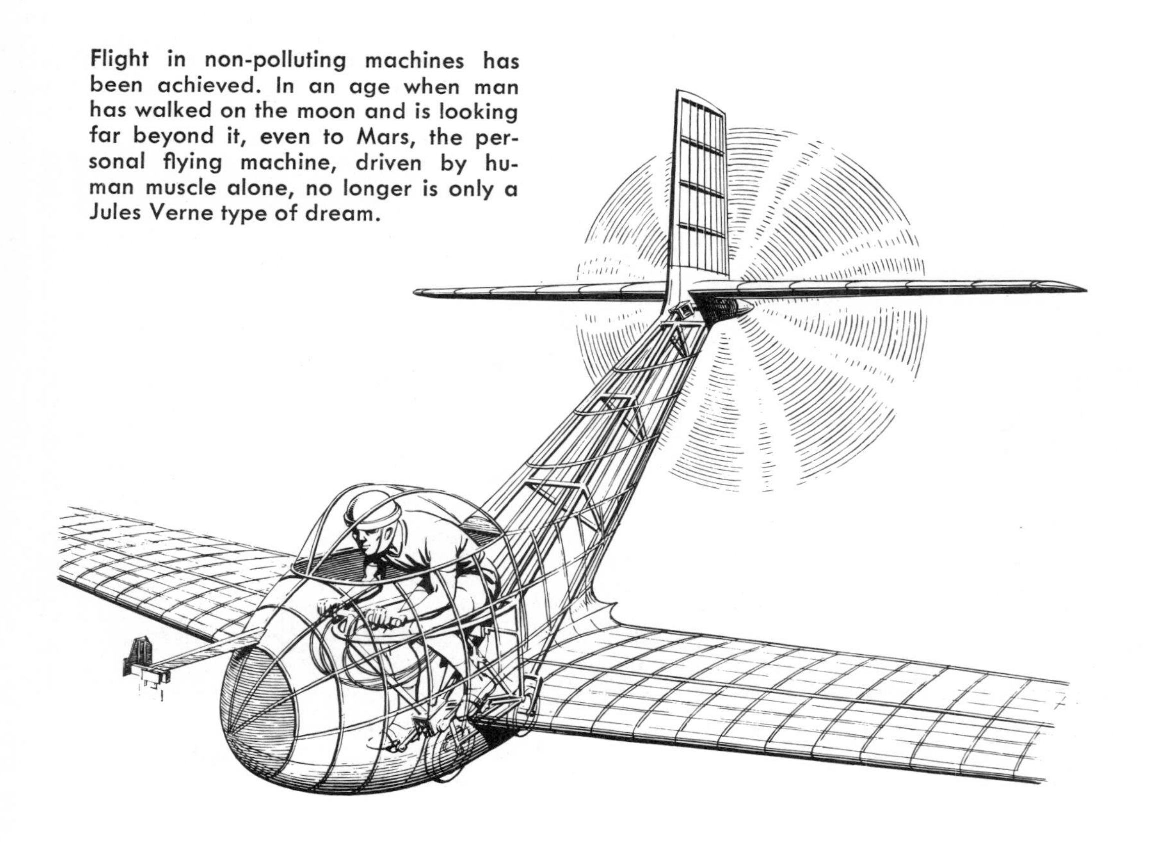

Flight in non-polluting machines has been achieved. In an age when man has walked on the moon and is looking far beyond it, even to Mars, the personal flying machine, driven by human muscle alone, no longer is only a Jules Verne type of dream.

This Piper Comanche 260C, with cruising speed of 185 mph and 1,300-mile range, is used by a private flier, island hopping in the Caribbean. An estimated 25,000 trips are made annually by private plane. You, too, can fly in one of many small high- and low-winged ships available. After your flying lessons and licensing you need aeronautical and nautical charts and a set of radio navigation aids with detailed airport information put out by Jeppesen in Denver.

Wacky Ways to Fly—*even without an aircraft*

Called "Spirit of Saint," a word play on Lindbergh's famous plane, this kooky flying flivver was built by aircraft dealer Russell St. Arnold. A ship's steering wheel guides it! It can carry 3, powered by its 100-hp Continental. Like the plane, the pilot is far out!

Then there's the whirlybird, the gyrocopter, also called "an agitated palm tree." The sustaining rotor blades, unlike a helicopter's, autorotate from the thrust of rear-mounted little engine and propeller. This amphibian, making 60 mph, can be home-built from a kit.

True helicopters, with power-driven rotors, also can be backyard-assembled, much to the distress of federal licensing authorities who fear amateur work. This one, cruising at 75 mph, costs about the same as a medium-priced car. The engine is not included in that cost.

Like automobiles, airplanes arouse nostalgia. Men remember when. Cole Palen of Rhinebeck, N.Y., owns a whole museum of World War I planes—all airworthy. The Fokker Triplane was made famous by Germany's ace, Baron von Richthofen.

Today man can fly even without benefit of an aircraft. He began it with a "rocket belt," a reaction device working on precisely the same principle as a skyrocket. He graduated to this miniature jet engine, with a range of 10 miles at 60 mph. The pilot maneuvers with twist-grips and control arms. Developer is Bell Aerosystems Company.

The Northrop F-5 Bantam supersonic jet is a 1,000-mph fighter that weighs only 12,000 pounds, compared to more than double that for a standard fighter. It has a wing span of just over 26 feet and takes half as much maintenance as the big jets. The infantry-support plane is shown above in a painting by Bob McCall for a *Popular Science* cover. At right, it is being serviced by mechanics, a quick and easy task because of its short legs. Born of war, the jet engine has been adopted by airlines the world over.

CHAPTER 7

The Telephone and How It Grew

"MR. WATSON, come here—I want you!" The words, first ever heard over a telephone, were Alexander Graham Bell's anguished cry for help to his assistant.

Bell, a black-whiskered Scot of twenty-nine, and his twenty-two-year-old technician were at opposite ends of a primitive experimental telephone line, in separate top-floor rooms of a Boston boardinghouse. Bell was preparing to try out a battery-powered transmitter that he hoped would convey intelligible speech. By mischance he upset one of his old-fashioned battery jars—and acid spilled over his workbench and his clothes. Dabbing ineffectually at the mess, he called to Watson for aid.

Watson put down his telephone and rushed to the other room. "Mr. Bell," he shouted, "I heard every word you said—distinctly!"

Forgotten in their elation was the spilled acid. The telephone worked.

The date of that historic moment was March 10, 1876. Months of experimenting had led up to it. Like many others, Bell had been trying to devise a multiplex telegraph—a kind that would send many messages at once over a single line. Knowing how valuable a device like this would be to the Western Union Telegraph Company spurred on the young inventor.

Bell called his version a "harmonic telegraph." Originally he intended to use tuning forks, so paired that each of a set at the sending end would make a certain one at the receiving end vibrate in resonance. When these proved disappointing, Bell turned to steel organ reeds. The reeds gave off a whine like a high note on an electric guitar. Bell magnetized them so that a reed's vibration induced a current in an electromagnetic coil at the sending end. The current, led to a similar coil at the receiving end, vibrated a matched reed there.

One June day in 1875, Bell and Watson, in different rooms, were tuning the reeds by tightening and loosening screws. One reed was screwed too tightly. Watson plucked it to free it. At the line's other end, Bell's ear caught a new sound—a twang rich in tones and overtones. He ran to his assistant exclaiming, "Watson, what did you do then? Don't change anything—let me see!"

What he saw was that the stuck reed behaved like a crude diaphragm. It generated current that varied in intensity just as did the sound waves from a plucked reed—an undulating electric current.

An inspiration came to Bell: If he could make a vibrating diaphragm translate spoken words into just such an *undulating electric current,* he could talk over a wire. He would have a telephone!

His first attempt, in 1875, produced an instrument with a diaphragm of goldbeater's skin. It was intended to serve as a transmitter or receiver. Disappointingly, although you could tell someone was talking into it, you couldn't make out the words.

Within a year Bell exhibited a telephone, acting as either transmitter or receiver, that *did* work. It was powered by the voice alone. To transmit, its vibrating diaphragm generated an undulating electric current; to receive, it worked in reverse, and current vibrated its diaphragm. But Bell's first success had come in the meantime with another kind of telephone.

Foreseeing the need for stronger current than a voice could generate, he contrived

Dramatic scene when telephone first talked, visualized by artist. Alexander Graham Bell had called to his assistant, "Mr. Watson, come here—I want you!" Tom Watson (right) ran in to report he had heard every word.

Clue to secret of telephone was twang of steel reed in Bell's "harmonic telegraph" (left). The first telephone (right), which Bell designed in 1875, was not successful.

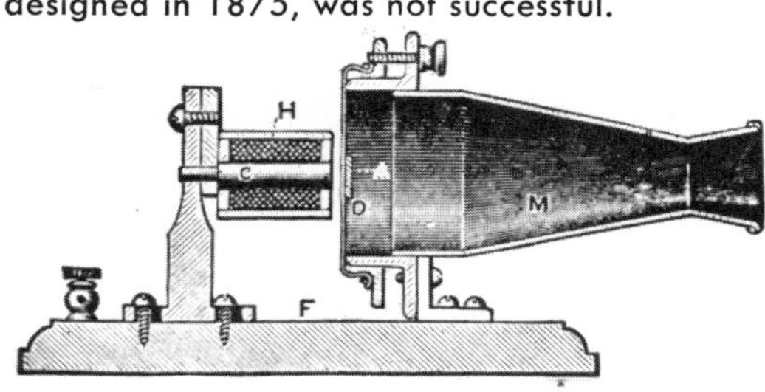

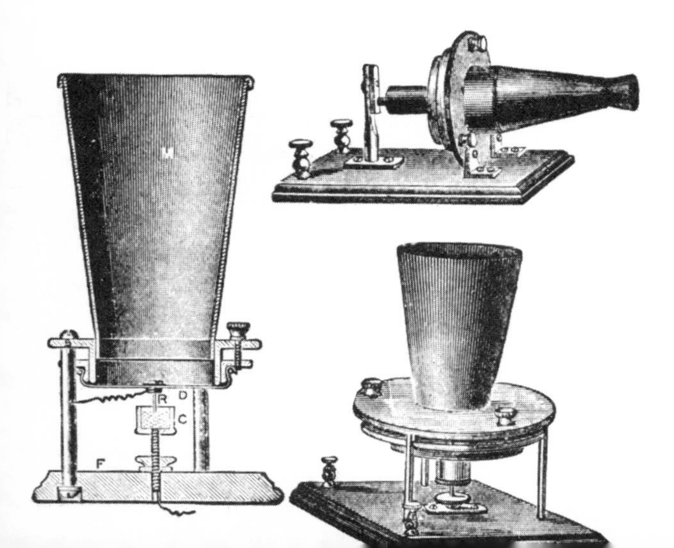

Bell's voice-powered telephone (upper 2 sketches, left) served as transmitter or receiver. Vibration of diaphragm induced current in electromagnet (H). Lower sketches show battery-powered "liquid transmitter." Here, vibration made wire (R) bob in cup of liquid (C), varying resistance to current.

a battery-powered "liquid transmitter." A platinum wire, dipping into a tiny cup of acidulated water, bobbed up and down as the diaphragm vibrated to a voice. Electrical resistance to current passing through the wire and fluid diminished as the wire plunged deeper, increased as it rose. This "variable-resistance" transmitter turned a steady battery current into an undulating one. A simple electromagnetic receiver completed the outfit. It was the liquid transmitter that conveyed, from Bell to Watson, the telephone's first words.

So Bell invented, not one, but *two* telephones: the batteryless voice-powered telephone and the battery-powered variable-resistance telephone.

The voice-powered version was actually used by the very first telephone subscribers. When *Popular Science* in 1878 told about "The Telephone and How It Works," this was the kind it described. Almost immediately it was outmoded and all but forgotten (until sound-powered military telephones in World War II revived the idea).

Bell's greater achievement was to give us the battery-powered telephone. Far as it has come from his pioneering design, it still applies his fundamental variable-resistance principle. And to this day your telephone operates, not on electricity from your house wiring, but on low-voltage direct current from the telephone company's storage batteries. Even during a power failure your phone keeps working.

When two of Bell's brothers had died of tuberculosis and it threatened him, too, the Bells moved from England to the more benign climate of Canada. At Brantford, Ontario, he thrived. In 1872 he moved to Boston, where he opened a school for the deaf and their tutors. Later he became professor of vocal physiology at Boston University.

Deafness loomed large in his life. His mother was deaf. So was his fiancee—Mabel Hubbard, whose father, Gardiner Greene Hubbard, was one of Bell's two financial backers. The other was Thomas Sanders, whose son, born deaf, was tutored by Bell.

Bell's practical familiarity with the anatomy of speech served him well in working on a telephone. He said to his future father-in-law: "If I can make a deaf-mute talk, I can make iron talk."

But when the venerable secretary of the Smithsonian Institution, Joseph Henry, viewed Bell's embryo apparatus in March of 1875 and encouraged him, Bell confessed a handicap: he lacked enough knowledge of electrical science. Laconically Henry replied, "Get it!" Taking the advice, Bell pored over books, to good effect.

The Centennial Exposition of 1876 in Philadelphia offered Bell a chance to exhibit his inventions. He was too busy with his school for the deaf to make the needed applications. It was Mabel Hubbard's father who arranged to display there, on a table labeled TELEGRAPHIC AND TELEPHONIC APPARATUS BY A. GRAHAM BELL, his voice-powered and battery-powered telephones and his harmonic telegraph.

What happened is told in an account of the telephone's early history that appeared in *Popular Science* in 1906:

From Philadelphia, Hubbard urged Bell by wire to come down from Boston to demonstrate his inventions during the judges' inspection the next day. Bell felt he couldn't go, just before final examinations at his school—but he took Mabel Hubbard, who was going, to a Saturday train.

She'd hoped to change his mind. When she realized he wouldn't be there to show off his work to the judges, she vanished, weeping, into the train. That was too much for Bell. Without ticket or baggage, he jumped aboard as the train was pulling out.

Judges at Centennial made Bell prove his telephone wasn't just a toylike "tension telephone," like this, with transmitter and receiver linked by a taut string or wire. This Civil War version, complete with bell to signal a call, was used to help escaping slaves.

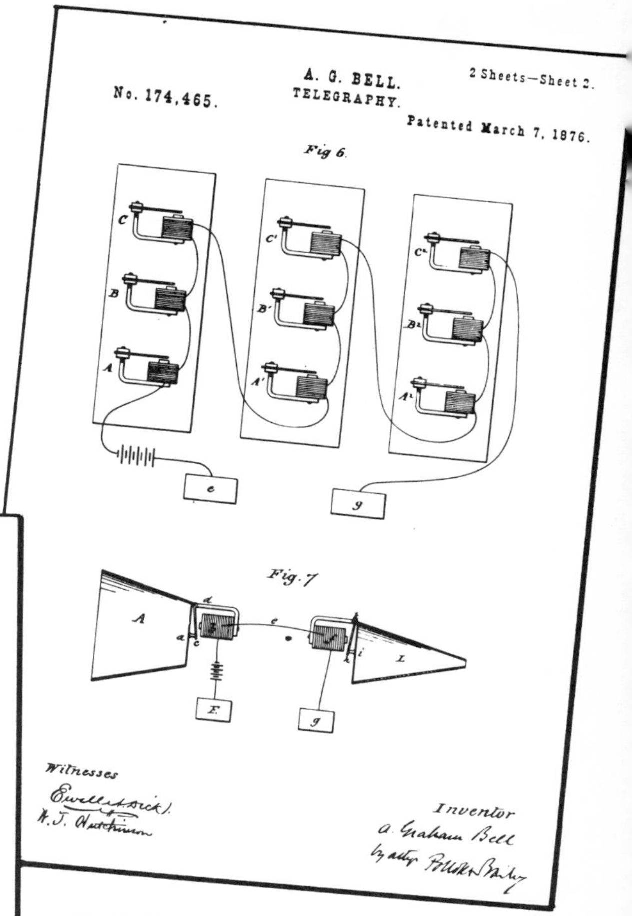

No. 174,465.
A. G. BELL.
TELEGRAPHY.
2 Sheets—Sheet 2.
Patented March 7, 1876.

Fig 6.

Fig. 7

Witnesses
Ewell A. Dick
N. J. Hutchinson

Inventor
A. Graham Bell
by atty. Pollok & Bailey

Bell's first telephone patent, March 7, 1876, used word "telegraphy." Fig. 6 shows Bell's harmonic-telegraph idea; his voice-powered telephone shows up in Fig. 7. Only claim to inventing a telephone is fifth and last item (below) summarizing what the patent covered.

174,465

lucting-wire
rhood of the
the conduct-
nagnet both
ion in each

dulations in
he vibration
nductive ac-
otion of the
hborhood of

dulations in
radually in-
stance of the

circuit, or by gradually increasing and diminishing the power of the battery, as set forth.

5. The method of, and apparatus for, transmitting vocal or other sounds telegraphically, as herein described, by causing electrical undulations, similar in form to the vibrations of the air accompanying the said vocal or other sound, substantially as set forth.

In testimony whereof I have hereunto signed my name this 20th day of January, A. D. 1876.

ALEX. GRAHAM BELL.

Witnesses:
THOMAS E. BARRY,
P. D. RICHARDS.

AT&T PHOTOS

First commercial telephone, 1877, with opening for transmitting and receiving.

Here was telephoning de luxe in 1878. Single instrument of time was shuffled between mouth and ear, for talking and listening—but a well-to-do subscriber could pay extra for 2 telephones and use them as pictured above.

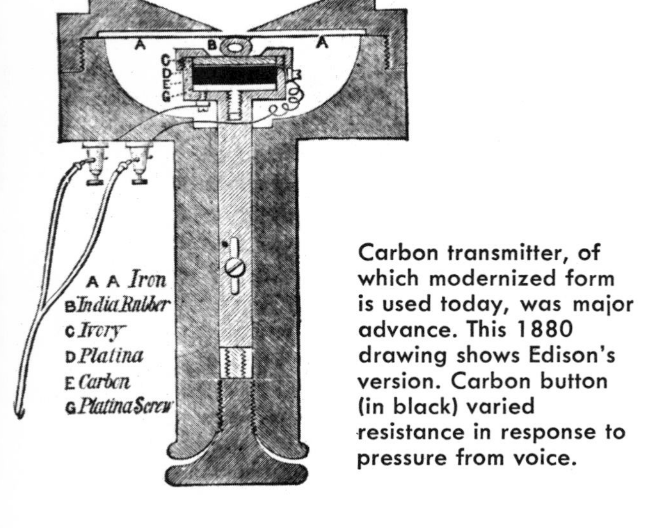

Carbon transmitter, of which modernized form is used today, was major advance. This 1880 drawing shows Edison's version. Carbon button (in black) varied resistance in response to pressure from voice.

It was a sweltering Sunday afternoon in July. Before the perspiring and weary judges reached Bell's exhibit, they were ready to quit for the day. Then a celebrity among them, Dom Pedro, Emperor of Brazil, recognized Bell and greeted him warmly. Dom Pedro had visited the Boston school, which he hoped to duplicate in Brazil. When Bell said he couldn't stay another day to demonstrate his inventions, the Emperor declared, "Well, then, let us look at them now."

Over the telephone, Bell recited the Hamlet soliloquy, "To be or not to be . . ." Listening at the receiving end, Dom Pedro shouted, "My God, it talks!"

Some of the judges insisted on proof they were not being tricked. They made Bell move his apparatus to another spot, to show it was really electrical—and not some form of the toylike old "tension telephone" which carried a voice mechanically over a taut string or wire. Bell worried over whether his delicate instruments would work after being moved, but they did—and earned him a Centennial prize.

Bell was married the next year. His wedding gift to his bride was all but a tiny share of his interest in his patents.

Bell insisted, too, that Hubbard and Sanders share in the profits from the telephone —though it was not mentioned in the written contract. The agreement had been that in return for their backing, they'd have a part of the profits from his harmonic telegraph, which was what they expected to be the real moneymaker. In 1876 Hubbard actually offered to sell Bell's telephone to Western Union for $100,000—and was turned down. Few then regarded the telephone as more than a scientific toy.

For some years *Popular Science* seemed uncertain who really invented the telephone. One early article credited Elisha Gray, another, Johann Philipp Reis. Both men *had* invented instruments they called "telephones," and lawsuits began.

A court decision of 1881 held Reis "a man of learning and ingenuity" but ruled that with his apparatus "articulate speech could not be sent and received . . . A century of Reis would never have produced a speaking telephone by mere improvement in construction."

Elisha Gray—by coincidence, one of Bell's judges at the Exposition—had the advan-

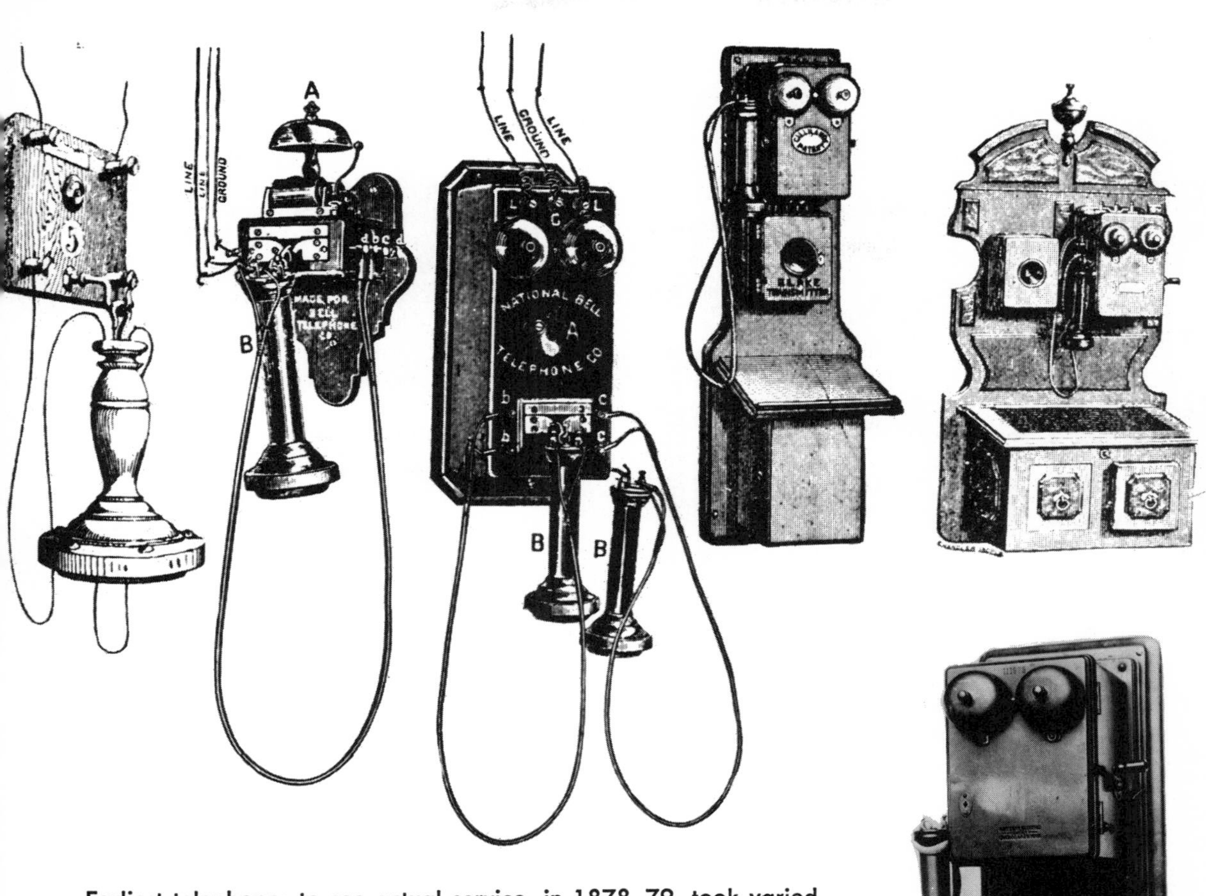

Earliest telephones to see actual service, in 1878–79, took varied forms above. From left: mahogany phone of subscriber to first exchange, in New Haven; phone with a bell (Central transmitted a squeal before); set with 2 duplicate phones, for talking and listening; and plain and fancy wall phones using Blake transmitter. Right, 1886-model phone, like some predecessors, had hand-cranked magneto that generated current to signal Central.

AT&T PHOTO

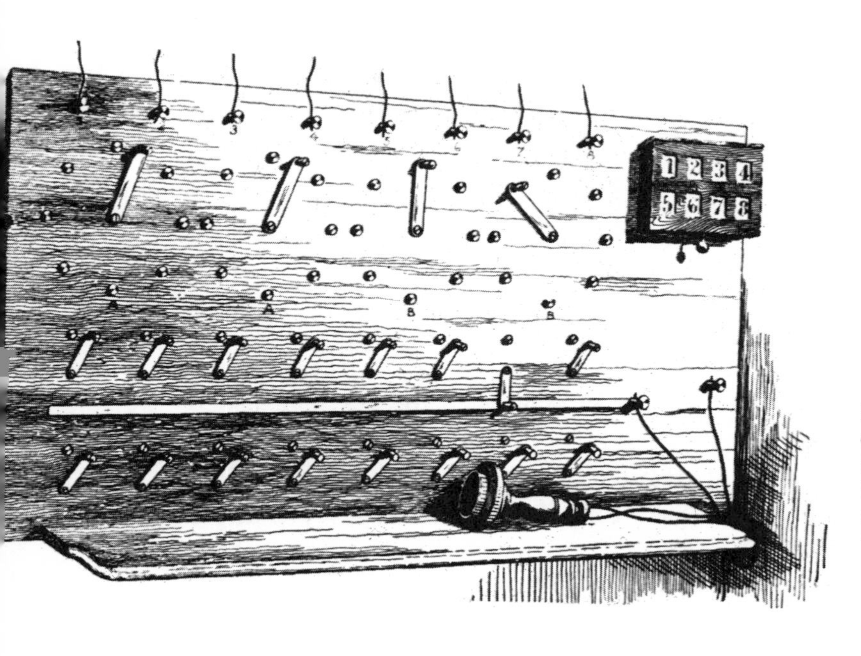

Boy at this switchboard of 1878, which serviced 30 subscribers in New Haven, responded to caller by demanding, "What do you want?"

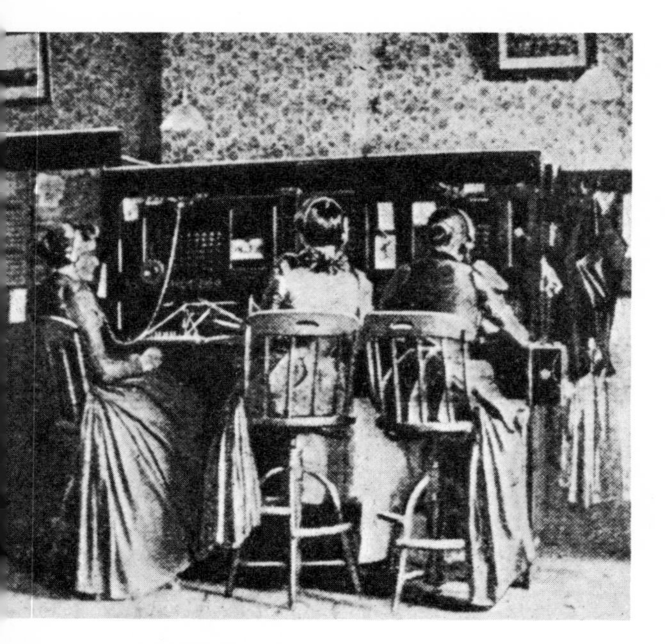

The telephone girl arrives. Early photograph shows Central operators at work on first crude multiple switchboard. It could service only a few hundred subscribers.

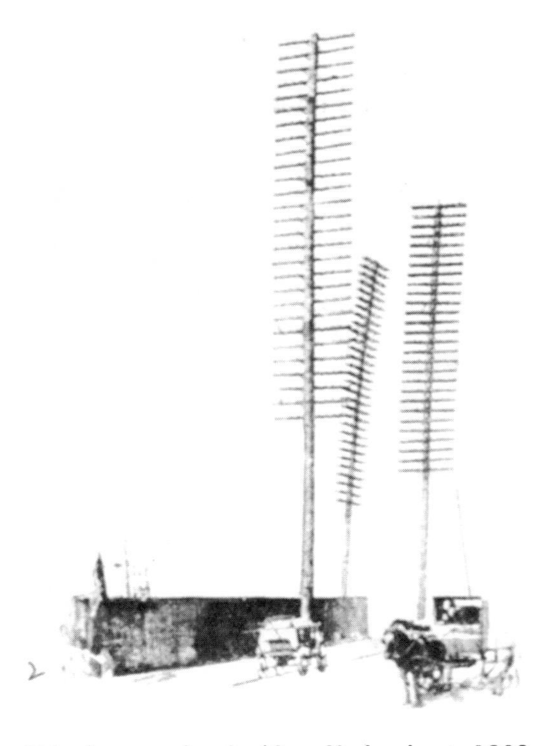

Telephone poles in New York about 1903, carrying some 280 wires, were the tallest ever erected. Their appearance brought public demand that lines go underground.

tage over him of being an expert electrician. By another coincidence, Gray submitted to the Patent Office a "caveat"—in those days a procedure indicating intention to invent —on the same day and just a few hours after Bell filed his patent application. Gray fared no better than Reis in the courts, which finally decided the telephone's inventor was Bell.

Bell's victory was clinched by his first telephone patent—granted three days before the telephone first talked and called the most valuable single patent ever issued. If you expect it to be all about the telephone, a look at it is a whopping surprise. Its title: "Improvement in Telegraphy." Mostly, its six pages describe ways to transmit an undulating electric current, and proposed applications—to Bell's harmonic telegraph.

Almost as an afterthought the text adds that the transmitter can be actuated "by the human voice"—and "a similar sound to that uttered" will then be heard at the receiver. Fifth and last of Bell's concluding claims of what he'd invented is a method and apparatus "for transmitting vocal or other sounds telegraphically." Only the last of seven drawings is of a telephone. It resembles Bell's voice-powered version. (Although pictured hitched to a battery, it didn't need one, Bell soon found.)

Just the bits and pieces on the telephone in this patent were enough. They were such conclusive evidence of Bell's priority as to withstand the most determined legal onslaughts against them.

Certainly, Bell's telephone was open to improvement. His liquid transmitter, which pioneered the key variable-resistance principle, was better suited to tender handling in a laboratory than to the rugged life of a household or business phone. It took the help of other inventors to bring the telephone's transmitter to practical form.

Thomas Edison conceived a major new idea for a variable-resistance transmitter. Instead of a wire dipping in liquid, he envisioned a solid "button," of a substance whose electrical resistance would vary with pressure from the vibrating diaphragm. In his tireless way he tried 2,000 chemicals.

One day an assistant brought him a broken lamp chimney, encrusted with lampblack. Edison scraped it off, pressed it into a little cake, put it between two plates of

Cross-country telephone lines went overland on poles, as pictured above, for many decades. Illustration from a 1930 AT&T advertisement in *Popular Science* depicts "An inter-city cable, part of the Bell System network that unifies the nation." Compare a later way of installing an overland telephone cable, at right.

First underground transcontinental phone line was laid in 1941 by this plow train. A caravan of machines dug a trench, lowered 2 cables in it, and filled in earth as it crawled across country. First underground coast-to-coast coaxial cable came soon after.

In October 1892 Dr. Alexander Graham Bell opened the first New York-Chicago long-distance line, 950 miles long. It consisted of a single circuit and could carry only 1 conversation at a time. Thick copper wire used weighed 400 tons and cost more than $100,000.

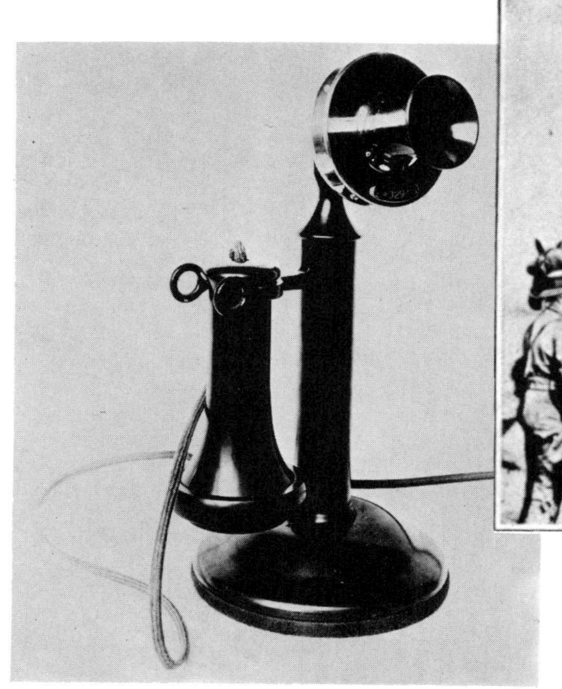

AT&T PHOTO

Horse-drawn rig, with earth-boring auger turned by a gasoline engine, dug 40 post holes a day for 1915 transcontinental line.

Upright phone held sway from 1915 to 1927, unchanged except for added dial at base.

metal—and had a successful "carbon-button" telephone transmitter.

Before claims of rival "inventors" were settled, there was lively competition between Bell's group and Western Union, now also seeking to establish a telephone business. Western Union acquired Edison's transmitter, to use with Gray's receiver.

Later Edison told *Popular Science* of being invited to the office of President Orton of Western Union to discuss payment for his transmitter. Edison hoped to get $25,000—but on the way, he recalled, "It seemed to me that what I had done was easily accomplished, and I felt shaky and uncertain about asking for so much."

Face to face with Orton, Edison proposed that he make an offer. Perhaps Orton remembered the Bell deal he'd refused. "One hundred thousand dollars," he said. Edison's reticence paid off handsomely.

Bell's group sought a transmitter able to compete with Edison's, and found it. First adopted was a type devised by Emile Berliner, pioneer of the microphone. Then came a transmitter invented by Francis J. Blake, an improved form of a carbon transmitter like Edison's. Granules of carbonized anthracite coal continue in use today.

In 1879 Western Union acknowledged the validity of Bell's patents and bowed out—leaving the field to the Bell enterprise.

At first, your phone looked like an oversize receiver. You could talk into it, or listen. When you finished speaking, you shifted the instrument from lips to ear to hear the reply. A lively conversation called for no little agility. Or you could spare yourself the exercise by indulging in the luxury of two telephones—one for talking and the other for listening, like the bearded gentleman pictured on page 132.

If you shouted at the top of your voice you might be heard, above squeaks, squawks, and howls. You got this noise because when two wires were strung side by side (say, a phone wire and a telegraph line), an effect called induction caused interference. At first, trolleys and arc lights in the streets caused interference too, because these and telephones all used the earth as a return path for the circuit; it was as if all were on the same wire. The solution was to add a return wire to a telephone line—a "full metallic circuit."

The first telephone wires, like telegraph wires, were of iron. Copper was a better conductor—but so soft then that a copper-

Popular Science Monthly

How Your New Automatic Telephone Will Work

Nearly every home in America is going to be its own exchange

By **Raymond Francis Yates**

WITHIN a few years telephone operators will be as scarce as horse-car drivers. With automatic telephony, you have only to manipulate a dial at the base of the telephone. All kinds of little "jiggers" at the telephone exchange dance around rapidly, going about their task of connecting your line with the number you are calling. The number called is recorded, an idle trunk-line between the exchanges is found, the proper line is picked out at the distant exchange and the bell of the distant telephone is rung.

In place of giving the number to the

letter N, and for the number. Only the first three letters of the exchange are "dialed." This makes seven times that the dial must be turned and allowed to return to its normal position.

When you lift the telephone receiver from its hook, a number of little busybodies at the exchange get into action, select your line, and automatically connect it with an idle "sender." When this is done, you are notified by a gentle

The dial of the automatic telephone. It works like this: If you wanted the number, Garfield 225, you find the number printed in the book, "GARfield." This means that the letters GAR would have to be "dialed." To do this, the finger is placed in the hole containing the letters GHI. The HI would have nothing to do with the call. The dial would then be turned until the finger came to the stop and then it would be allowed to return. The same operation would be fol-

The years-long task of converting New York City's telephones to dial operation was about to begin when this November 1921 article introduced *Popular Science* readers to dial telephoning. There had been dial telephones before the turn of the century, but their wide use awaited automatic exchanges. A caption explained: "When the subscriber operates his dial, the electrical impulses are submitted to a 'sender' which receives and registers them and in turn translates them . . . Each automatic exchange will be equipped with hundreds of these little senders, which are a vital part of the system."

wire line would stretch and break of its own weight. When hard-drawn copper wire was perfected around the end of the 1870s, it replaced iron telephone wire—and, *Popular Science* reported, became "one of the fundamental contributions to modern communication."

Bell and his partners had made two important decisions: that the telephone should be leased, never sold, and that it would have to have a "central" to interconnect subscribers. The first telephone exchanges were in Connecticut.

George W. Coy mailed a circular to 1,000 prominent people in New Haven, offering to install a telephone connected to an exchange. He expected at least 50 replies; he got 1, from a pastor. Persevering, he rounded up about 30 more subscribers, and opened the first commercial telephone exchange in January of 1878. Others sprang up rapidly.

An early caller was greeted by Central with a brusque "What do you want?" Boys served as operators. The telephone girl's dulcet "Number, please?" came later. Then dial telephoning arrived, to connect you with your party without human intervention.

There were dial telephones before the turn of the century, but wide use awaited automatic exchanges. When *Popular Science* introduced readers to dial telephoning in 1921, it had only just begun—in Norfolk, Dallas, and Omaha. New York City had still to commence the huge conversion.

Undreamed of then were some of today's refinements of dial telephoning. Latest is ESS, the electronic switching system—a computer-controlled exchange. In contrast with the earlier electrical-and-mechanical equipment, ESS is virtually all-electronic and has practically no moving parts.

The things it can do! You're going to a friend's home for the day. You pick up your phone and dial his number, along with a brief code number meaning "transfer." Your incoming calls will then be switched automatically to your friend's house.

Then there's "camp on." If you're phoning and get a busy signal, you just dial a code number, then the number you want. The electronic exchange dials your party as soon as he hangs up, and calls you back.

Another wrinkle lets you make calls with only three or four twirls of the dial. You give the phone installer a list of numbers you call most often, paired with identifying numbers like 1 to 30. ESS stores the paired numbers in its electronic memory. From then on, you need only dial an "11" and the identifying number to get your party—say, "114" if you want 818-4244.

The first long-distance lines came slowly. Electrical undulations carrying your voice over ordinary telephone wires weakened with the miles. Copper wire as thick as your finger, the only known answer in 1892, cost $130,000 for the 950-mile New York-Chicago line opened that year.

Here entered Michael I. Pupin, who as a boy had sold his schoolbooks and a watch to get to America from Serbia and ended up as a Columbia physics professor. In 1900 he invented a loading coil which, when inserted at intervals in a telephone line, helped to carry conversations farther. To deter electrical undulations' weakening with distance, it employed induction—the former enemy of telephone men, which oddly now became their ally. The Pupin coil "made transcontinental telephony possible," AT&T's research director J. J. Carty later recalled in *Popular Science*.

In 1915 the first coast-to-coast telephone line was opened, between New York and San Francisco. *Popular Science* reported that it was necessary to dig 13,900 postholes for the telephone poles.

Into the story of the telephone now came Lee De Forest's 3-electrode vacuum tube—the Audion, or triode tube, which already had revolutionized radio. Being an amplifier, the Audion could periodically boost the strength of voice-carrying electrical undulations on a long line—instead of merely preventing weakening. Vacuum-tube "repeaters" began to supplant Pupin coils.

Radiophone experiments, too, could begin with De Forest's tube. Trials in 1915 transmitted speech between the U.S. and Hawaii, from the Eiffel Tower in Paris to Arlington, Virginia. Land telephone lines joined radiophone links to remote cities.

Finally, the new tube accomplished the feat of "multiplexing" the telephone—sending more than one conversation at once over a wire. It provided radio waves that traveled along a phone line to carry additional conversations, in what was called a "carrier system," or wired radio. "Five telephone conversations are carried on today simultaneously over one toll-line circuit," said a 1919 AT&T advertisement.

That was just the opener. Successfully tested between New York and Philadelphia in 1936 was a major innovation, the coaxial cable, which carried 240 conversations at once. Less than an inch in diameter, the tubular cable did the work of a 480-wire bundle as thick as a man's arm. (Today some coaxial-cable installations carry more than 30,000 simultaneous conversations.)

Meanwhile the radiophone experiments of earlier years bore fruit. In 1927 transatlantic telephone service via radio began between New York and London. Candidly AT&T admitted in a *Popular Science* advertisement that the service operated only "at certain hours daily" and "static will at times cause breaks in the ether circuit."

There had to be a more reliable way for America to talk with Europe—and there was. By 1955 the first transatlantic telephone cable linked New York and London. It added 36 phone circuits by submarine cable to the 12 then available by radio.

Actually it was a pair of coaxial cables; one eastbound, one westbound. At 40-mile intervals, little repeater stations—102 in all—were capable of amplifying voices a million-fold. The heart of each repeater was a trio of electron tubes rugged enough for 20 years of constant service—since it would cost $250,000 to replace one!

At about the same time, telephoning overland without wires made its debut: Beams of microwaves appeared on the scene, offering tremendous carrying capacity to aid in coping with a phenomenal increase in long-distance phone calls.

Microwave beams consist of radio waves of extremely high frequencies (around 4,000 to 11,000 megacycles) focused into a narrow bundle like a searchlight's. They also travel only in a straight line, to points within range of sight. So they are relayed across country between towers, spaced at intervals of, say, 30 miles apart.

In a story on microwaves in 1962, *Popular Science* prophesied: "If someday we wish to talk with men on the moon, existing microwave systems can handle the job."

And they did.

The oldest telephone some of us remember, the long-lived upright telephone of

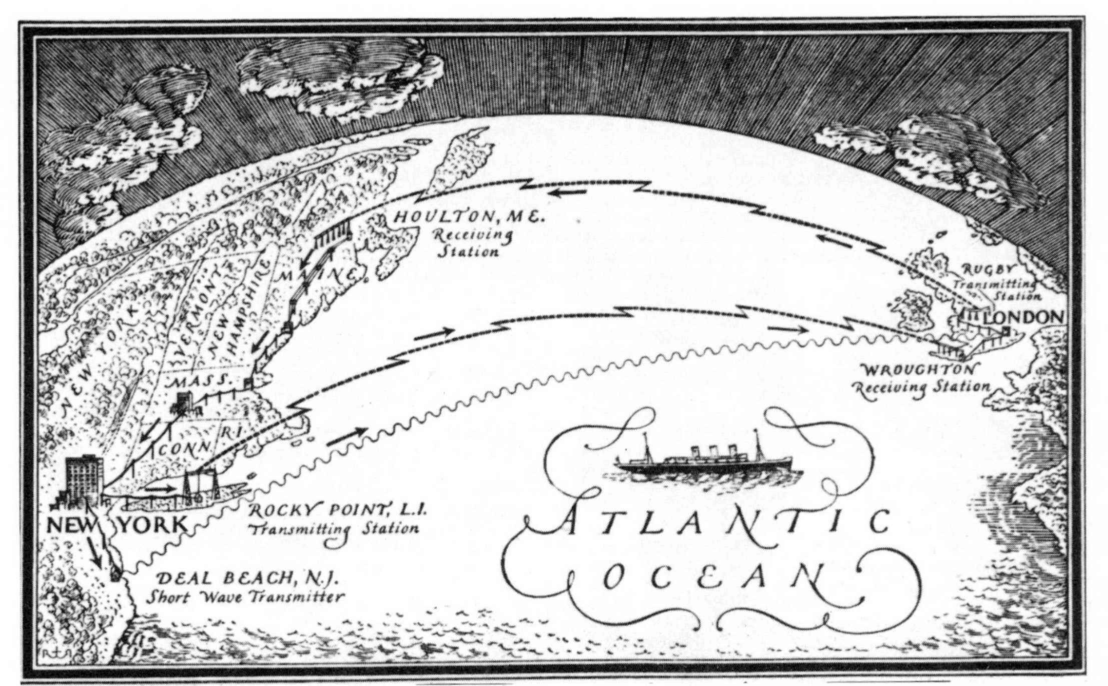

Transatlantic radiophone service began in January 1927. Map showing radio paths and **stations is from AT&T ad in** *Popular Science* **that frankly admitted service was limited to** "certain hours" and at times marred by static. In bad magnetic years, circuits often **failed.**

"French" telephone (or handset) of 1928 conveniently combined transmitter and receiver in one piece for ease of handling. It wasn't until 1954 that phones departed from staid black and became available in 8 colors.

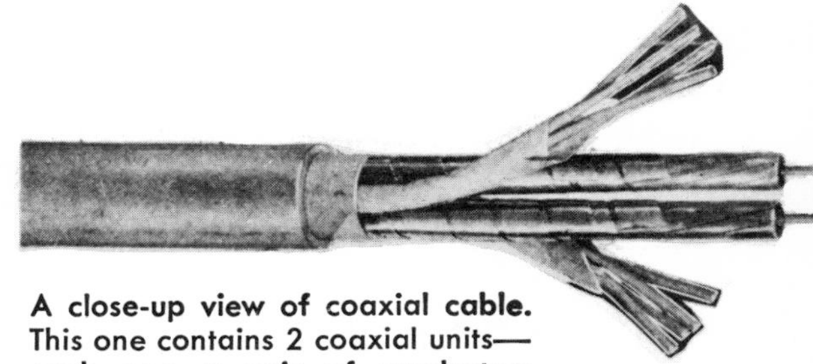

A close-up view of coaxial cable. This one contains 2 coaxial units—each one a pair of conductors made up of a copper tube and a wire running along its center—as well as 8 other wires.

Experimental coaxial cable laid between New York and Philadelphia proved able to carry 240 telephone conversations simultaneously, doing work of standard 480-wire cable. This 1937 picture shows an engineer at panel that controlled messages on the coaxial cable.

Before Morse's telegraph, observers with telescopes worked semaphore arms on towers atop hills to relay messages from one station to the next. Title of this picture in the magazine was "Aerial Telegraphy, 1793."

Microwave relay towers bring back hilltop-to-hilltop communication in sophisticated new guise, to carry myriads of telephone conversations. Chain of 7 linking New York and Boston (below), and detailed breakaway of a tower (facing page) are pictured in these 1947 illustrations from *Popular Science*.

1915, remained unchanged (except for adding a dial at its base) until 1927. That year saw transmitter and receiver combined in the more convenient handset, as the telephone company calls it. Telephones in a choice of hues struck a colorful note in 1954. Pushbutton calling came along in 1960 with the Touch-Tone phone. Now on the way is an electronic telephone, smaller than your present phone and weighing half as much.

Meanwhile has come the marriage of telephone and TV in the Picturephone, over which people can talk and see each other at the same time. When introduced in 1965, it had a little 2-by-3-inch screen. Now a screen about twice as large each way gives a wide-angle view, a long-range one, or a close-up of a picture that you want to transmit. On the drawing boards are color and 3-D Picturephones.

For the telephone, the sky is literally the limit. Where microwave towers cannot go—across the seas—satellites now serve as microwave relay stations for telephone and TV circuits. In synchronous orbits 22,300 miles high, they hover motionless with respect to the rotating earth at their mid-ocean stations.

A single one of the latest drum-shaped 17½-foot-high Intelsat IV telephone satellites, writes Dr. Wernher von Braun in a recent issue of *Popular Science,* transmits an average of 5,000 simultaneous overseas conversations—compared to a capacity of 750 via the fifth and latest transatlantic telephone cable.

Soon we should see overland telephone links via satellite, too. At this writing the Federal Communications Commission was expected presently to give its nod to one of seven rival proposals, submitted by leading communications firms, for establishing

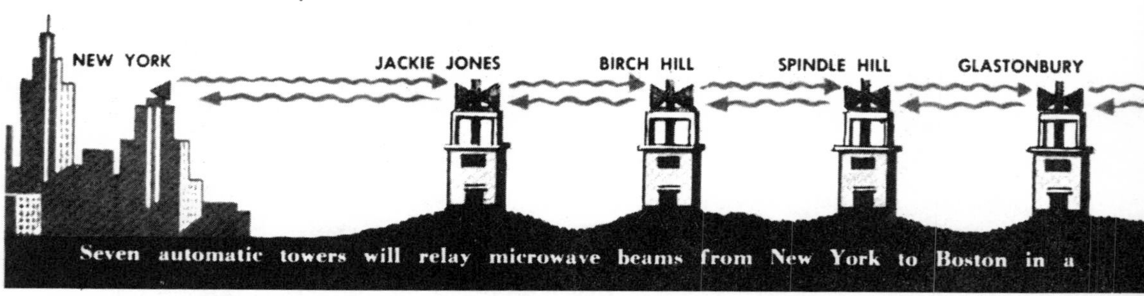

telephone system, soon to begin operation. Large picture shows layout of one tower.

Bell's early batteryless version of the telephone was forgotten until voice-powered one reappeared on U. S. Navy warships in World War II. Photo shows use by seaman at range finder to report range to gunner.

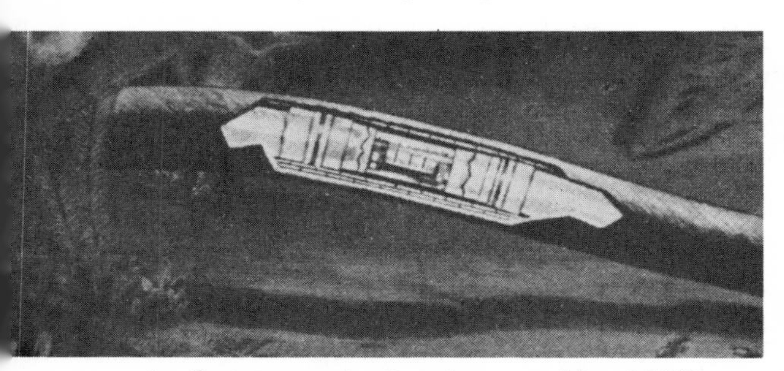

In first transatlantic phone cable, 1955, repeaters at 40-mile intervals contained 3 electron tubes able to last twenty years.

a domestic satellite microwave relay system. By some estimates it could be in operation by 1974 or 1975.

What lies beyond? Foreseen for the late 1970s are fantastic underground "pipelines" for telephone messages between distant cities and coast-to-coast. A copper-lined steel pipe no bigger around than your wrist, buried about 4 feet deep, will carry an almost incredible quarter-million conversations simultaneously—as many as if the entire population of Richmond or Dayton were talking at once.

The pipe will convey its torrent of speech on "millimeter" radio waves, so called because only a few millimeters in length. Their corresponding high frequency of 40 to 110 billion cycles, much higher than for microwaves in current use, gives them more carrying capacity than all lower radio frequencies combined.

Like a high-speed railroad, early studies indicate, the millimeter-wave pipe must follow a virtually straightaway path and have only the most gradual curves. To work out its unique features, the Bell System plans a 20-mile-long experimental installation in northern New Jersey in 1974.

That is a glimpse of the telephone's past, present, and future. Today it seems astounding that men eminently successful in industrial and commercial circles once ridiculed Bell as "the man who is trying to make peo-

World's largest cable-laying ship, the 8,000-ton *Monarch,* pictured off Newfoundland where it began in 1955 to lay first transatlantic telephone cable to Scotland. Below, right: unbroken miles of cable, painstakingly coiled by hand in ship's huge tanks. Cable added 36 phone circuits under the sea to the 12 radio circuits already available.

ple believe you can talk through a wire" and were even shortsighted enough to declare that "the telephone could never be of any practical use in business affairs."

Actually the amazing instrument that Alexander Graham Bell first exhibited on a sultry July afternoon in 1876, at the Centennial Exposition in Philadelphia, has since then made over the nation's habits. In business it has separated a company's office from its factory and enabled far-flung activities to be directed from a central headquarters. At home, it has shattered the barrier of distance to put you in instant touch with family, friends, and all whose help you need in anything from everyday chores to illness or emergency.

And its story is not yet done.

Touch-Tone telephone was introduced in 1960 as alternative to dialing and *Popular Science* showed this new version in 1964. Note use in Picturephone at bottom of page.

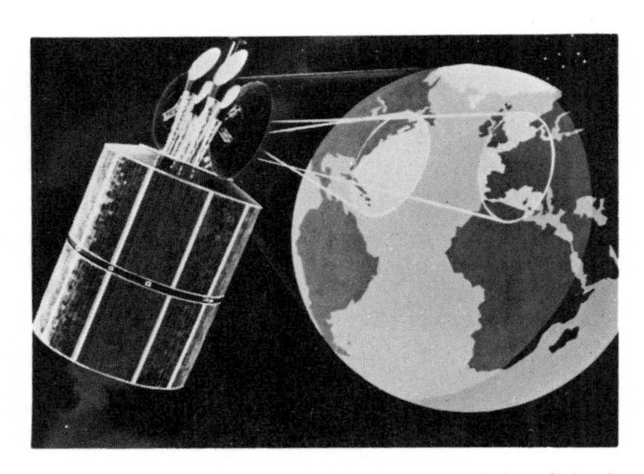

Drawings compare satellites for transocean telephoning, from Early Bird of 1965 to Intelsat IV of 1971.

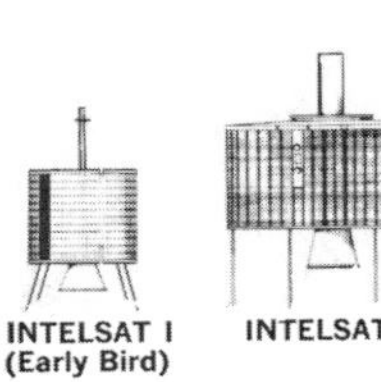

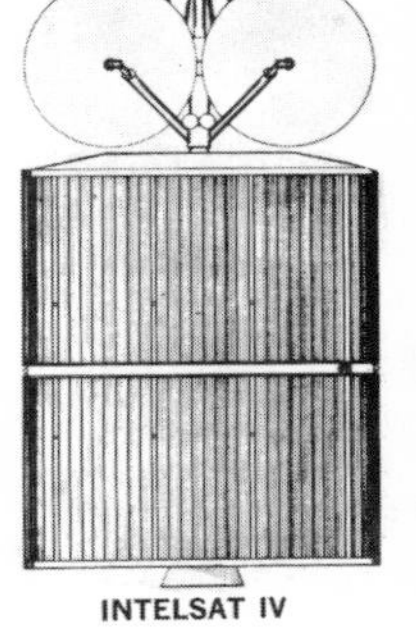

First Picturephone, 1965, had a 2-by-3-inch screen. Improved 1968 version (below, left) gives choice of bigger wide-angle picture, long-range shot, or electronic close-up. Control unit regulates camera height, electronic zoom, image brightness, and speaker volume. Below, right, "graphic mode" for sending charts or documents is shown.

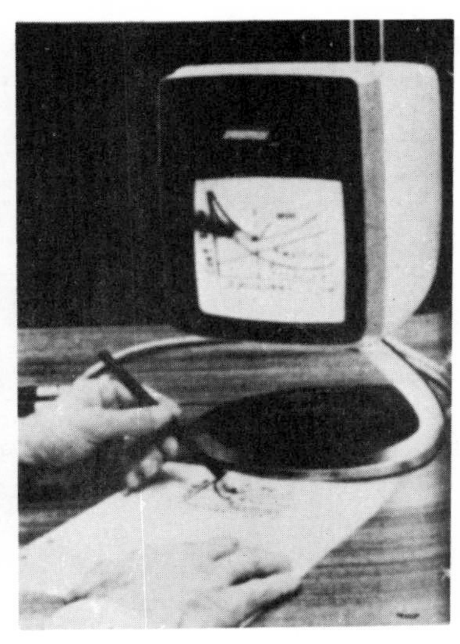

CANNONBALLS

SHELLS

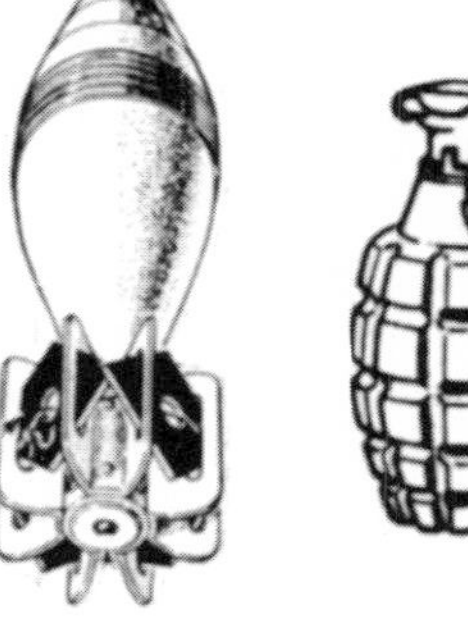

MORTAR SHELLS

HAND GRENADES

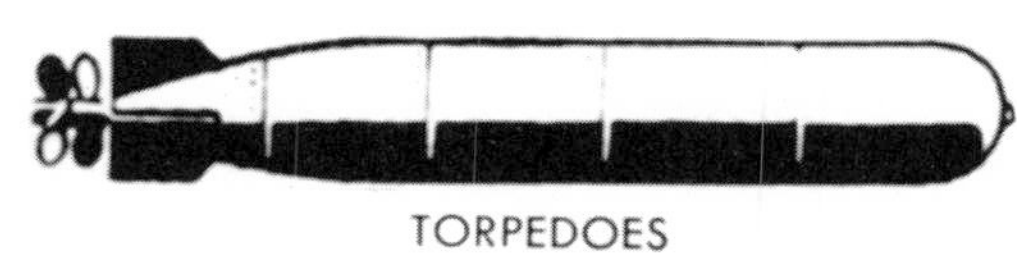

TORPEDOES

BARRAGE ROCKETS

CHAPTER 8

Military Science— From Cannonballs to ICBMs

By 1872 the maximum range of a cannonball had just about reached 3 miles: the wishful figure that had led the U.S. in 1793 to proclaim its sovereignty out to a 3-mile limit beyond its shores.

In 1972 intercontinental ballistic missiles could strike an adversary's heartland 8,000 miles away—an apt example of the spectacular changes wrought by a hundred years in military science.

Until the world can police itself effectively, self-defense still forces a nation to lavish its inventive talent upon instruments of warfare. Do they belong in this chronicle of a century's accomplishments, along with more constructive products of man's ingenuity? We believe they do, for compelling reasons:

Military science has been the innovator —first to introduce jet planes, nuclear-propelled ships, many another radical and costly new development. Through its pioneering the world has often gained peaceful advances that would otherwise have come far later, if ever: atomic power, space rockets, radar, for example.

And it is military science that calls forth the fantastic extremes of machines' performance. For life-or-death struggles it drives ships and aircraft beyond all normal limits of range, speed, endurance.

No wonder, then, that the special fascination of military gear makes armchair generals and naval buffs of the least bloodthirsty of laymen. To them, as to historians, the clash of these mighty engines of war in battle offers drama transcending all peacetime experience.

Popular Science has mirrored that interest. How thoroughly it covered military science may be gauged from the fact that every one of the pictures in this chapter is from its pages; there was no need to look further to illustrate a single important detail. In 1945 the magazine could trace back World War II's bazooka to a novel ex-

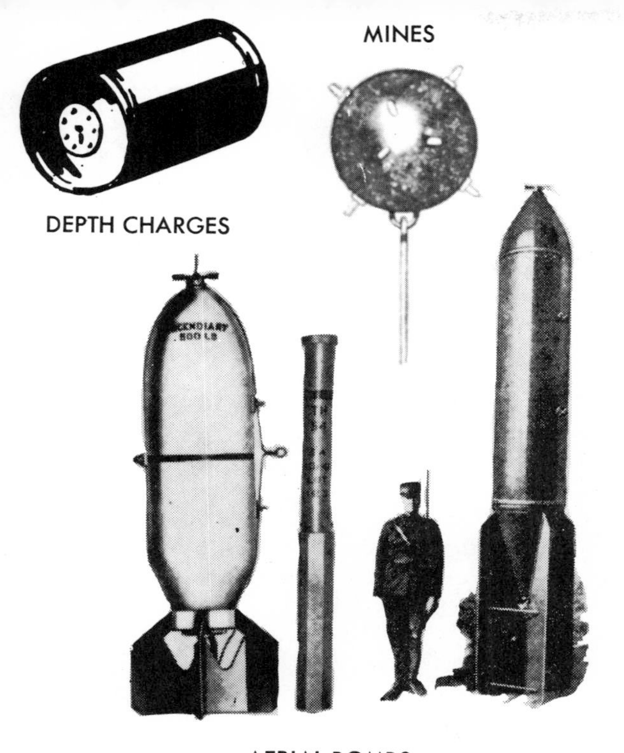

Munitions of military science, over a century, have taken forms as varied as these. Sampling of *Popular Science* pictures shows kinds from post-Civil War guns' cannonballs to the jet and rocket missiles of recent years.

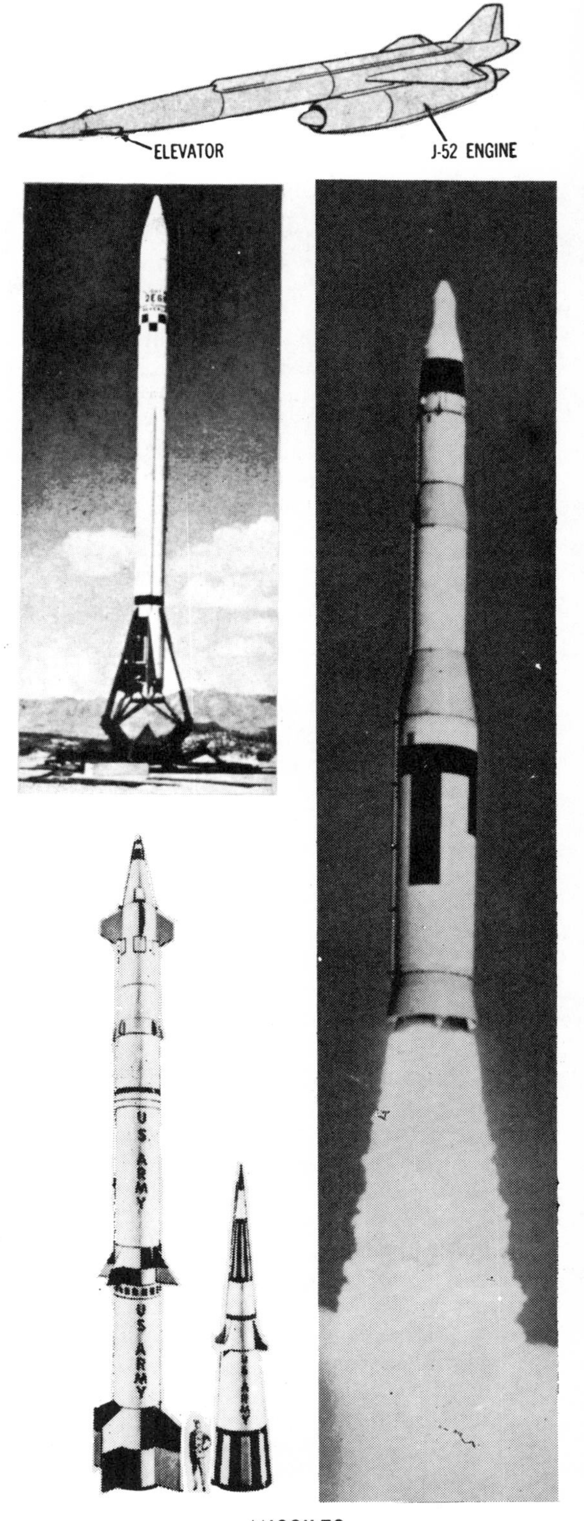

plosive phenomenon, the "Munroe effect," announced by the discoverer in *Popular Science* in 1900. The Navy Department's Director of Naval History wrote in 1962, "We often refer to your magazine for events of the last century."

What it saw and described in a hundred years, a "pictorial scrapbook" forming most of this chapter can tell better than words. For an introduction, only the briefest summary of highlights will suffice:

• *Land forces* gained a formidable weapon in Maxim's machine gun of 1883 and its successors. An attempt to break the deadlock of World War I's trench warfare brought the surprise innovation of military tanks—forerunners of potent modern versions. Star shells and searchlights introduced night warfare. A German "mystery gun" shelled Paris from the amazing range of more than 60 miles; its secret was an exceptionally long barrel, *Popular Science* later revealed.

Between two world wars, mounted cavalry gave way to motorized troops—and horse-drawn gun carriages, to self-propelled artillery. France built the last of the world's great fixed fortifications, the Maginot Line—fated to fall in short order to a German blitzkrieg.

In World War II recoilless guns and bazookas gave infantry the firepower of field artillery. Rocket missiles began to vie with shells. The revolutionary proximity fuse for shells and missiles turned misses into hits.

• *On the sea,* steam warships of 1872 and later still had auxiliary sails. From the Civil War's *Monitor* and *Merrimac* evolved U.S. fighting ships that won the Spanish-American War. The submarine *Holland,* which in 1900 became the U. S. Navy's first one, foreshadowed the undersea warfare to come in the decades that lay ahead. All-big-gun battleships arrived with H.M.S. *Dreadnought* and U.S.S. *Michigan.* German U-boats of World War II sank the *Lusitania* and worked havoc with Allied shipping.

Between world wars a bitter controversy over air power versus sea power followed inconclusive trials of bombs against warships—and led to court-martial of a too-outspoken champion of aviation, Brigadier General "Billy" Mitchell. Navies, luckily hedging, adopted aircraft carriers.

In World War II, U.S. carrier forces won the decisive Battle of Midway, first ever fought by fleets beyond sight of each other. Amphibious landing craft spearheaded the invasion and liberation of Nazi-conquered Europe and the island-by-island recapture of Pacific strongholds from the Japanese. Battleships slugged it out for the last time, at Surigao Strait in the Philippines, in a great naval battle that crushed Japanese sea power. By war's end the U.S. had succeeded hard-hit Britain as ruler of the seas.

Since then the world's battleships have gone—the last, U.S.S. *New Jersey,* after twice emerging from retirement, to contribute its big-gun fire off Korea and then Vietnam. Replacement of guns by missiles has been remaking other warships' armament; and nuclear propulsion gives them unlimited range at top speed.

• *In the air,* except for Civil War observation balloons, military aviation began with U.S. purchase of a Wright airplane in 1909. First to use airplanes in warfare was Italy, in 1911, for reconnaissance and bombing over Libya.

World War I airplanes flew observation missions, strafed trenches—and, like raiding zeppelins, dropped bombs. Air combats began. Lowering a brick on a rope served one pilot to splinter an enemy craft's wooden propeller. Then military airplanes were armed with machine guns and small cannon.

Primitive guided missiles were being developed as early as the twenties.

In World War II's desperate Battle of Britain, England's fighter pilots beat off Hitler's bombers. Japanese carrier planes sneak-attacked Pearl Harbor. In turn, Germany reeled under long-range strategic bombing, and incendiary bombs ravaged Japanese cities. Germany struck back at England with the first long-range missiles—the V-1 "buzz bomb" and the V-2 rocket. A new military era began in 1945 when atom bombs—the only two used in warfare to date—hit Hiroshima and Nagasaki in Japan and ended World War II. (Another chapter, "The Atomic Era Arrives," tells their story.)

Later have come hydrogen bombs of still more fearsome power and intercontinental missiles to carry them across the seas. As of now, the two superpowers, the U.S. and Soviet Russia, both possess awesome "deterrent" arsenals of these nuclear weapons; neither could loose them at the other without incurring a catastrophic blow in return. Knowing it has seemed to put a damper on reckless "rocket-rattling."

Soft-pedaled, too, in recent years are even the scaled-down "tactical" nuclear weapons being touted for battlefield use a decade or two ago—for example, an "atomic cannon" unveiled for pictures to *Popular Science*'s own photographer—but conspicuously unused in combat to date. Is the distinction too close for comfort between nuclear weapons that wouldn't, or just possibly might, risk provoking escalation toward all-out nuclear war?

Perhaps, after all, if wars there must be, they may continue to be fought with "conventional" weapons—those of the past, plus fantastic new non-nuclear ones such as laser weapons, reported on the way. And meanwhile a review like this one of military science's past still has valid lessons.

The Early Days

Pot-bellied Rodman guns, dating from mid-1850s, were still being provided for coast defense up to about 1885. These and naval Dahlgren guns of same period had smooth bores and fired cannonballs.

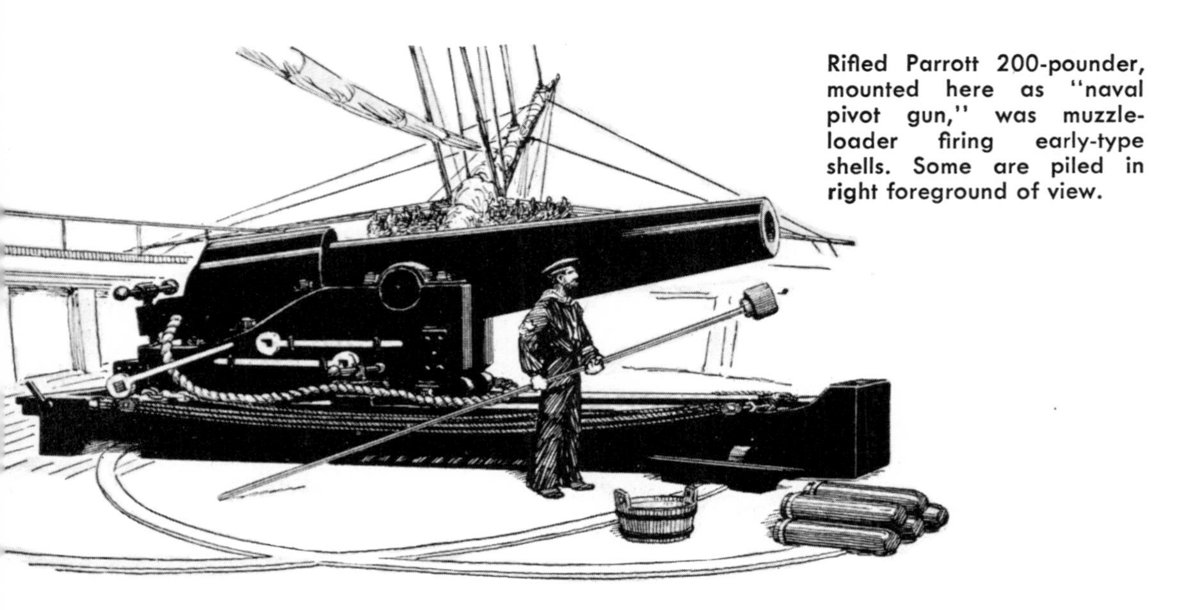

Rifled Parrott 200-pounder, mounted here as "naval pivot gun," was muzzle-loader firing early-type shells. Some are piled in right foreground of view.

Navies clung to auxiliary sails long after arrival of steam warships. Picture from 1893 *Popular Science* article, "Modern War Vessels of the United States Navy," shows cruiser *Cincinnati* decked with sails.

Enter the ironclads. Capt. John Ericsson's *Monitor* of Civil War fame set the style for about 50 more monitors built by U.S., the last completed in 1903. A monitor's revolving turret and low freeboard gave it the curious "cheesebox-on-a-raft" appearance above.

Confederates' iron-sheathed *Merrimac* was on rampage of sinking wooden Union warships by gunfire and ramming when, in 1862 battle pictured, *Monitor*'s 11-inch guns drove it out of war. Some of later ironclads were armored like *Merrimac*.

Nothing can more thoroughly awaken a feeling of awe than the sight of immense structures like the great modern iron-clads (Fig. 65), vessels having a total weight of 8,000 to 10,000 tons, and propelled by steam-engines of 8,000 or 10,000 horse-power, carrying guns whose shot penetrate solid iron fifteen inches thick, and having a power of impact, when steaming moderately, sufficient to raise 35,000 tons a foot high.

View of "modern iron-clads" and text are from a *Popular Science* issue of 1878.

First true machine gun, using recoil of cartridge to load and fire next one, was lethal creation of Hiram Maxim in 1883. It attained unheard-of firing rate of 600 shots a minute.

Sixteen-inch gun on coast carriage was product of renowned Krupp works in Germany by 1897, when picture of it above appeared in *Popular Science*. Weight was 71 tons.

First U.S. battleships, of 1895–97, included *Iowa*—previewed by *Popular Science* picture above. In 1898 *Iowa*'s blazing guns helped win Spanish-American War.

U. S. Navy's first submarine, the *Holland,* has a successful trial. Photos, top to bottom, show it in cruising trim, diving, and rising after the dive.

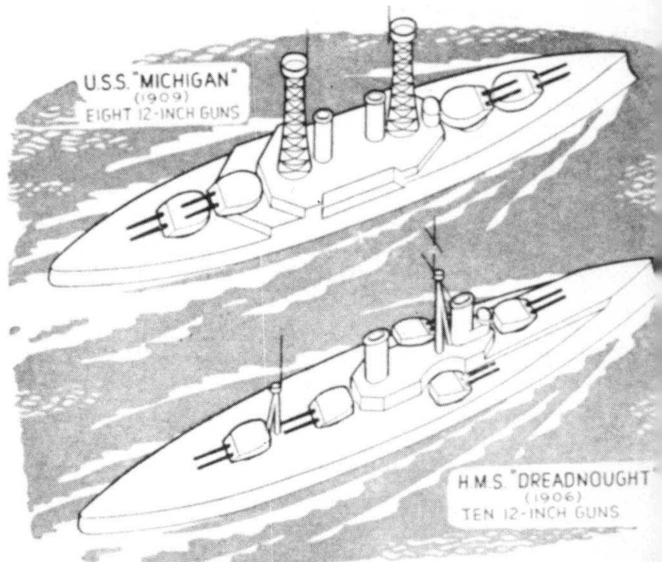

Two naval milestones: world's first all-big-gun battleship was Britain's *Dreadnought* of 1906. Placing all battleship's guns on center line, for unhindered fire to either side, was pioneered by America's *Michigan* of 1909.

World War I

Early role of airplanes in World War I was to swoop low over trenches, risking bullets from ground, to strafe enemy troops with machine-gun fire, as pictured here in *Popular Science*.

Aerial combat, plus forward-placed propellers, brought machine guns that fired through the whirling props. At first air fighters, like this French pilot, reinforced wood propeller with steel against chance hits. Later came machine gun synchronized with propeller so all bullets went between blades without hitting them.

Zeppelin raid! Searchlights spot a marauder over England. Great airships' bombing attacks lost terror as craft proved vulnerable to antiaircraft guns and planes' incendiary bullets—and their twisted aluminum skeletons began littering earth.

Aerial bombs of 1918: kinds dropped by German zeppelins and Gotha warplanes, above, ranged up to a 660-pound airship bomb. Sizes were puny by later decades' standards. Thirty years after, air bombs weighed up to more than 20 tons.

Tanks were surprise British weapon to break deadlock of World War I's entrenched armies. Bullets, trenches, barbed wire were no obstacles to them. Land warfare has been mobile ever since.

Soldier listens for sappers—whose laborious tunneling under enemy to blow him up bespoke the static nature of trench warfare.

Allied shipping ran gantlet of German submarines like one in 1916 photo below, armed with 10 to 12 torpedoes and able to cruise 2,000 miles before refueling. Wake of dreaded torpedo, right, often was threatened vessel's first warning. Ships had to sail in convoys protected by warcraft. One early anti-sub weapon: drum-shaped depth charges ("ash cans") triggered by pressure to explode at preset depth.

Between Two Wars

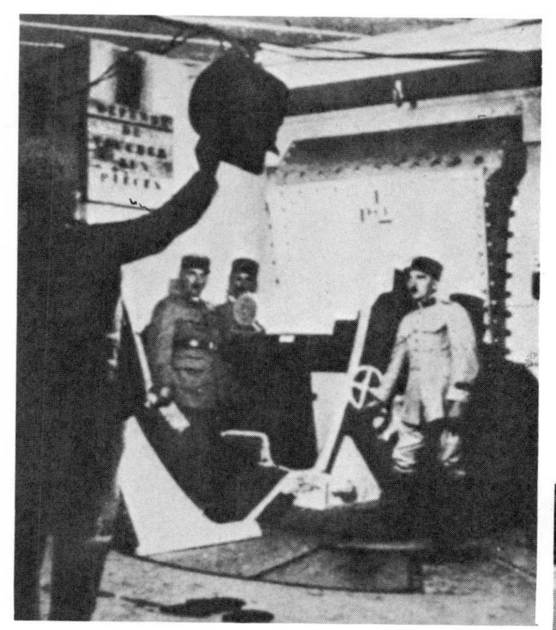

Self-propelled guns, like this 75-mm one, superseded artillery drawn by horses or tractors. Motorized pieces ranged from 37-mm up to 155-mm size.

Maginot Line, built by France in 1930s, linked border fortresses like one above by fantastic system of tunnels, ammunition magazines, and underground electric railways (portal, right). It was to fall when Germans overran weak spot at its northern end.

First U.S. aircraft carrier, 1922, was the *Langley* (right), converted from a collier and nicknamed the "Covered Wagon." U.S. carrier *Saratoga* (below) and sister ship *Lexington,* both completed in 1927, were 888-foot giants of 33,000 tons.

World War II

Recoilless 57-mm and 75-mm guns—the first a shoulder weapon (above)—gave infantrymen the firepower of field artillery. To banish kick, special shell cases vented gas rearward.

Military gliders, towed by planes to scene of action and released, repeatedly proved effective to land troops, guns, and vehicles behind the lines of a surprised enemy.

Carriers lead battleships and cruisers of a U.S. naval formation in Pacific. In World War II, for the first time, aircraft carriers gained ascendancy over battlewagons as capital ships of the world's navies.

British motor torpedo boat of 1939, launching torpedoes below, set style for U. S. Navy's PT (patrol torpedo) boats like famous Kennedy-skippered one. Cockleshell craft rely on speed to dodge gunfire of warship attacked.

"Best all-around American-built fighter of World War II," Air Force called propeller-driven P-51 Mustang—faster, more maneuverable, and longer-ranged than any other it then had. Even after the advent of jet airplanes, Mustangs still found use in Korean War.

War's storied Flying Fortress, the B-17 at left, was workhorse of strategic-bombing missions over embattled Europe. Design varied with successive refinements; B-17E version pictured had power-operated turrets.

Paratroops, making practice jump above, came into own in World War II. In one exploit they dropped on Philippine fortress of Corregidor to rout Japanese from caves.

"Bombs away!" Target is a Messerschmitt warplane factory in Germany, shortly before Allied invasion of Europe.

B-29 Superfortresses, built to carry large bomb loads to distances beyond reach of the B-17, made possible the strategic bombing of Japan. Their first raid on its homeland was in mid-1944.

THE APPLICATIONS OF EXPLOSIVES. 453

Among experiments made to demonstrate the resistance of structures to attack by a mob was one upon a safe twenty-nine inches cube, with walls four inches and three quarters thick, made up of plates of iron and steel, which were re-enforced on each edge so as to make it highly resisting, yet when a hollow charge of dynamite nine pounds and a half in weight and untamped was detonated on it a hole three inches in diameter was blown clear through the wall, though a solid cartridge of the same weight and of the same material produced no material effect. The hollow cartridge was made by tying the sticks of dynamite around a tin can, the open mouth of the latter being placed downward, and I was led to construct such hollow cartridge for use where a penetrating effect is desired by the following observations:

In molding the gun cotton at the torpedo station, as stated above, a vertical hole was formed in each cylinder or block in which

The Bazooka Story

Top to bottom: Prof. C. E. Munroe tells in 1900 issue of discovering power of hollow charge pictured; 1945 issue shows how similar charge applies Munroe effect to pierce tank armor; diagram shows charge built into projectile launched from hand-held tube.

IT MAKES STEEL FLOW LIKE MUD!

THE BAZOOKA'S GRANDFATHER

How Prof. Charles E. Munroe's amazing discovery, reported in Popular Science Monthly 45 years ago, is now blasting enemy tanks and fortifications.

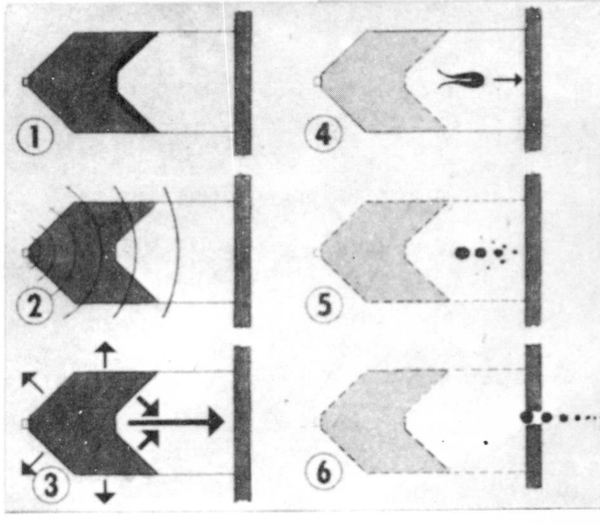

Explosion of hollow charge (1) creates detonation waves (2). In hollowed-out portion, force is concentrated to form a jet (3). Metal torn off the cavity lining is turned inside out (4) or forms pellets (5) that join the jet to plow through armor (6)

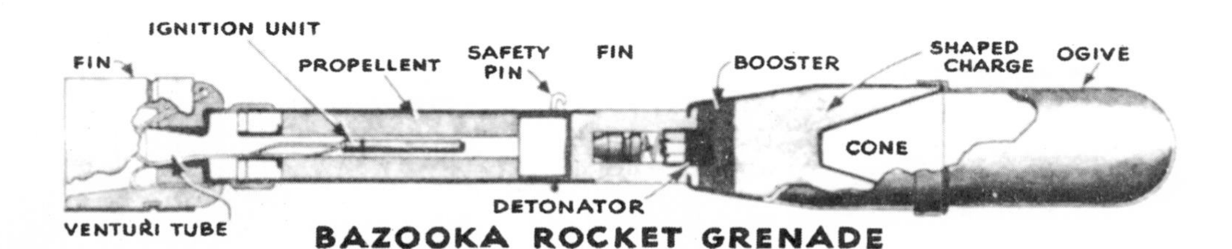

BAZOOKA ROCKET GRENADE

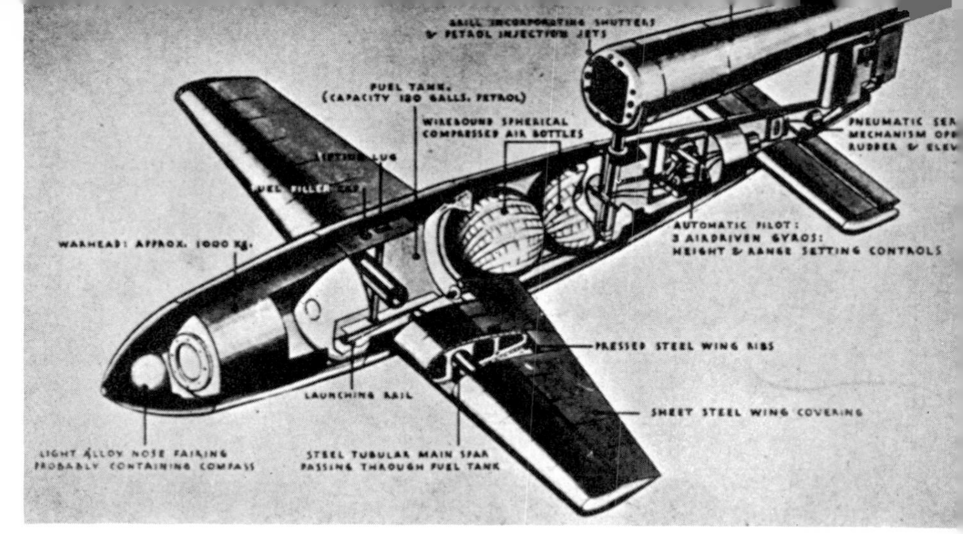

Germans' V-1 "buzz bomb," a guided missile propelled by a pulse-jet engine, bombarded England from across Channel. Steered by compass, it had range of up to 150 miles; then it dived to earth and 1-ton warhead exploded. More than 8,000 V-1s struck at London and vicinity.

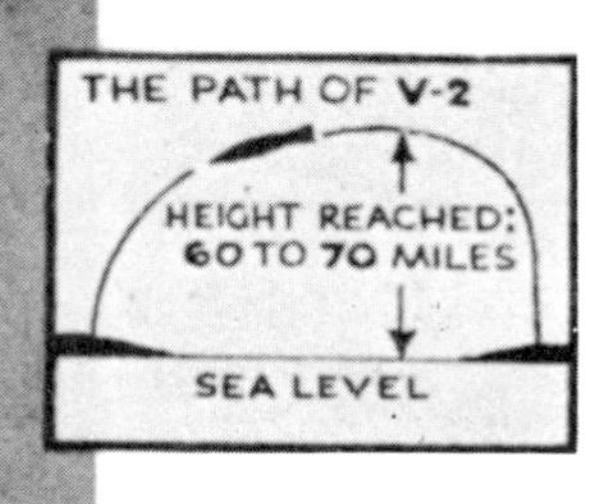

German V-2 rocket missile, 46 feet tall and of 200-mile range, came next. Allied invasion of Europe ended its use. Captured V-2s (one pictured in flight) aided postwar U.S. trials of rockets to explore upper atmosphere.

Tanks that swam led U.S. landing forces ashore to storm Japanese-held Pacific islands. Our first armored amphibian, above—an 18-tonner armed with a 37-mm cannon and 2 machine guns—went into action at Kwajalein in 1944.

The U.S. won the race to get the atom bomb—and the first one in warfare, right, blasts Hiroshima in Japan with force of 20,000 tons of TNT. A second hits Nagasaki, and war is over. Japan surrenders to irresistible new weapon.

Up to the Present

Gigantic "hot rod" of the seas, 85,000-ton U.S. carrier *Enterprise,* speeds at fantastic 40 mph under power of 8 atomic reactors. It climaxed naval application of nuclear propulsion that began in 1955 with U.S. submarine *Nautilus* (see chapter: The Atomic Era Arrives).

Atomic cannon of 1952, which could hurl 11-inch nuclear shells, was dramatic example of "tactical" atomic weapons designed for use on a battlefield

Robot helicopter called Dash, armed with homing torpedo, has become one of sophisticated new anti-submarine weapons.

World's last battleship in action, U.S.S. *New Jersey,* fires 16-inch-gun broadside off Korea. Final appearance was at Vietnam.

B-52 Stratofortress, our "hydrogen bomber," first flew in 1954. Besides dropping bombs of conventional explosive, it could launch hydrogen-bomb-carrying missiles 1,000 miles from a target: jet-engined Hound Dog and then rocket-propelled ballistic Skybolt.

Controversial F-111 of 1964, too heavy for Navy, gave Air Force the world's fastest fighter-bomber. With swing wings retracted (photo), double-sonic plane can pass 16-inch shells in flight.

U. S. Army got biggest helicopter, Skycrane, in 1964. Easily it lifts 6½-ton truck (picture at left), or 155-mm howitzer, or 90 troops.

Largest airplane ever built, C-5A Galaxy of late 1960s, flies typical cargo pictured—tank, trucks, Jeeps, troops—at 500 mph.

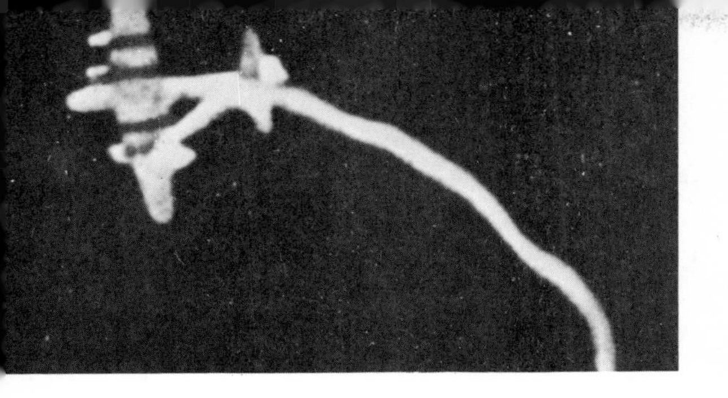

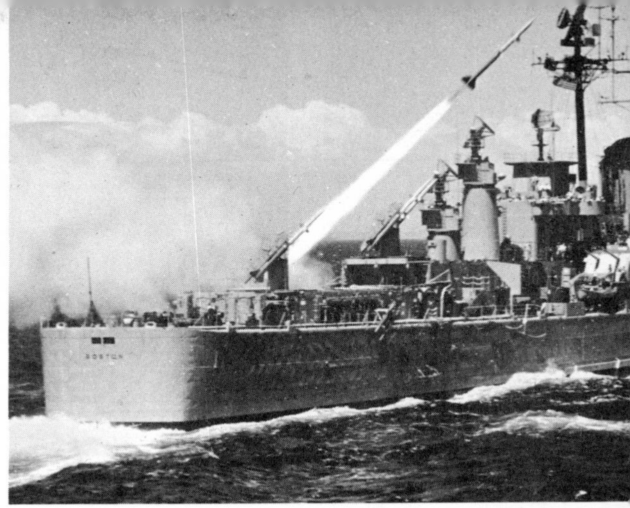

Falcon anti-aircraft rocket missile pursues drone B-17 to a kill. Trail shows how homing missile, guided by radar or heat sensor, closes on prey despite its evasive maneuvers. Making debut late in World War II, guided missiles took on wide variety in the 1950s.

U.S. missile cruiser *Boston,* with Terrier anti-aircraft weapons on twin launcher above, pioneered Navy's change-over from guns to missiles. At top, *Boston* launches a Terrier.

Combat-tested Sidewinder missile has been highly successful aircraft-to-aircraft rocket weapon since 1953. Using infrared heat sensor, it homes on tailpipe of an enemy plane.

First U.S. intercontinental missile, in 1958, was 5,000-mile, hydrogen-warhead Snark—a 69-foot jet craft taking off (right) with rocket boosters. ICBMs outmoded it.

Evolution of U.S. intercontinental ballistic missile (ICBM) has progressed from liquid-fuel Atlas and Titan rockets to instant-ready solid-fuel Minuteman, left, with hydrogen warhead in nose cone, above. A 1970 innovation puts 3 independently targetable warheads in a single ICBM.

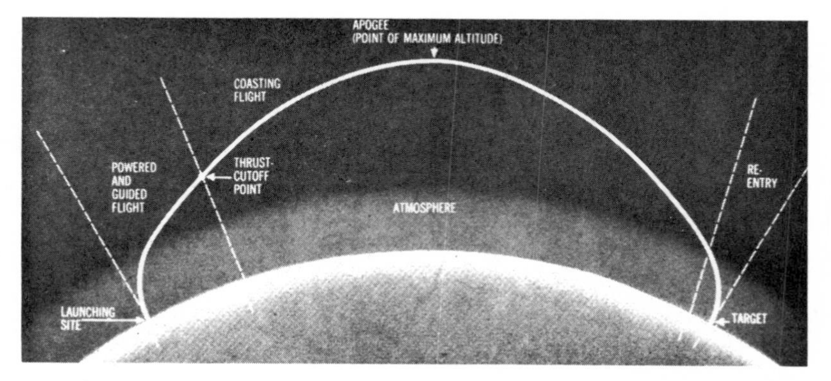

Flight of a long-range ballistic missile is shown above. Under rocket power, missile rises vertically and inclines in direction of target. Then rocket engine cuts off and missile coasts on, in ballistic trajectory like that of shell from a gun. Course takes an ICBM into space and back, to re-enter atmosphere near target.

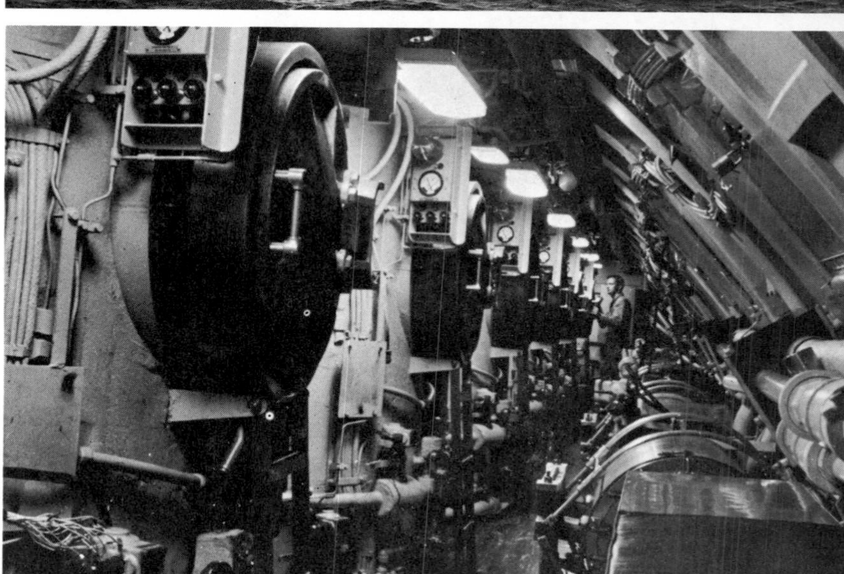

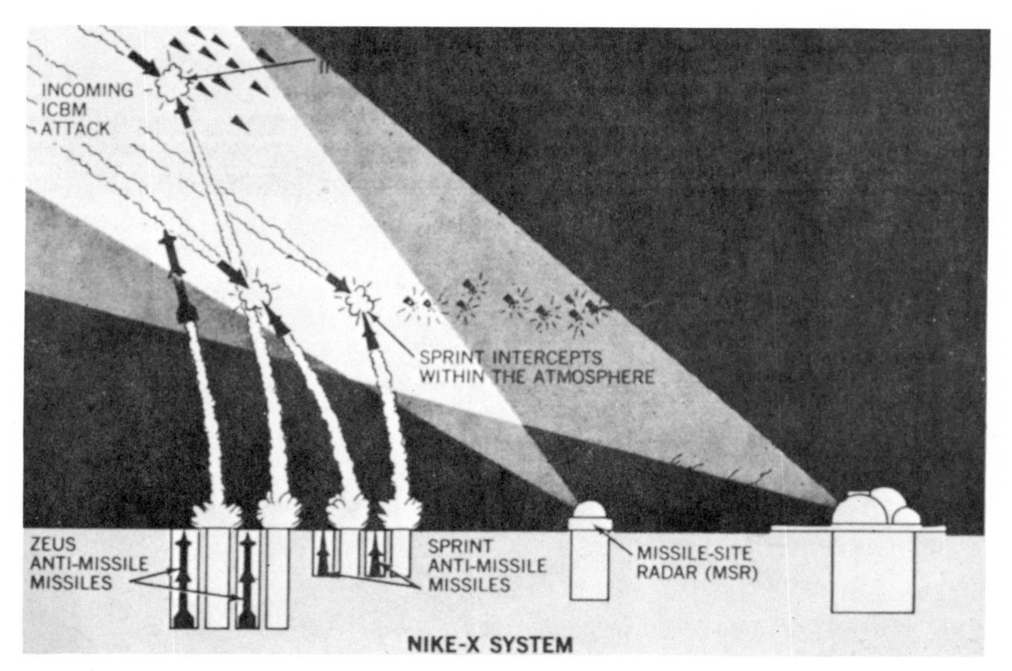

Even the ICBM has its counterweapon: anti-missile missiles. First U.S. system, above, combines 2 radar-guided ones: 100-mile-range Zeus, backed up by shorter-range Sprint. Photos, right: Sprint in flight; and, below, both missiles. Due to replace Zeus by 1974 is Spartan, of 400-mile range. Missiles' hydrogen warheads disarm incoming ones with X-rays.

Polaris nuclear missiles (and later Poseidons) are launched under water, below, by U.S. Polaris submarines. Opposite page: missile tubes in submarine *Edison;* next above, *Edison* at surface. Polaris subs could strike back even if ICBMs on land were destroyed.

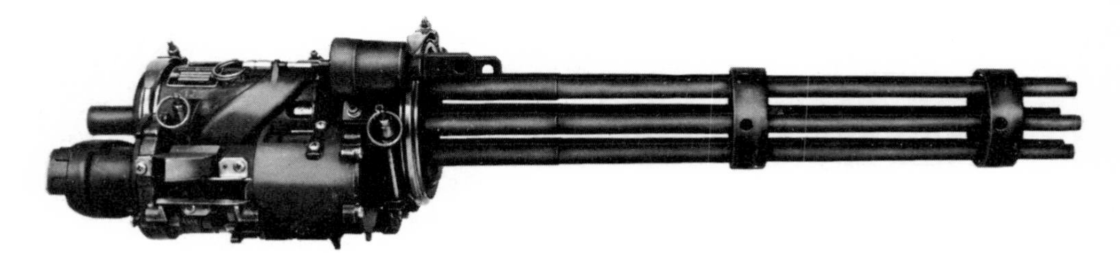

Machine guns reached new peak of rate of fire with 100-bullets-a-second Minigun above. Mounted in cargo planes and helicopters, the 6-barreled 7.62-mm (.30-inch) guns saturated the air with bullets to protect road vehicles in Vietnam from attack, as at right.

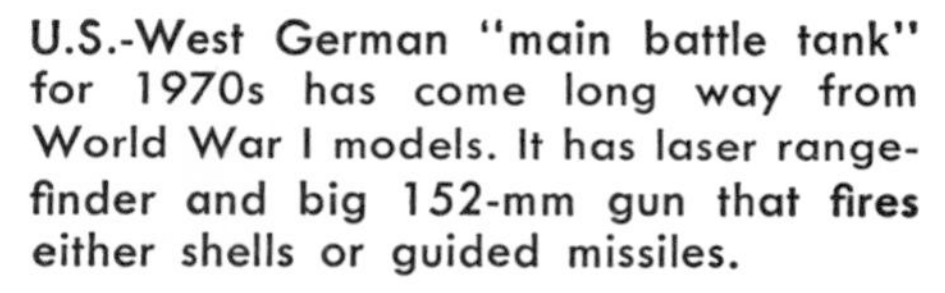

U.S.-West German "main battle tank" for 1970s has come long way from World War I models. It has laser range-finder and big 152-mm gun that fires either shells or guided missiles.

B-1 supersonic bomber (right), scheduled to fly in 1974, will be Air Force's swing-wing successor to subsonic B-52. Approaching target, B-1 will hug ground to escape radar detection.

Navy of future ma
have "surface-effe
ships" like
experimental 100-
tonner, left, curren
being built. To red
drag for speed, it
rides on bubble of
pressurized air
between underwat
keels along sides.

CHAPTER 9

Conquest of Space

ASTRONOMERS from all over the world met at Brighton, England, in August of 1970, for a congenial task made possible by the Space Age. They gave the official approval of the International Astronomical Union to a list of names proposed for more than 500 features on the hidden far side of the moon, discovered by U.S. and Soviet spacecraft. For one large lunar crater, they decided, the name on future moon maps will be Tsiolkovsky.

Thus honored with a perpetual memorial in space was the man now given worldwide credit for being first to point out how to get there.

The idea of propelling a spaceship with rockets came to Konstantin Tsiolkovsky, a Russian schoolteacher, in 1883. In the airless void of space, with nothing for a propeller to bear against, a rocket offered the only practicable form of motive power he could see. From his mathematical studies, possible designs for space rockets eventually took shape on paper, beginning in 1903 with a rocket burning a forward-looking combination of liquefied oxygen and hydrogen.

Although Tsiolkovsky experimented in a home laboratory, there is no record that he ever tested a rocket or even built one. His contributions were purely theoretical—and, after noting them, the story of space must shift to America.

An audacious attempt to fly in a man-carrying rocket made the front page of the New York *Times* in 1913. The would-be pilot was Rodman Law, daredevil stunt man. His 44-foot-tall rocket, pictured in *Popular Science,* looked like a grotesquely magnified Fourth of July skyrocket—even to a huge stick trailing behind. A thin steel cylinder held power—and, above it, a passenger in a compartment with a hinged conical cap. The chosen launching site was a hillock surrounded by marshland in Jersey City, New Jersey.

About the destination of his trial flight, Law was uncommunicative, but made no claim to being space bound. Rumors had it that he meant to parachute to earth at Elizabeth, 12 miles away.

From a platform reached by a ladder, Law clambered aboard. An assistant, identified only as "Sam," lit a sputtering fuse. As spectators readied binoculars, the spark reached the powder chamber. With a mighty bang, the skyrocket blew up.

Law landed 30 feet away in a clump of rushes. His clothes were torn and powder-blackened, his face and hands slightly singed, but he luckily escaped serious injury. He did not figure in the annals of rocketry again.

The world's first practical rocket scientist, America's Robert H. Goddard, began his experiments on a much more modest scale. Near Worcester, Massachusetts, where he was professor of physics at Clark University, he first test-fired simple powder rockets that flew a few hundred feet high, under the power of miniature engines about a foot long. In a Smithsonian Institution paper of 1919, he demonstrated mathematically that a rocket could be fired to the moon—a possibility dramatized by a *Popular Science* artist in 1920—and was dismayed by newspaper stories expecting him to do it right away.

In an article he wrote for the magazine in 1924, he corrected the popular misconception that a rocket propels itself by pushing against the air. What really drives it forward, he pointed out, is the recoil from the gases of combustion that it ejects

An imaginary scene that came true: this close-up view of the moon, with the earth in the sky above, appeared in *Popular Science* in 1873. Men saw it by 1968.

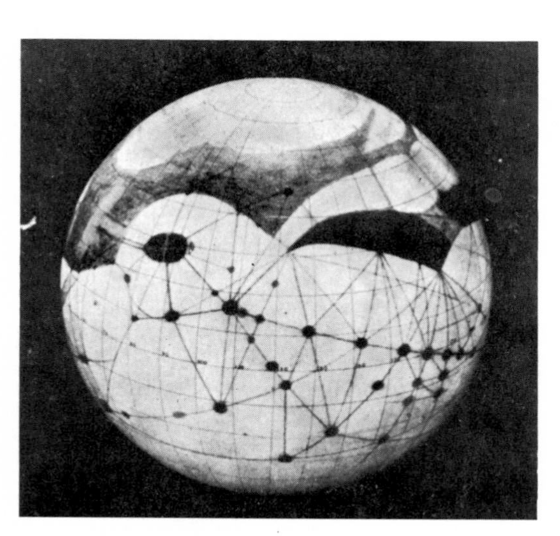

"Canals" of Mars, discovered in 1877 by Schiaparelli, were work of intelligent beings to astronomer Percival Lowell. He charted them on globes like the one above, around the turn of the century.

A manned rocket flight in this 44-foot-tall version of a Fourth-of-July skyrocket was attempted in 1913 by daredevil stunt man Rodman Law. He miraculously escaped serious injury when rocket blew up.

rearward. Since the presence of air actually hinders their ejection, a rocket engine works even better in a vacuum than in the earth's atmosphere.

The world's first liquid-fuel rocket, burning gasoline and liquid oxygen and ignited with a blowtorch, was test-fired by Goddard in 1926. Again a tiny one, it flew for only 2½ seconds and 220 feet—but it was the ancestor of today's mightiest rockets.

Much to the relief of Goddard's Massachusetts neighbors, plagued and sometimes frightened by the noise of his rocket projectiles, he moved his experiments in 1930 to a large ranch near Roswell, New Mexico. Now at the stage of big-league rocketry, he shot aloft liquid-fuel rockets 10 to 22 feet long, reaching a maximum altitude of 8,000 or 9,000 feet. They could be heard 8 miles away.

There were other pioneering rocket experimenters—notably including Hermann Oberth in Germany, who built and tested small rocket engines, wrote books on future space travel, and indulged in picturing fanciful designs for the spaceships he saw ahead. (A young lad destined to become famous in rocketry, Wernher von Braun, assisted Oberth in his engine trials.)

With World War II the scene of big-scale rocketry shifted abroad. At the Peenemünde Experimental Center on the Baltic coast of Germany, the mightiest rocket ever built had taken shape.

The missile stood 46 feet tall and measured 65 inches in diameter. It weighed more

Moon-rocket conception of 1920 in *Popular Science* embodied ideas of Prof. Robert H. Goddard of Clark University.

Prof. Goddard exhibits a modest-sized rocket engine of his early design, in illustration at right. His experimental rockets then used smokeless powder, an improvement over the old-fashioned gunpowder in earlier ones.

Hermann Oberth, German rocket pioneer, envisioned in 1927 the passenger-carrying moon ship below. Working on rockets with him was a young lad named Von Braun.

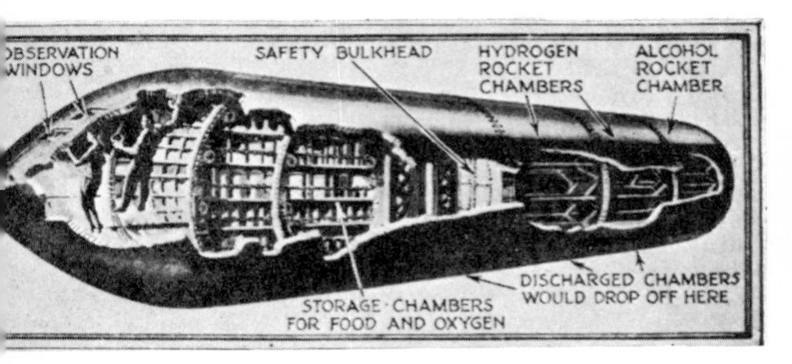

than 13 tons. Goddard's biggest "birds" were pygmies compared to it. The designers were Captain Walter Dornberger of the German Army and Wernher von Braun, Oberth's former assistant.

To Von Braun the great rocket, made possible by army funds and facilities, was the forerunner of the spaceships of his dreams. To the German Army it was a formidable new weapon. Burning alcohol and liquid oxygen, it could soar to a 60-mile height or more and hurtle down with an explosive warhead upon a target 200 miles away. It made World War II history under its military name, the V-2.

Thousands of V-2s were built. Beginning in 1944, they rocketed down upon targets across the Channel in England and on the Continent—until invading Allied armies overran their launching sites and finally captured the principal V-2 manufacturing plant in the Harz Mountains.

To America fell 100 captured V-2s—and a far bigger prize. By great good fortune its war booty included Von Braun himself. The brilliant rocket scientist, seeing America as the land where his ideas for space flight could best be pursued, had planned it that way. Evading the Russians approaching from the east, he and his staff traveled south to meet oncoming U.S. forces and surrender to them.

Bringing his welcome talents to America, Von Braun became a U.S. citizen in 1955. He headed the Army Ballistic Missile Agency (ABMA) formed in 1956 at Huntsville, Alabama.

The captured V-2s, fired from the White Sands Proving Ground in New Mexico, inaugurated a modest postwar U.S. program to explore the fringe of the earth's atmosphere. Their flights to heights up to 133 miles had little to do with getting into space. During the same period, however, several noteworthy events did mark steps toward that end.

An advanced 2-stage rocket called Bumper was fashioned by adding a small Wac Corporal rocket to the nose of the big single-stage V-2. Multistage rockets with solid-fuel boosters had arrived earlier, but Bumper appears to have been the first all-liquid-fuel one. In 1949 it reached an altitude of 252 miles and a speed of 5,150 mph—records that stood in the United States for seven years. It was a significant step toward space rockets—which, along a course leveling off to horizontal, would have to reach a speed exceeding 17,000 mph to put a man or a payload in orbit.

Launching of world's first liquid-fuel rocket by Goddard, pictured standing beside it, made history in 1926. Engine was at top.
SMITHSONIAN INSTITUTION PHOTO

The rocket that finally outperformed Bumper was Jupiter C, designed by Von Braun's ABMA team to test nose cones for a Jupiter ballistic missile. Its first flight took it to a record 682-mile height and 3,400-mile distance. Then it made news when a nose cone it carried withstood white heat from air friction in re-entering the atmosphere and was recovered intact.

The nose cone was made of revolutionary new "ablative" material, which absorbs heat by melting or charring and vaporizing at its surface without being entirely consumed. Not only did it show a missile warhead could survive re-entry; it also proved that future manned spaceships could return to earth without burning up.

A more immediate way to explore space could be with what *Popular Science* in 1954 called a "poor man's space station"—a small unmanned artificial moon to be rocketed into earth-circling orbit. The young American physicist who proposed

Big-league rocketry had arrived by 1935 when Goddard's trial rockets reached this respectable size. To the vast relief of dwellers near his former Worcester, Mass., test site, he and his alarming noisemakers had moved to new quarters on a large range near Roswell, N.M. In tests there, lasting into 1941, he launched a series of liquid-fuel rockets ranging in length from 10 to 22 feet.

this concept, S. Fred Singer, named it MOUSE for "*m*inimum *o*rbital *u*nmanned *s*atellite of the *e*arth."

"Can We Live Without Weight?" was a *Popular Science* headline of 1952. When spaceships coasted in orbit or to destinations beyond the earth, passengers would become weightless—and the article told of experiments just begun to see if living creatures in this strange state of "zero gravity" would suffer ill effects.

Small high-flying Aerobee rockets carried mice and monkeys to a 37-mile height and released the animal compartment to plummet to earth. The compartment's free fall simulated zero gravity for nearly 2 minutes, until a parachute opened to land its live cargo gently and safely. The animals were unharmed by their wild ride.

Front-page newspaper headlines of July 30, 1955, proclaimed the electrifying news that America was taking its first step into space. President Eisenhower had just approved a plan to launch a basketball-sized satellite, with instruments that would report scientific data to earth by radio. It would be a U.S. contribution to the Inter-

Big German V-2 war rockets, 46 feet tall, find peaceful use as World War II booty: at White Sands, N.M., they rise to heights of more than 100 miles and begin program to explore fringe of earth's atmosphere.

national Geophysical Year (IGY)—actually an 18-month period of worldwide earth-science observations, scheduled to begin in July of 1957 and to extend through all of 1958. *Popular Science* drawings of Singer's similar MOUSE served the New York *Times* to visualize what the coming satellite might look like.

Actually, at that moment a U. S. Defense Department advisory committee was still weighing a choice between two rival satellite-and-launch-rocket plans. One, called Project Orbiter, proposed to put up a satellite with a 4-stage combination of already-existing rockets: a Redstone, a direct descendant of the V-2, and clusters of smaller Sergeants or Lokis. This plan was advanced by the Von Braun group at the ABMA. The alternative was a Navy plan, Project Vanguard. Although based on the technology of successful Viking and Aerobee sounding rockets, it called for designing and building a wholly new 3-stage launching rocket, virtually from scratch—with all the problems, delays, and mishaps that that risked.

The committee's unhappy choice was Project Vanguard. Only later would America find that with that decision, in the fall of 1955, it had thrown away its chance to lead the way into space. At that time, no one doubted that the Vanguard satellite would be the first in the world.

An actual man-made earth satellite was *Popular Science*'s meat, of the juiciest kind. Seizing upon first available details of the Vanguard "moon," Associate Editor Herbert R. Pfister produced a striking plastic-encased model of it to illustrate in the magazine. So much admired was his scientific showpiece that he was commissioned to make two duplicates, one for the Hayden Planetarium in New York and the other for the Washington headquarters of the U. S. National Committee for the IGY.

The magazine's pages detailed the launching plan stage by stage; counseled readers on how to watch for the satellite (it would take binoculars or a small telescope to see it); and presented all but the actual blueprints of the launch rocket.

Then thunderstruck Americans got the startling news:

Sputnik 1, the world's first earth satellite, was successfully launched by the Russians on October 4, 1957. A 23-inch globe with trailing antennas, it was sailing flauntingly over the U.S. and the rest of the world, beep-beeping a "Here I am" as it went by. It circled the earth once every hour and 35 minutes, at an altitude ranging from about 140 to 600 miles, with its better-visible carrier rocket tumbling along behind. Its weight was announced by the Russians to be an incredible 184 pounds, 7 or 8 times the weight of the projected Vanguard satellites.

While Project Vanguard was still getting around to trying to launch a satellite, Russia did it again, on November 3, 1957. Sputnik 2 was an astounding 1,121-pounder, with a dog named Laika aboard.

By early December the first complete Vanguard rocket stood ready for launch at last, from Cape Canaveral, Florida (later renamed Cape Kennedy). Technically a "test vehicle" for operational Vanguards still to come, it carried a baby 6-inch test satellite, entitling it to be billed as "a satellite-bearing vehicle" in a public announcement of the impending launch.

Thousands flocked to the cape to watch Vanguard recover U.S. prestige. Before their eyes, and the world's, the rocket rose about 4 feet. It fell back, toppled—and spectacularly exploded.

At the United Nations headquarters in New York, Soviet delegates asked American ones if the U.S. would like to receive aid under the Reds' program of technical assistance to backward nations.

In Washington, ever since Sputnik 1, Von Braun and his team had been pressing for a chance to come to the rescue. Their spurned Project Orbiter had become an improved Project Explorer; their new and successful Jupiter C, they urged, could quickly be adapted to put up a satellite. Finally they got the nod, from President Eisenhower himself.

On January 31, 1958, a Jupiter C rocket launched the first U.S. satellite—Explorer 1. A cylinder 80 inches long and 6 inches in diameter, weighing 18 pounds, it was built around a fourth rocket stage for the Jupiter C. At last an orbiting U.S. spacecraft could beep back.

It was a crowning anticlimax when Project Vanguard (now being unkindly nicknamed Rearguard) did succeed in orbiting a grapefruit-sized "baby" satellite, Vanguard 1, in March of 1958. Nine days later, Jupiter C lofted another Explorer.

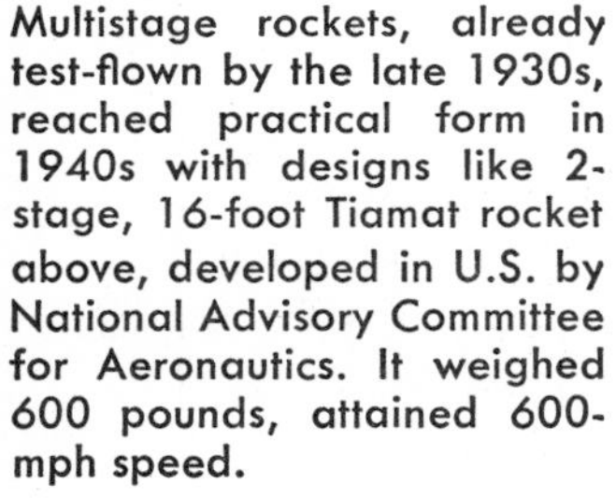

Multistage rockets, already test-flown by the late 1930s, reached practical form in 1940s with designs like 2-stage, 16-foot Tiamat rocket above, developed in U.S. by National Advisory Committee for Aeronautics. It weighed 600 pounds, attained 600-mph speed.

First large 2-stage U.S. rocket, Bumper, was created by adding a second stage to a big V-2. Its altitude record of 252 miles, set in 1949, stood in the U.S. for seven years—as also did its speed record of more than 5,000 mph. It was a step toward space rockets.

To see how weightlessness would affect living creatures in space flight, trials of rocketing animals aloft began with Aerobee rocket above. From a 37-mile height it dropped animal compartment for free fall, until slowed by parachute for a gentle landing.

Recovery intact of Jupiter C nose cone in 1957 proved that spacemen and their craft could survive the heat of re-entry into atmosphere to return to earth.

SMITHSONIAN INSTITUTION PHOTO

Weightless mice are photographed as they fall in compartment from Aerobee rocket. They were found unharmed by experience.

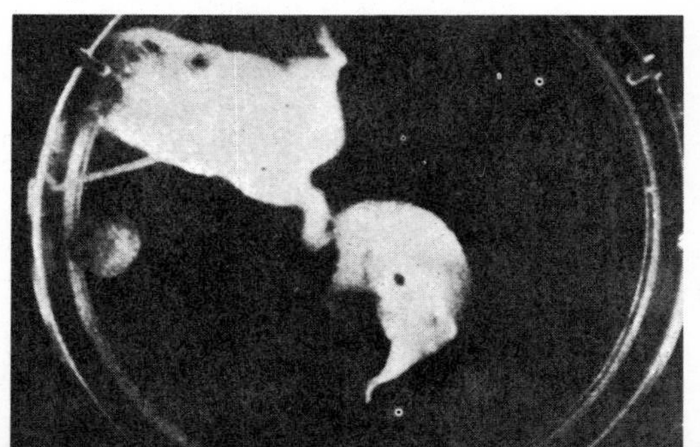

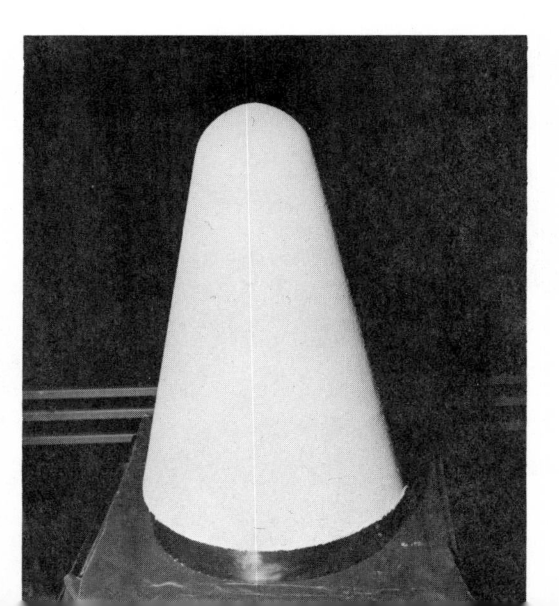

The New York Times.

VOL. CIV. No. 35,616. Entered as Second-Class Matter, Post Office, New York, N. Y.

Copyright, 1955, by The New York Times Company.

NEW YORK, SATURDAY, JULY 30, 1955.

LATE CITY EDITION
Fair and seasonably warm today. Fair, quite warm tomorrow.

FIVE CENTS

U.S. TO LAUNCH EARTH SATELLITE 200-300 MILES INTO OUTER SPACE; WORLD WILL GET SCIENTIFIC DATA

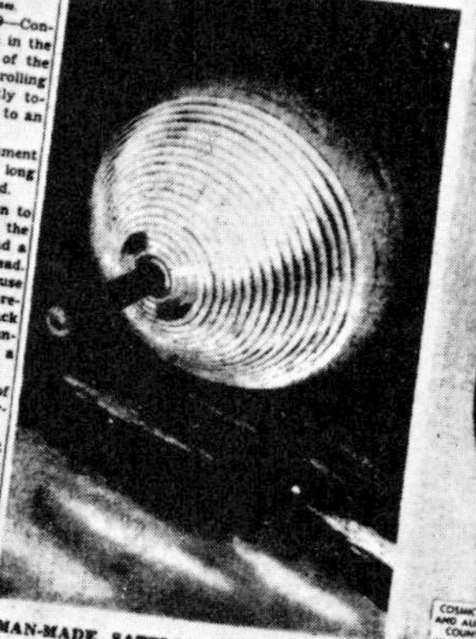

MAN-MADE SATELLITE: Artist's renditions of the earth-circling satellite, based on a concept of Prof. S. F. Singer of the University of Maryland. Professor Singer's specifications—diameter of about two feet, weight 100 pounds and speed 17,280 miles an hour—conform closely with those of the announcement from the White House.

PACE 18,000 M.P.H.

Rocket to Start Object Size of a Basketball in 1957 cr 1958

Texts of press conference and documents, Pages 8 and 9.

By RUSSELL BAKER
Special to The New York Times.

WASHINGTON, July 29—This country plans to launch history's first man-made, earth-circling satellite into space during 1957 or 1958.

Tentative plans envision an unmanned globular object about the size of a basketball. The satellite will flash around the earth about once every ninety minutes at a speed of 18,000 miles an hour in a fixed path 200 to 300 miles above the ground.

These plans were announced this afternoon at an extraordinary White House news conference attended by a battery of prominent scientists.

James C. Hagerty, White House press secretary, joined the scientists in stressing the

Court Rejects Plea To Deport Bridges

Special to The New York Times.

SAN FRANCISCO, July 29—Federal Judge Louis E. Goodman today refused to strip Harry Renton Bridges, Pacific Coast labor leader, of his United States citizenship.

Judge Goodman, in the Government's fourth attempt to deport Mr. Bridges to his native Australia, ruled the prosecution had not proved its charges that the longshore leader had been a member of the Communist party before he was naturalized Sept. 17, 1945.

United States Attorney Lloyd H. Burke and Lynn J. Gillard and Robert H. Schnacke, assistant Federal attorneys, prosecuted the case. They said a decision to appeal would depend on the outcome of consultation with Department of Justice officials in Wash-

Continued on Page 10, Column 6

CONGRESS CHIEFS ABANDON PLANS TO ADJOURN TODAY

House to Meet on Monday —Fuel Gas Bill Sidetracked —Public Housing Set Back

Special to The New York Times.

WASHINGTON, July 29—Congress was caught tonight in the traditional minor frenzy of the eleventh hour as the controlling Democrats labored urgently toward bringing this session to an end.

All hope for an adjournment by tomorrow night, as had long been planned, was abandoned.

The Senate was in position to finish its work by then but the House of Representatives had a solid docket of work still ahead.

Late in the day the House Democratic floor leader, Representative John W. McCormack of Massachusetts, officially announced that there would be a House meeting on Monday.

He prepared a calendar of business that will result in carrying Congress into next week.

The House quit for the night at 6:17 P. M. and will reassemble at 10:30 tomorrow morning.

Pressing to meet the original quitting date as nearly as they could, the Senate leaders officially cast aside until next year the most controversial single measure remaining on that side.

Defeat for President

This was a bill, passed 209 to 203 last night by the House, to exempt natural gas producers from Federal price control.

The decision was in a modified sense a blow to the President, who had expressed support in principle for the bill.

It was a much heavier blow, however, to the leading Democrats of Congress, including ... Rayburn of Texas, ... the House of Repre... and the ailing Sena... cratic chieftain, Senator ... B. Johnson of Texas.

J. E. HOOVER SHUNS CITY POLICE POST

Declines Mayor's Bid to Be Commissioner—Wagner Is Said to Seek Outsider

By PAUL CROWELL

J. Edgar Hoover, director of the Federal Bureau of Investigation, has declined an invitation by Mayor Wagner to become the city's next Police Commissioner.

The offer of appointment to the $25,000 post now held by Francis W. H. Adams was made by the Mayor early this week through an unidentified emissary described at City Hall as a close friend of the F. B. I. chief.

Mr. Adams announced his resignation last Sunday but is remaining at his post until the Mayor appoints a successor.

R.A.F. IS RETURNING 400 U.S. SABRE JETS

Russians Already Striving ...

...ree Ex-G.I. Turncoats Land in San...

...R. Leahy, right, reads a summary of court-martial ... into custody. The prisoners are, from left, Otho G. ...

Almost too late for anyone to care, the first full-sized 20-inch, 24-pound Vanguard satellite was orbited in February 1959; and an oversize 52-pounder in September 1959. That ended the ill-fated Vanguard project.

America's wounded pride at being beaten into space by the Sputniks could be salved by the accomplishments of its own first two satellites. To Explorer 1 went the credit for discovering the hitherto unknown Van Allen belts of radiation that surround the earth. Even pipsqueak Vanguard 1 scored a triumph, when observations of its orbit led to the discovery that the earth was pear-shaped rather than round.

Satellites of every description followed the Sputniks, Explorers, and Vanguards into space. Beyond a doubt the most satisfying of early ones, to the general public,

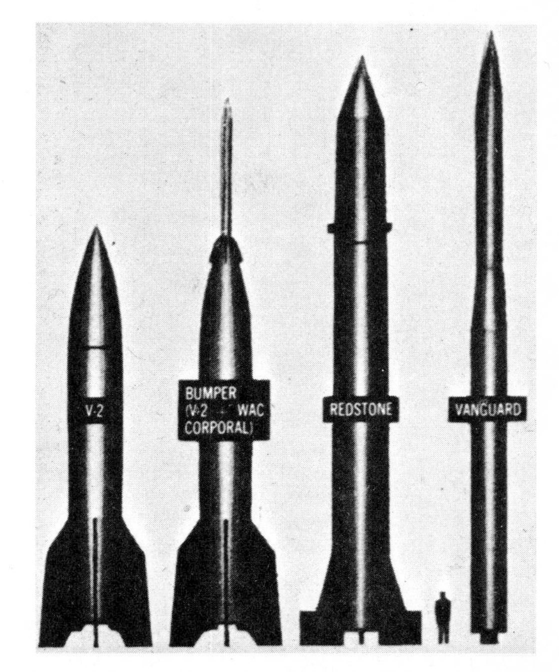

Launching rocket for Vanguard satellite, at far right above, was new 3-stage one designed virtually from scratch for purpose. Picture compares its height and bulk with those of V-2, Bumper, and Army's Redstone.

Facing page: White House announcement that U.S. would orbit a satellite made headlines in 1955. New York *Times* illustrated front-page story with *Popular Science* drawings of similar satellite proposed by Prof. S. F. Singer. When first details became available of new satellite, Vanguard, *Popular Science* built elaborate model of it (lower view). Hayden Planetarium and International Geophysical Year committee got duplicates.

Plan for orbiting Vanguard satellite 200 to 400 miles above earth is shown below. First 2 stages of rocket lift it beyond atmosphere; it coasts on; and third stage accelerates it to required 25,000 feet a second (about 17,000 mph) to keep on circling earth.

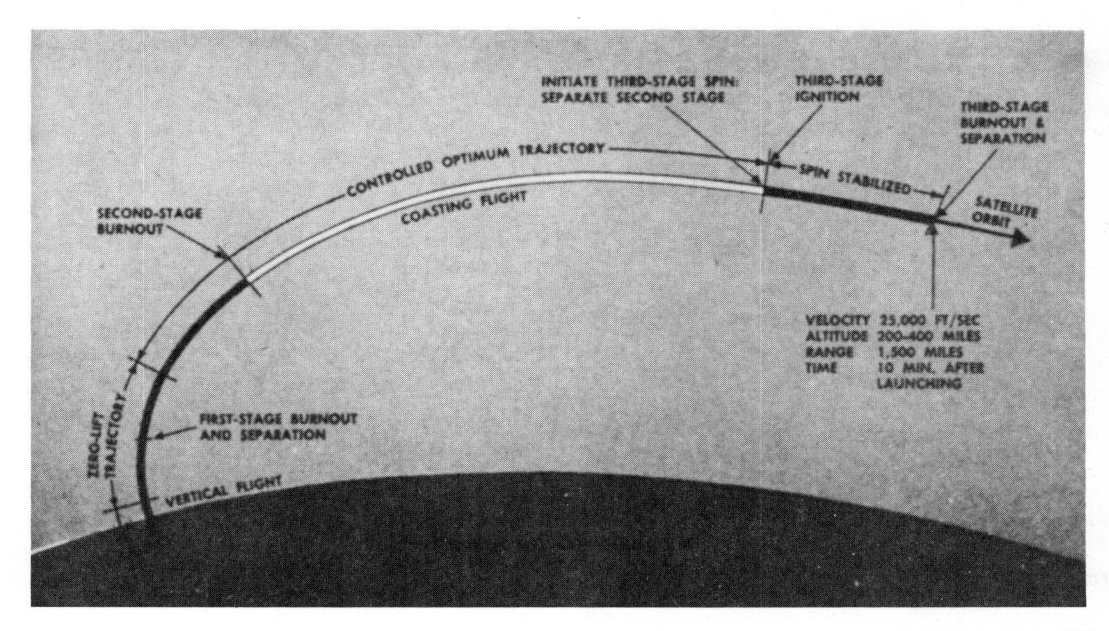

Moonwatch team of amateur observers, one of many organized to help track a Vanguard satellite, rehearses sighting and timing it. The satellite they saw first was a surprise.

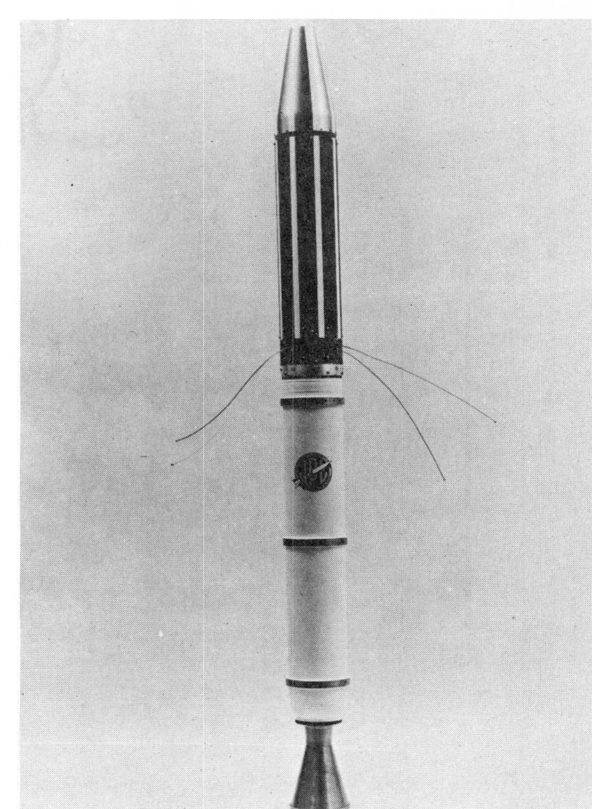

First U.S. satellite, 80-inch-long Explorer 1, was successfully orbited with an army rocket combination in 1958, as the original Vanguard program floundered in failures.

Russia beat U.S. to launching a satellite in 1957 with Sputnik 1, above, a 23-inch globe with trailing antennas. A telescopic camera photographed the trail of its rocket carrier, below, across the U.S. sky.

was America's Echo 1, launched in 1960. It was the first you could see with the unaided eye. Vying with the brightest stars in brilliance, it sailed across the night sky like a high-flying airplane—right on time, by a schedule you could find in your daily paper. Its record size explained its brightness: it was a balloon-like globe 100 feet in diameter, made of aluminum-coated plastic film and inflated in orbit. Using it as a primitive communications satellite, experimenters bounced radio signals off it and back to a distant destination on earth.

Meanwhile all altitude records went by the board and the sky became the limit, as the speed of rocket launching vehicles reached a magical figure just beyond 25,000 mph. That was the "escape velocity" sufficient to break free of the earth's gravity—and whiz off into deepest space, to the moon or destinations beyond. A Soviet series of unmanned Luna space probes began it in 1959 and U.S. versions immediately followed.

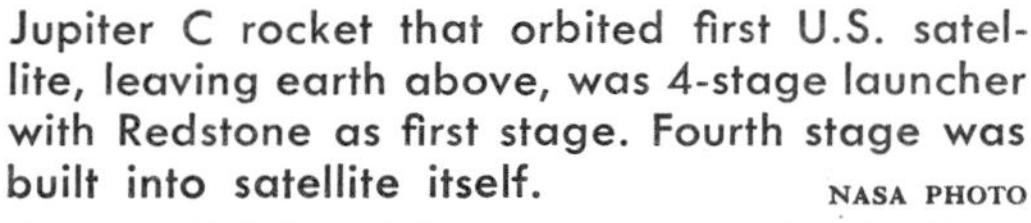

Jupiter C rocket that orbited first U.S. satellite, leaving earth above, was 4-stage launcher with Redstone as first stage. Fourth stage was built into satellite itself. NASA PHOTO

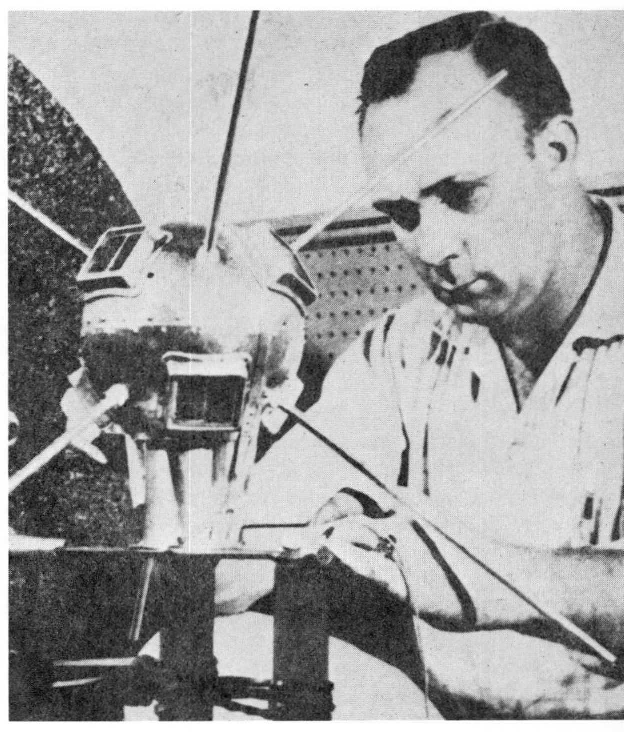

Vanguard project finally launched its first satellite—a baby one of grapefruit size, above. Only later did it manage to orbit a full-size, 20-inch Vanguard satellite.

Most satisfying of early satellites to a layman was huge 100-foot-diameter Echo 1 of 1960—the first one visible to unaided eye, as it crossed night sky 1,000 miles high. Sphere of metalized plastic was inflated after it had been rocketed into orbit.

Time-exposure photo catches luminous trail of Echo 1 during 40 seconds. Brighter than most stars, it sailed past at the apparent pace of a high-flying airplane. Even in a brilliantly lighted city, a watcher could spot it clearly. Newspapers gave timetable.

An early star of this celestial show was the Soviets' Luna 3. As it flew past the moon at about 40,000-mile range in 1959, a camera aboard it took the first pictures ever made of the moon's hidden far side—and radio equipment transmitted them to earth. True, the photos were so fuzzy that only a few of the largest and most conspicuous far-side lunar features could be made out—but it was the first time man had ever seen that mysterious side of the moon, always turned away from the earth.

To *Popular Science* the feat was "a dazzling exhibition of rocketry and instrumentation" and ushered in a new "go-there-and-see" kind of astronomy. Succeeding years would bring ever more exciting go-there-and-see views from distant worlds.

A civilian U.S. aerospace agency—NASA, the National Aeronautics and Space Administration—came into being in 1958 to take over non-military U.S. space projects. Once more, Von Braun was "captured" when NASA commandeered him and his ABMA team and made him director of NASA's new Marshall Space Flight Center at Hunstville. Transfer to the Air Force of responsibility for long-range missiles had opened the way for NASA's fortunate coup.

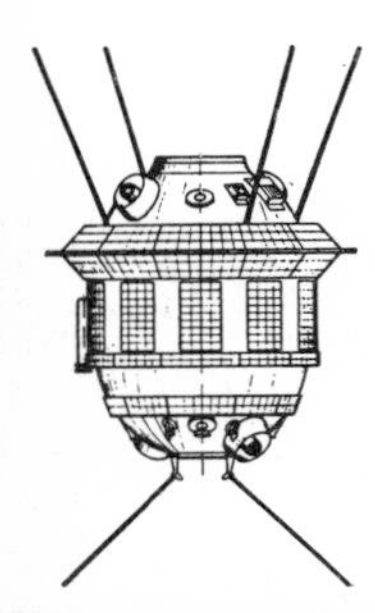

Man's first view of far side of moon was Soviet triumph of 1959. Unmanned "Lunik" spacecraft (left) radioed photo below of features never seen before. One was huge crater forming dark spot at upper right—now named Tsiolkovsky, after space pioneer.

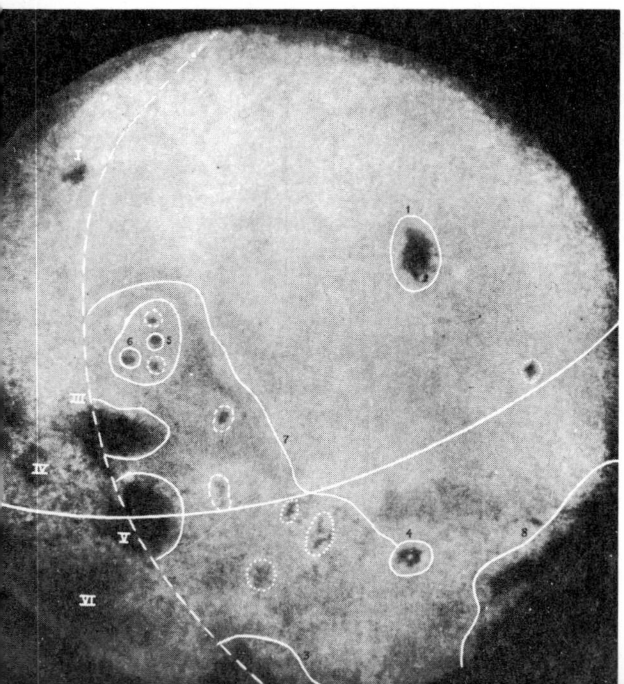

And now the time had come to put a man in space. America's first intercontinental ballistic missile, Atlas, had streaked 6,300 miles down the Atlantic Missile Range in a successful trial in November 1958. For the first time, the U.S. had a rocket vehicle powerful enough to orbit a manned spacecraft.

A 1-man satellite called Mercury took shape. First trials, boosted by a Redstone rocket, would be "suborbital" flights arcing into space and right back again, ending with landing the capsule by parachute in the sea. Then the Atlas rocket would launch it on orbital flights.

A chimpanzee named Ham, the first suborbital passenger, rode the Mercury to a safe landing in January 1961. That was as far as the program had got when the Russians pulled off another of their surprise "space spectaculars":

Their cosmonaut Yuri Gagarin was in orbit. He made one complete circuit of the earth in a Vostok 1 spacecraft and landed safely on April 12, 1961. The United States was "skunked" again.

This time its answer was to set itself an utterly fantastic new space goal. Even before it had matched Gagarin's feat, President Kennedy told Congress: "I believe that this nation should commit itself to achieving the goal, before this decade is out, of landing a man on the moon."

He was proposing to accomplish before 1970 an exploit that Dr. I. M. Levitt, director of the Fels Planetarium of the Franklin Institute, had only lately predicted in *Popular Science* would take place about the year 2000.

The U.S. did commit itself. Mercury and subsequent manned space flights were to be steppingstones to the moon.

Astronaut Alan B. Shephard rode the Mercury spacecraft to a 116-mile height in a 19-minute suborbital flight, and "Gus" Grissom did it again, in 1961.

First American in orbit was John H. Glenn, Jr., who circled the earth 3 times in February 1962. So did Scott Carpenter in May. Walter M. Schirra, Jr., orbited 6 times around in October, and the Mercury project was climaxed with a final 22-orbit mission, lasting 34½ hours, by L. Gordon Cooper, Jr., in May 1963.

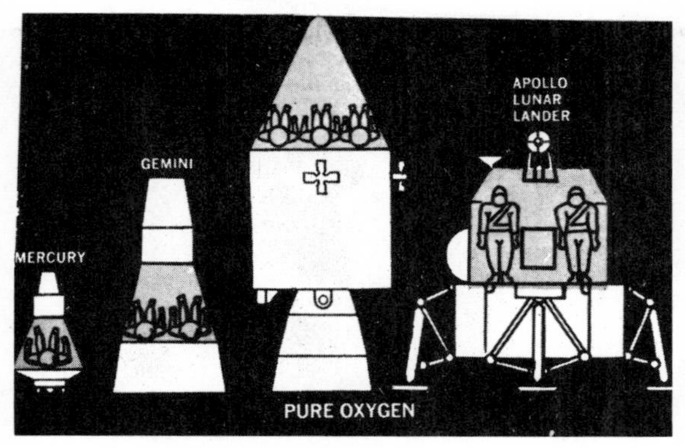

Manned spacecraft followed unmanned ones. Drawing compares successive American kinds: 1-man Mercury, 2-man Gemini, 3-man Apollo, and the lunar lander that carries 2 of its crew to surface of moon and back.

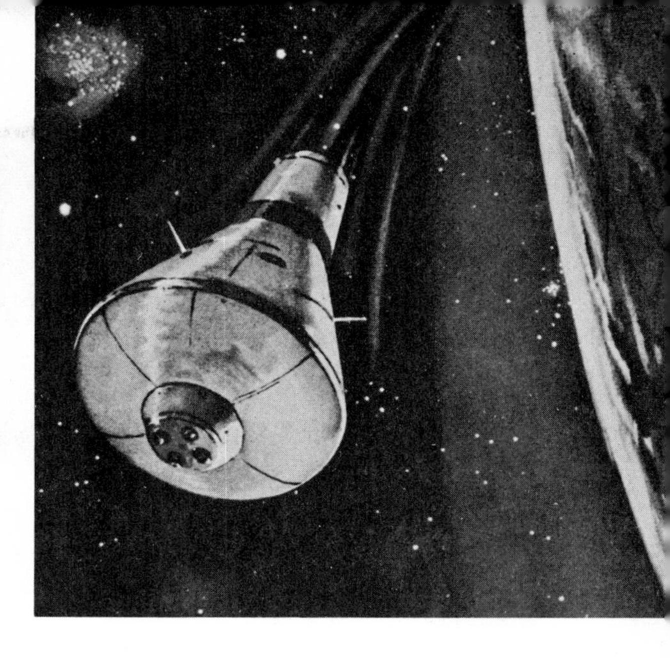

Two-man Gemini spacecraft below, contrasted with Mercury in background, served for ambitious 1965–66 series of missions. Crews practiced rendezvous and docking, "spacewalking" for work outside craft—and proved that weightlessness for the record time of 2 weeks in space brought no ill effects.

First American to orbit earth, in Mercury spacecraft, was John H. Glenn, Jr., in 1962. *Popular Science* previewed the feat three years earlier with the drawing above. Rockets at wide end slowed craft to descend.

Man conquers space! A Gemini astronaut ventures forth from his speeding craft to float beside it in the void beyond the earth and to maneuver himself about with the aid of a hand-held jet gun. Only his spacesuit, fed with oxygen by a trailing hose, shields him from rigors of this airless, hostile realm.

A rendezvous in space: artist portrays approach of an earth-orbiting Gemini (at lower left of picture) to dock with an unmanned Agena spacecraft (at upper right of view).

After Mercury came America's 2-man Gemini spaceships—put into earth orbit by a modified version of the Air Force's potent new ICBM rocket, Titan 2, with a booster thrust of 430,000 pounds. Ten strikingly successful manned Gemini missions were flown during 1965–66.

Keeping at their one-upmanship, the Russians had already launched the world's first multi-man spacecraft—with *three* aboard—in 1964. A new Russian "space spectacular" of March 1965 was the first "spacewalk," in which cosmonaut Alexei Leonov emerged from a 2-man spacecraft in orbit and floated on a tether for 10 minutes. But now the Americans were scoring some notable "firsts" themselves.

Put in orbit simultaneously, Gemini 6 and 7 achieved the first rendezvous in space, when they maneuvered to within 1 foot of each other.

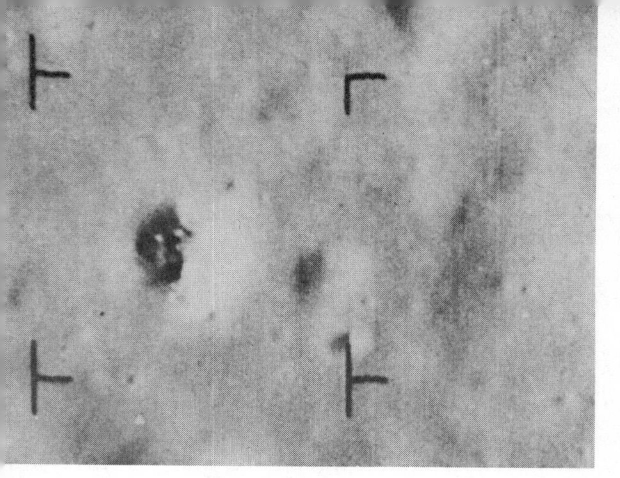

Blazing trail to moon, Ranger TV craft spots a loose rock lying in lunar crater—proof that dust wouldn't swallow up a landing ship.

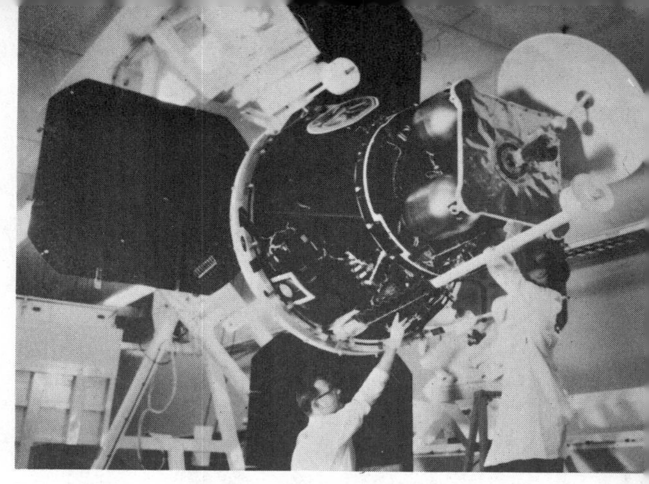

Lunar Orbiter above is one of series that surveyed moon-landing sites—and took the spectacular moon photos below.

Oblique view of majestic 56-mile-wide crater Copernicus from a moon-circling Lunar Orbiter was called "the picture of the century."

Spacecraft's camera peers straight down into great crater Tycho—nearly 3 miles deep, with a towering pinnacle rising near center.

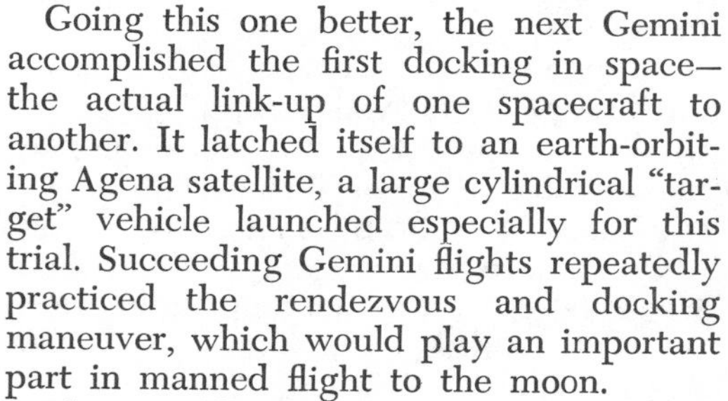

Going this one better, the next Gemini accomplished the first docking in space—the actual link-up of one spacecraft to another. It latched itself to an earth-orbiting Agena satellite, a large cylindrical "target" vehicle launched especially for this trial. Succeeding Gemini flights repeatedly practiced the rendezvous and docking maneuver, which would play an important part in manned flight to the moon.

The world's record for duration of a space flight was set at nearly 8 days by Gemini 5 and then boosted to 2 weeks by Gemini 7. Untroubled by weightlessness for so prolonged a time, the astronauts proved that there need be no worry on that score about a mission as long as a voyage to the moon.

Gemini astronauts sallied forth from their orbiting craft to turn the innovation of spacewalking into a practical part of a

Fantastic bull's-eye formation of Orientale basin, mostly on moon's far side, is seen for first time in its entirety by a Lunar Orbiter.

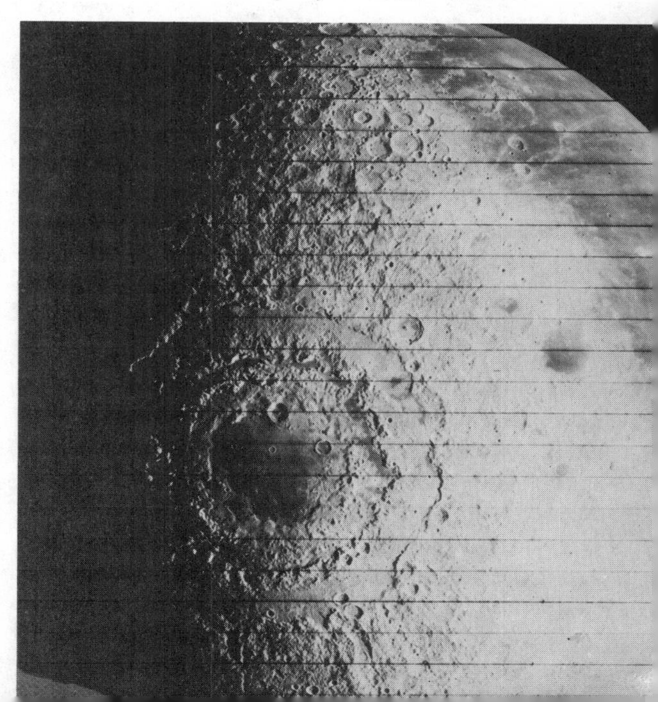

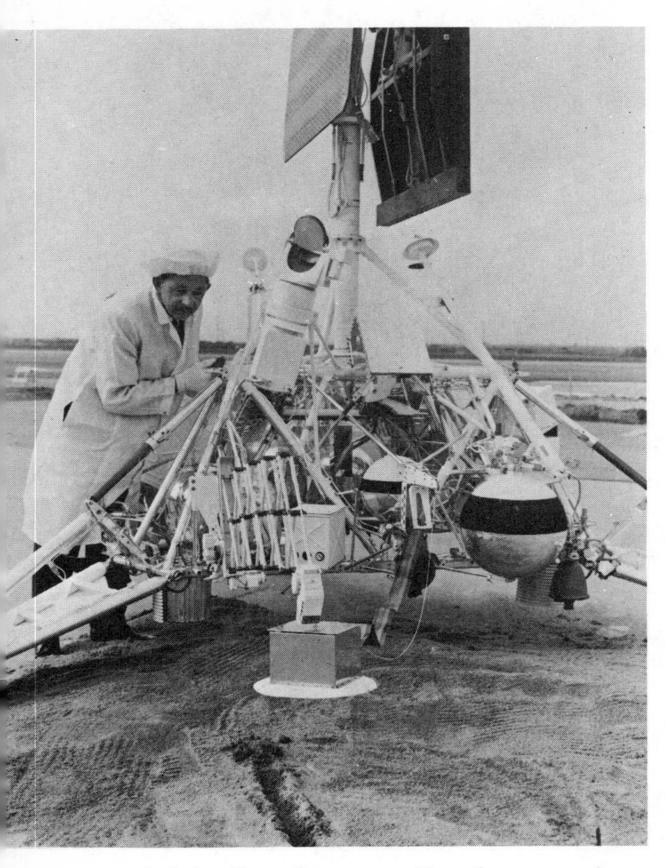

Soft-landing Surveyors like the one above made final tests of moon's surface. Wielding remote-controlled lazy-tongs scoops, they dug miniature trenches to gauge soil's consistency, as pictured below by a craft's TV camera—and even made rough chemical analysis of it with radioactive "chem kit" in a lowered box, visible in the foreground of the upper view.

spaceman's repertoire. With a hand-held gun that squirted gas jets at the touch of a trigger, Edward H. White II became the first spacewalker to maneuver himself about at will. Now man was truly the master of that eerie and hostile new environment, the forbidding void of space.

America turned to designing the 3-man moonship it would call Apollo—and the race for the moon was on.

Asked what our astronauts would find on the moon, a U.S. space expert is said to have quipped, "Russians." It was no idle jest. Our modest Mercurys and Geminis were far outranked in size and weight by the Reds' Vostoks and Voskhods. They had the big booster rockets; ours, so far, were no match.

And they were doing precisely what we were—blazing the trail to the moon by exploring it with unmanned craft. For before you can design a manned lunar lander, you first must learn what sort of terrain it's going to land on. Some seriously feared, for example, that the moon was covered with deep dust in which a landing craft would instantly sink and vanish forever.

First to reassure us on that score was our Ranger 7 spacecraft, one of a series designed to send back television pictures as they hurtled to a crash landing on the moon. From 3-mile height, one of the beautifully clear pictures spotted a loose rock lying in a lunar crater made by its impact—visible proof that it rested upon a firm surface.

Writing on the Ranger 7 findings in *Popular Science*, Von Braun opined, "I think a man walking on a lunar *mare* will make visible footprints." How right he was, man's first steps on the moon would show.

On the airless moon, wings or parachutes would be useless for landing. Riding down the shaft of flame of a braking rocket offers the only way to a gentle touchdown on the lunar surface. First to accomplish it, after 4 or 5 unsuccessful tries, was the Soviets' television spacecraft Luna 9. Beginning 4 months later, the U.S. soft-landed 5 out of 7 Surveyor spacecraft.

Not only did the Surveyors radio back thousands of TV pictures of surrounding moonscapes. Under remote control from earth, some of them wielded tiny scoops on lazy-tongs booms to dig miniature trenches, test the bearing strength of lunar

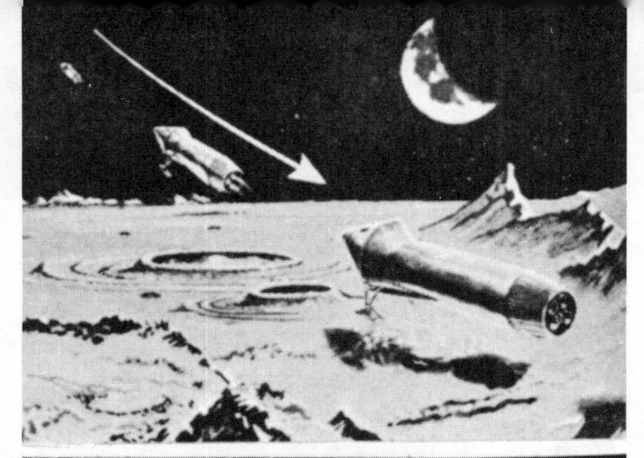

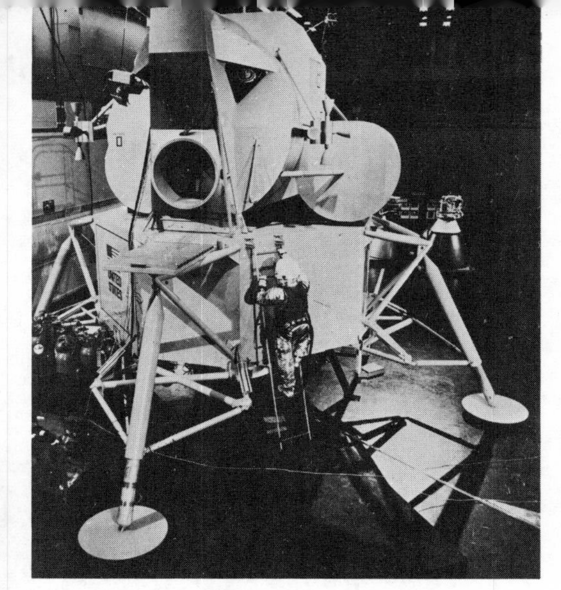

With lunar-rendezvous plan adopted, the landing craft—Apollo's Lunar Module (LM)—takes shape in mock-up above. Crew of 2 are to ride it standing up, as below.

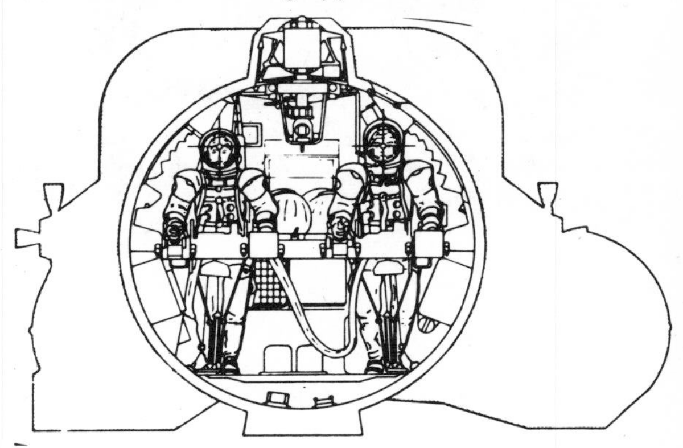

Early plan for manned lunar landing envisioned direct earth-moon flight by giant Nova rocket ship, touching down at near-horizontal angle (top view). Returning part's takeoff would barely clear surface (above). Apollo's lunar-rendezvous plan replaced this one.

Separating from mother ship in lunar orbit and coasting downward, LM fires braking rocket and tilts to upright attitude to make a soft landing on surface of moon.

For return from lunar surface, LM leaves behind its lower portion. Ascent rocket boosts manned cabin skyward to a rendezvous with mother ship circling the moon.

Anatomy of a docking: probe on Command Module (CM) enters cone-shaped member of LM, latches, and retracts to draw craft together. A ring of latches then secures them.

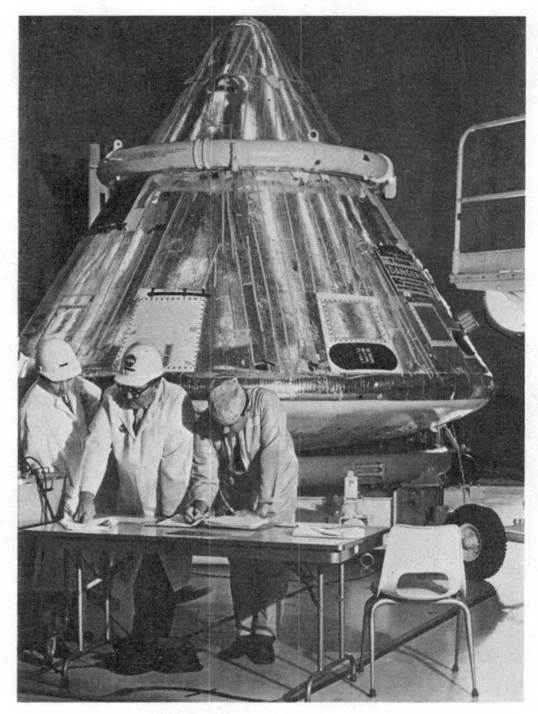

Apollo's conical CM carries crew from earth to moon and back. A cylindrical service module with a rocket engine, and the LM, complete the moonship.

soil, and heft scoopfuls of it. "If we could have chosen ideal soil conditions for man's first landing on another heavenly body," commented Von Braun in *Popular Science,* "we would have come pretty close to writing down the specifications actually met by the moon."

While the Surveyors were doing their thing, U.S. photo spacecraft called Lunar Orbiters swooped low overhead, mapping the moon for likely manned-landing sites. The project had interesting offshoots. It yielded the most spectacular photographs of famous lunar features that the world had ever seen. And it completed a sharply detailed photo map of the hind side of the moon that Luna 3 and later Soviet craft had hardly more than begun. That ended what had been a Russian monopoly, by right of discovery, of proposing names for the features of the moon's far side. (Amicably the U.S. and U.S.S.R. divided the honors about equally in submitting the names that the world's astronomers approved in 1970.)

How the 3-man Apollo spacecraft would get to the moon—and therefore, what it and its launching rocket would look like—were questions undecided until late 1962.

First favored was a "direct ascent" flight plan—straight from the surface of the earth to a moon landing, non-stop. That would be a brute-force approach. The launching rocket, called Nova, would have to be of such fantastic size and power—12 million pounds of thrust at takeoff—that it looked purely visionary, but experts believed it could untimately be built.

Given such a rocket, a direct flight plan seemed simplest and least risky to a layman's eyes—until details took shape on drawing boards.

Still 50 feet long when it reached the moon's surface, the manned ship would be likely to topple if it tried to land vertically. So it would come in for a landing almost

Dr. Wernher von Braun, dean of rocket designers, headed task of building mightiest space vehicle the world had ever seen—the 364-foot-high Saturn V moon rocket.

horizontally, tail first. Its touchdown would be a near belly-whopper on skids—with jets on its bottom hopefully cushioning the shock. If this worked, the astronauts' return journey in the conical front end would begin with a nearly level takeoff, barely clearing the moon's jutting crater rims and peaks. In printing illustrations of these hair-raising maneuvers, *Popular Science* took care to emphasize they were official NASA drawings—and no wild imagining of its own.

An "earth-orbit rendezvous" flight plan—the one reportedly favored by the Russians—was an alternate one under NASA consideration. Rockets from earth would travel only as far as to an earth-circling orbit, where a moonship would be assembled and start on its way. The 2 rockets needed to deliver materials, fuel, and crew to orbit would have to be of formidable size, but nearer within reason than the Nova.

Actually NASA's choice fell on neither of

First Saturn V thunders aloft, in unmanned 1967 trial. Bold "all-up" gamble tests all 3 stages at once instead of separately. Maiden flight of enormous rocket is flawless.

Daring trial of first manned Saturn V is a voyage around moon, by Apollo 8 in late 1968. Drawing of moon-circling craft is superimposed upon a moon photo it took.

Apollo 11, moonborne by Saturn V, sets forth on man's greatest adventure. On July 20, 1969, 2 of the 3 astronauts aboard became the first men to set foot upon the moon.

these flight plans. It went instead, in November 1962, to an appealing new "lunar-orbit rendezvous" plan proposed by John C. Houbolt of NASA's Langley Research Center at Hampton, Virginia:

The manned Apollo moonship would fly from the earth's surface to earth orbit and thence directly to orbit around the moon. There, a 2-man lunar module (LM) would detach itself and descend to the moon's surface. Returning, it would rendezvous and dock with its mother craft in lunar orbit, and the recombined moonship would head homeward to earth.

Engineers reached for their slide rules. A single launch rocket, with 7,680,000 pounds' thrust at takeoff, would do it. An early convert was Von Braun, who accepted the challenge of building the mightiest rocket in history—Saturn V, the moon rocket.

As if this were not enough for one man to do, Von Braun somehow found time and energy to become a prolific author on his favorite subject, space flight. Books, technical papers, popular articles flowed from his gifted pen. Beginning in 1963 he became a regular contributor to *Popular Science* of answers to readers' questions on space and later of full-length articles on current space developments.

And now, key decisions made, the pace quickened. Into being came the Apollo moonship's command module (CM), its 3-man crew compartment; the service module, its "engine room"; and the LM, its spidery 4-legged moon-landing craft.

The 1,640,000-pound-thrust Saturn IB, one of a series of ever-larger Saturn rockets created by Von Braun and his team, served in 1966 for initial unmanned tests of such Apollo components as its heat shield. They were successful.

Nevertheless, a London bookmaker's odds still favored the Russians to land a man on the moon before the Americans.

Before the end of 1967, the first Saturn V

"I think a man walking on a lunar *mare* w
make a visible footprint," predicted V
Braun in 1964. Five years later, imprint
Apollo 11 astronaut's boot on the surface
the moon (facing page) shows him so rig
NASA PH

stood poised on its launch pad for its maiden flight. It was an eyeful.

Taller than the Statue of Liberty, the stupendous 3-stage rocket towered 364 feet high. Five mighty engines in its 33-foot-diameter first stage would gulp more than 3,000 gallons every second of kerosene-like RP-1 fuel and liquid oxygen. Liquid hydrogen and oxygen would feed the 5 engines of its second stage and the single one of its third.

Urgency had telescoped its trials—and its first flight of November 9, 1967, was an "all-up" test. Instead of being tried out separately, the normal cautious procedure, all of the stages were getting their initial tryout at once. Even Von Braun, confident as he was of his baby, was taken aback by the magnitude of the gamble.

Ignition! The fires of Hades erupted from the great booster engines. The ground shook as if rocked by an earthquake. "Go, baby, go!" spectators screamed above the thunderous sound of the rising rocket's blast.

The fateful flight was flawless, perfect, almost too good to be true. America had its moon rocket.

Belatedly, in its second and last unmanned trial of April 1968, trouble mysteriously appeared. Two second-stage engines shut down prematurely; the third-stage engine failed to re-ignite in space as it would have to do in a moon mission; and undue vibration had followed the launching. Intensive detective work brought speedy remedies.

So sure were experts of having fixed the difficulties, that Saturn V's next flight would be manned—and for what a flight! Although it would carry no LM for a landing, this Apollo 8 mission would be man's first voyage to the moon—to circle it and then to return to earth.

Just before Christmas of 1968, a white Saturn V put Apollo 8 in a 24,200-mph trajectory to the moon—faster than man had ever traveled before. Past the leading

First men on the moon: "Buzz" Aldrin, standing on lunar surface above, and Neil Armstrong, mirrored in his visor. "Kangaroo hop," below, works well in weak lunar gravity.
NASA PHOTO

Precious cargo, boxful of first moon rocks, arrives on earth. Security guard shoos photographer aside as 2 men bear priceless booty in triumph to Houston laboratory.

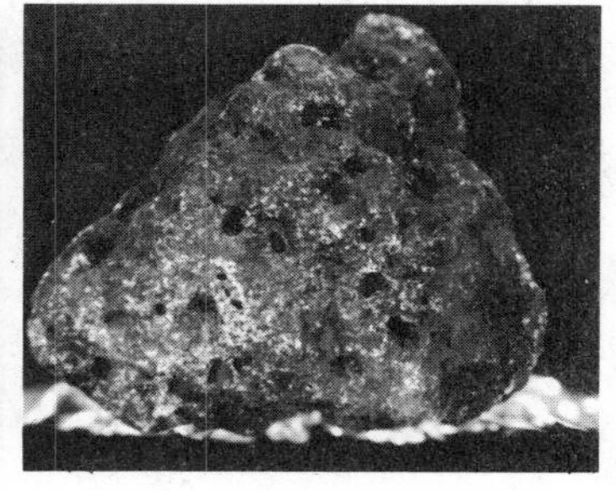

Three moon rocks from Apollo 11's haul give rockhounds something to feast their eyes upon. Fabulous age of moon rocks, some older than any found on earth, promises that they will provide clues to the earth's own mysterious early history.

Glassy "moon beads," highly magnified below, were surprise find in lunar soil.

Pinpoint touchdown allows Apollo 12 crew to visit Surveyor 3, landed on moon in 1967. NASA PHOTO

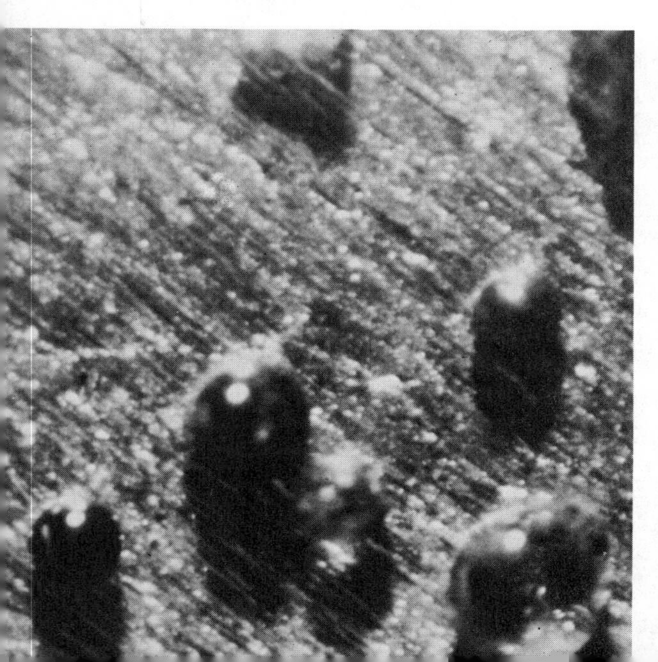

Left on moon, robot device at left detected water vapor; the one above, moonquakes.

edge of the speeding moon scooted the ship, "like a car beating a locomotive to a railroad crossing," wrote Von Braun in *Popular Science*. Rehearsing a coming lunar-landing maneuver, it rocket-braked itself to be captured in lunar orbit. Ten times it circled the moon, as the crew thrilled to a close-up view of lunar landmarks seen only through telescopes before, and took magnificent photos. Then Apollo 8's rocket kicked it out of lunar orbit and it streaked for home—to hurtle like a shooting star into the earth's atmosphere at record re-entry speed of nearly 25,000 mph.

There was scant margin for error in the angle at which it hit the earth's air. Too steep, and it would burn up; too shallow, and it would rebound into space like a stone skipping on water, a fatal calamity too. Right on the beam, Apollo 8 made it.

Two succeeding Apollo missions practiced releasing the LM and docking again with it—first in earth orbit, and then in lunar orbit—and the stage was set for man's greatest adventure.

The fantastic event took place on July 20, 1969. Casting off from their mother ship, in the landing craft they had named *Eagle*, Neil A. Armstrong and Edwin ("Buzz") E. Aldrin, Jr., swooped down from lunar orbit toward the smooth gray surface of the moon's Sea of Tranquillity. As they neared it, a blast of their descent engine checked their speed and braked their fall.

Down the pillar of flame they rode. Blue lights flashed on in the cabin, as feelers extending from *Eagle*'s bottom made contact with lunar soil. Across a quarter of a million miles to earth came Armstrong's radioed voice: "Tranquillity Base here. The *Eagle* has landed."

Never before had so many on earth watched a happening at once. All over the world, 530 million people at television sets saw space-suited Armstrong lead the way

First shipwreck in space: oxygen-tank explosion ruptured Apollo 13's service module, above. Crew barely got home safely.

Spectacular crash of rocket stage on moon to register man-made moonquake was accomplished despite Apollo 13 accident.

Weird unmanned vehicle called Lunokhod 1, looking like something out of science fiction, was put on moon by Russians in 1970. Remotely controlled by radio from earth, 8-wheeled moon car lumbered over lunar surface for 10 months, transmitting pictures from TV eyes at 2 camera ports (left end in photo). Solar panels supplied power to propel electric vehicle, at leisurely pace of less than one tenth of a mile per hour, during lunar day; it parked during lunar night. In all it covered a distance of about 6 ½ miles.

out the opened hatch, descend the rungs of a ladder, and plant a foot upon the moon. "That's one small step for a man, one giant leap for mankind," came his now-historic words.

Before the eyes of the world, Armstrong and Aldrin learned to walk on the moon. Some scientists had feared they would find it difficult to move about in the weak gravity of the moon, one-sixth of the force on earth—but they mastered the knack in minutes. A "kangaroo hop" proved one effective way, they found.

For 2 drama-packed hours they explored their eerie surroundings, collected 54 pounds of moon rock and soil to bring home, and emplaced a robot moonquake detector to be left in operation upon the lunar surface.

Back aboard *Eagle,* they fired the ascent rocket built into its upper half—and soared aloft to rejoin the mother ship and its pilot, Michael Collins, in a perfect rendezvous and docking. Their homeward trip was a repeat of Apollo 8's. The safe splashdown touched off joyous celebrations of America's success. People of Huntsville hoisted Von Braun to their shoulders and paraded with him about the courthouse square.

"Worth more than all the gold in Fort Knox," a NASA official called Apollo 11's moon-rock samples, borne in triumph to Houston for the most intensive study any minerals ever had. Most intriguing feature of these and later missions' lunar samples was their amazing age. Some antedated any rocks found today on earth, where the elements have obliterated the record of this planet's beginning. Scientists hoped the moon rocks would help to fill in the blank pages of the story.

Not just one, but two manned moon landings in 1969 fulfilled the goal set by President Kennedy.

Before the year was over, Apollo 12 touched down to a pinpoint landing within easy 600-foot walking distance of its target—Surveyor 3, an unmanned spacecraft

NASA official receives speck of moon soil from Russians in first U.S.-Soviet swap of specimens. Exchange followed remarkable feat of Russians' unmanned Luna 16 craft in landing on moon and returning to earth with 3½-ounce sample from Sea of Fertility.

NASA PHOTO

First manned car on moon, lunar roving vehicle brought by Apollo 15, carried lunar explorers at up to 8 mph on extended tours of mountainous Hadley Rille region. Radio on car maintained direct communication with earth and transmitted TV views of scenes.

NASA PHOTO

soft-landed in the moon's Ocean of Storms in April 1967. Retrieved were Surveyor's TV camera and other parts, to be added as trophies to a rich new 75-pound haul of moon-rock and soil samples. On earth, analysis of the Surveyor fittings would tell how they fared in more than two years of exposure to the lunar environment—and aid in choosing the materials of future spacecraft for long-duration missions.

The first shipwreck in space cut short the next lunar-landing try—Apollo 13 in April 1970.

Nearly halfway to the moon, an explosion shook the craft. One of the 2 main oxygen tanks in the service module had blown up—letting the precious life-giving gas escape. The only way home was the long one: to loop around the moon and return. There was perilously little oxygen left to breathe—and to supply power in the craft's fuel cells—for the 4 days it would take. The crew thought they were goners, they later reported.

Emergency consultations with the Houston Control Center brought them through. The LM, with its own limited oxygen supply, became their lifeboat until the time came to return to the CM for re-entry. A safe splashdown relieved the anxious world.

To allow time to find the cause of the accident and remedy it, further Apollo flights were postponed until 1971.

Two striking feats of unmanned Russian lunar spacecraft were other memorable accomplishments of 1970.

Luna 16 alighted on the moon, collected some 3½ ounces of lunar soil with an electric drill, took off again—and brought home Russia's first moon sample, from a hitherto unexplored region. The U.S.S.R. swapped bits of it for Apollo samples. (Luna 20 repeated Luna 16's feat in 1972.)

From another Russian spacecraft soft-landed on the moon, Luna 17, rolled a weird 8-wheeled vehicle with television eyes, Lunokhod 1. Under remote control from earth, the lumbering unmanned car

Coming in 1973: Skylab, first U.S. manned space station, is pictured with Apollo craft docked to it at lower left. Windmill-like solar panels, just above, power sun observatory. Saturn rocket stage forms 2-story lab at right. For interior views, see next page.

explored the moon for 10 months, traveling 6½ miles in all.

After another successful moon-landing expedition, to the Fra Mauro area in the lunar highlands—Apollo 14 in January-February 1971—the U.S. had a moon car of its own. A manned one, it was a hot rod compared to Lunokhod.

Folded into the Apollo spacecraft, the 2-man, battery-powered vehicle went to the moon with the mid-1971 Apollo 15 expedition to the Hadley Rille region, where it was unpacked and put to good use. The 8-mph electric car enabled the astronauts to explore an area the size of New York's Manhattan Island, during their extended 3-day stay on the moon's surface. It was one of the most rewarding expeditions yet.

Two more moon-landing flights in 1972, again taking electric cars along, were scheduled to conclude the current Apollo program. What then?

Next in U.S. manned space flight is to be an earth-orbiting space station called Skylab, due to be launched in 1973 to a 270-mile height. As big as a 3-room house, it will measure 83 feet long over-all.

Its main section, built into an empty upper stage of a Saturn V, will be a 2-story experimental workshop. In addition it will include an astronomical observatory for solar studies; an airlock for spacewalking; and a "multiple docking adapter" to which one or more man-carrying Apollo spacecraft can dock.

After Skylab has been launched unmanned, successive 3-man crews will arrive in Apollo craft, and set up housekeeping inside—for an initial mission of 28 days and two succeeding ones that each may last up to 56 days if all goes well. Observations of the effects on humans of these unprecedented periods of weightlessness will be among a vast number of zero-gravity tests to be performed in the orbital laboratory.

After Skylab will come a revolutionary new kind of manned space vehicle—the "space shuttle"—if Congress approves a $5½ billion, six-year program to develop it that received the backing of President Nixon in 1972. It would be the biggest U.S. space project since Apollo.

The space shuttle will be a reusable spacecraft with wings, to ferry men, satellites, and supplies between the ground and earth orbit. It will take off like a rocket, fly in orbit like a spaceship, and return like an airplane. Comparable in size

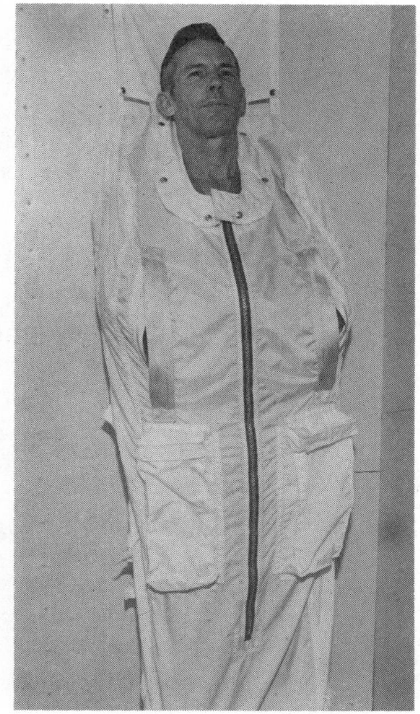

Inside Skylab space station: left, weightless crewman walks erect by grasping overhead handrail; center, zipped-up nightie anchors a sleeper to keep him from floating about; right, flying this self-propulsion gear inside Skylab is one of 47 experiments planned.

to a DC-9 aircraft, this "spaceplane," as *Popular Science* calls it, will have a length of 120 feet and a wing span of 75 feet. It can carry up to 14 passengers besides its crew of 2.

Planners expect it can be used over and over again for as many as 100 missions, and will reduce the cost of orbiting a pound of payload to $100, from the present figure of $600 to $700.

By 1971, only fourteen years after Sputnik 1, 4,900 man-made satellites had been put up and 2,148 were still in orbit. No longer were satellite-launching nations an exclusive club; even Communist China had orbited its second satellite.

Although the glamour and drama of manned space flights had held the spotlight upon them, the unmanned earth orbiters ranked high for their practical benefits to the world.

Among outstanding ones were the U.S. weather satellites that began with little Tiros 1 in 1960 and had grown to 682-pound Itos versions by 1970. In that time they returned more than a million pictures of the earth's cloud cover, and other data, for weather forecasting and research. In another ten years, Von Braun then predicted

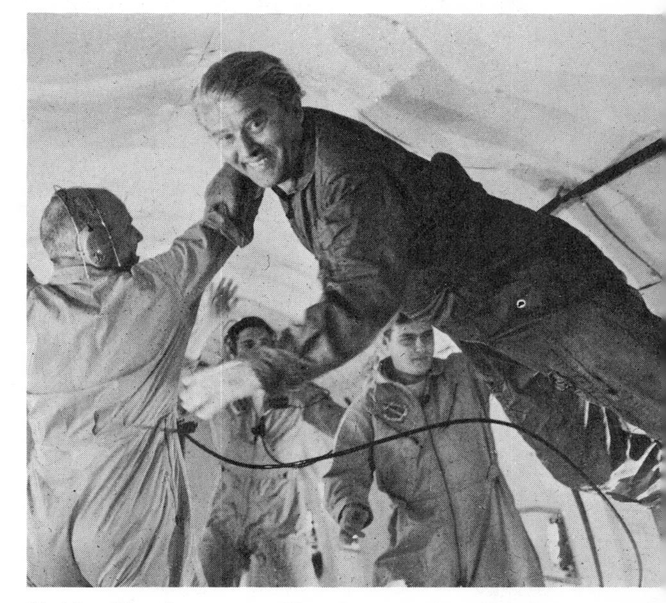

Making like Superman, Von Braun enjoys brief 30-second taste of floating weightless, in padded cabin of airplane flying ballistic trajectory. Such flights on earth serve to train astronauts and to test designs of space equipment for use under zero gravity.

Notable U.S. satellites: top, first weather satellite, Tiros 1; above, Syncom prototype of "stationary" satellite; below, latest-type transocean phone-and-TV satellite, Intelsat 4.

in *Popular Science,* "Weather satellites will tell you whether you will have a rainy vacation if you leave two weeks from now."

Telephone-and-TV satellites formed a worldwide net to relay voices and pictures across the oceans. Since 1964 they have applied a trick pioneered by that year's Syncom 2 satellite, to hang seemingly stationary over a selected point on earth. It could be done, Syncom 2 demonstrated, by orbiting a satellite with a speed that just matches that of the earth's rotation. Such a "synchronous" satellite hangs at an altitude of 22,300 miles above the earth.

Interplanetary exploration, a fantasy of science fiction before, began for real in 1962 when the U.S. spacecraft Mariner 2 flew past Venus at less than 22,000-mile range. Ogling the planet, its instruments reported an intensely hot surface temperature and other data. Pictures would have been pointless; belying its name, Venus is anything but photogenic. Completely hidden beneath dense clouds, it looks, *Popular Science* observed, like "a fuzzy white tennis ball."

Mars was another story. In 1965 our Mariner 4 transmitted the first close-up TV views of the planet, from about 6,000 miles away. They staggered astronomers. Pocked with craters, the Red Planet's surface unexpectedly looked more like the moon than the earth.

Subsequent U.S. and Soviet spacecraft have outdone these "firsts." In 1969 Mariners 6 and 7 got closer and far-more-revealing pictures of Mars. In 1970, after repeated tries, Russia succeeded in parachuting down a capsule that transmitted data for 23 minutes from the burning-hot surface of Venus. The end of 1971 saw 3 spacecraft orbiting Mars at once—our Mariner 9 and the Russians' Mars 2 and 3. Their picture-taking was being hampered by the Red Planet's most intense dust storm in memory. Perhaps for the same reason, a TV capsule parachuted to the surface by Mars 3 transmitted 20 seconds of "pictures" that were virtually blank and then quit working.

Early in January 1972 the dust cleared, and Mariner 9 began transmitting Mars photos of unprecedented detail. They include the first close-ups of markings at all resembling the Martian "canals" that Schia-

Finding Mars cratered like moon (right) was stunning surprise of first close-up TV views, transmitted in 1965 by U.S. interplanetary spacecraft Mariner 4. But views from later spacecraft show variety of other features setting Mars apart from either moon or earth.

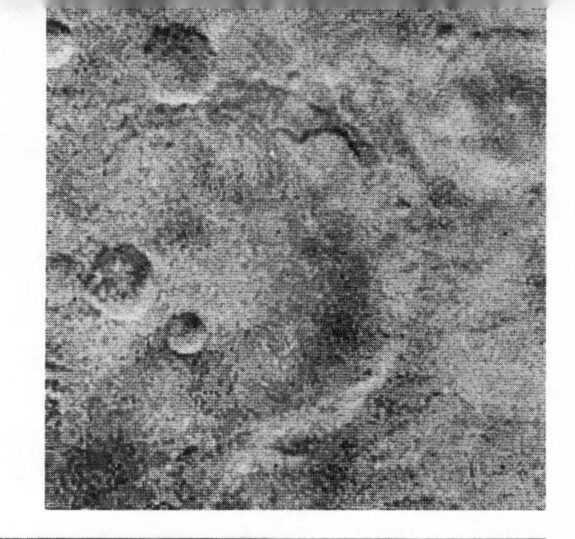

ight of Mariner 4 past Red Planet in 1965 visualized by *Popular Science* artist, with ajor features of spacecraft indicated by bels. Later visitors to Mars have been Marers 6 and 7 in 1969; and Mariner 9 and Soviet craft, Mars 2 and 3, in 1971.

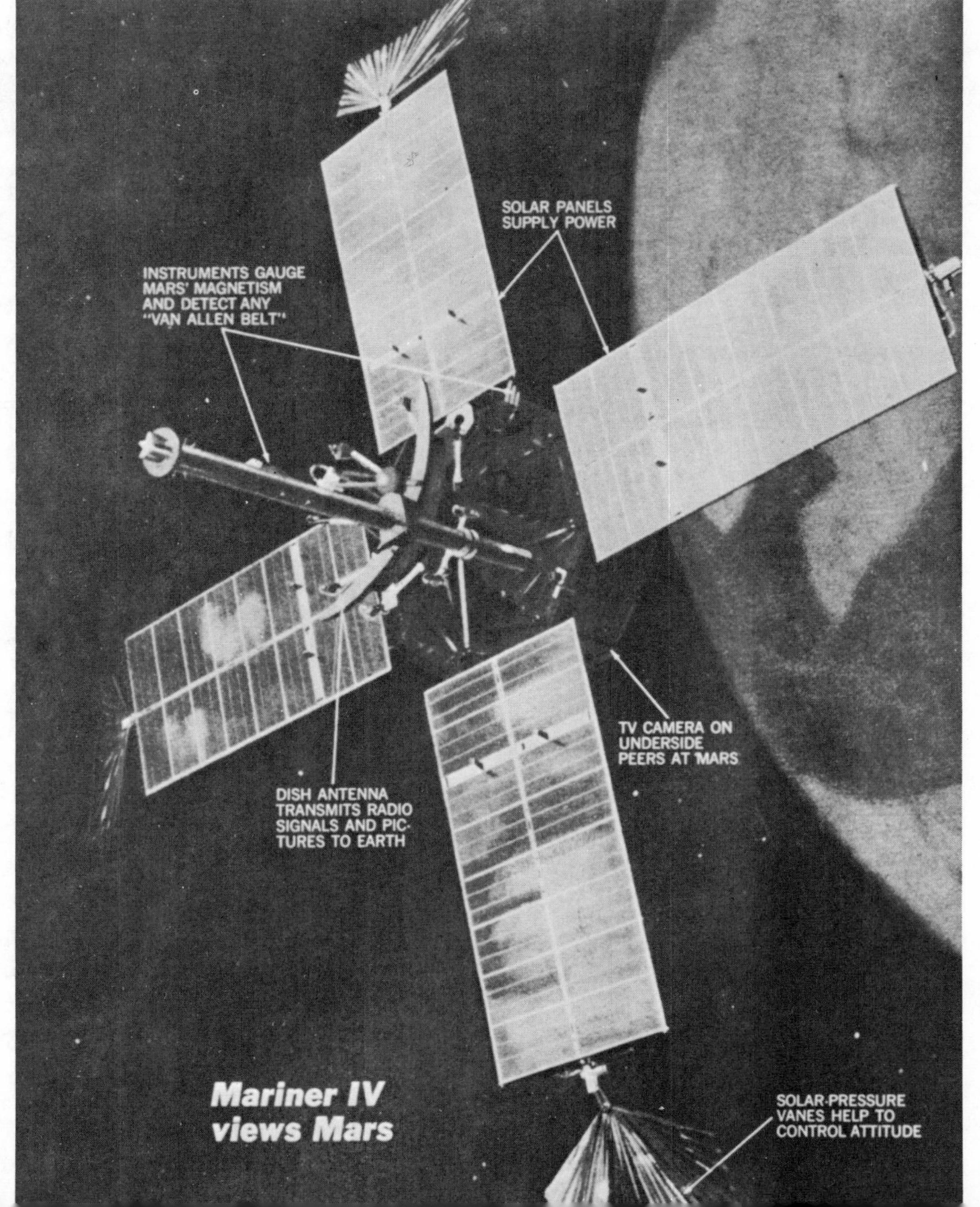
SOLAR PANELS SUPPLY POWER
INSTRUMENTS GAUGE MARS' MAGNETISM AND DETECT ANY "VAN ALLEN BELT"
TV CAMERA ON UNDERSIDE PEERS AT MARS
DISH ANTENNA TRANSMITS RADIO SIGNALS AND PICTURES TO EARTH
Mariner IV views Mars
SOLAR-PRESSURE VANES HELP TO CONTROL ATTITUDE

Is there life on Mars? Long-awaited answer may come in 1976 when landing craft above descends to Martian surface from U.S. Viking orbiter. Long arm at lower left will gather soil sample for chemical tests, to detect any sign of living organisms, by sophisticated analytical instruments inside craft. Findings will be transmitted from the landing craft by radio to earth. Discovering even the lowest form of life on another planet would be major event. This picture and the space-flight illustrations that follow on the next 3 color pages were painted especially for *Popular Science* by leading illustrator Bob McCall.

parelli discovered in 1877 and that American astronomer Percival Lowell believed to be the work of intelligent beings. One photo shows an almost straight canal-like feature bordered by dark lines—which look to geologists like natural "tension cracks," and not artificial handiwork.

Exploring other planets would follow. In 1972, the U.S. Pioneer 10 was on its way to a TV rendezvous with far-off Jupiter in December 1973.

Soon, too, interplanetary craft promise to solve the most fascinating mystery of space: Is there life on other planets?

Recent years' findings, to be sure, have dimmed the prospects of finding "little green men," or people of any other hue, elsewhere in our solar system. For example, the rigors of Mars' climate and scarcity of water make it look like an unlikely habitat for anything resembling human beings—or higher animals.

Far from ruled out, however, is the possibility of Martian life of simpler sort, conceivably primitive plants or micro-organisms. And the discovery of extraterrestrial life, even of such lowly forms, would be a momentous event to science. It could come as early as 1976, if not before.

In that year a life-detecting capsule was scheduled to descend to the Martian surface from a U.S. Viking spacecraft orbiting the planet. The capsule would reach out an arm, scoop up a sample of the soil, and take it inside for sophisticated chemical tests that would reveal the presence of any living thing. A positive response would open new vistas to the imagination, with proof at last that life does indeed exist in other worlds.

Braking rocket lowers 2-man Apollo landing craft to a gentle touchdown on airless moon—where wings or chutes are useless. During approach (lower half of composite view), craft tilts to vertical, and target area looms ever larger in triangular, scale-marked windows.

PAINTING BY BOB McCALL

"Space shuttle," a winged spacecraft due by 1978 or soon after, will ferry up to 14 men into orbit—and return to earth like an airplane, for further missions. Uses include taking aboard an

out-of-order satellite for repairs in orbit (right), and carrying crews to and from a space station of the future (left). Spacemen propelled by jets from backpacks (foreground) aid in operations.

PAINTING BY BOB McCALL

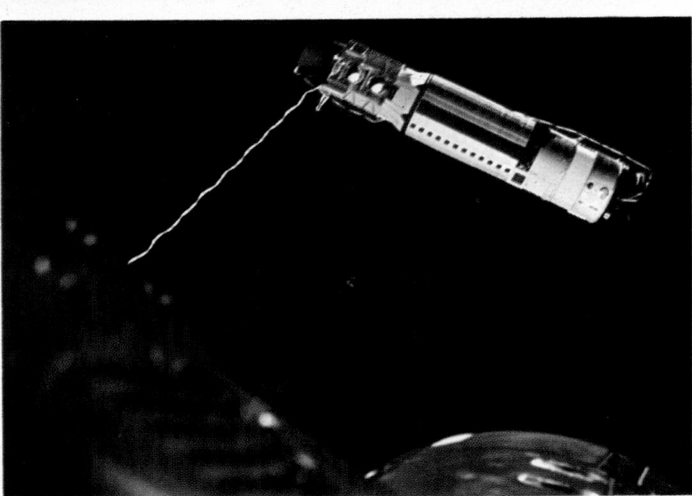

Crew of Gemini 11, making rendezvous in earth orbit with unmanned Agena spacecraft in 1966, show how future astronauts could provide artificial gravity to avoid problems of weightlessness aboard space stations and interplanetary spacecraft. Above, Gemini has attached 100-foot tether to Agena (center of picture) and is backing away to take up slack.

Gemini fires control jets sideward, slowly spinning the two craft like ends of a dumbbell. Centrifugal force creates artificial gravity—only 1/1000 as strong as on earth, but enough to settle crew in seats.

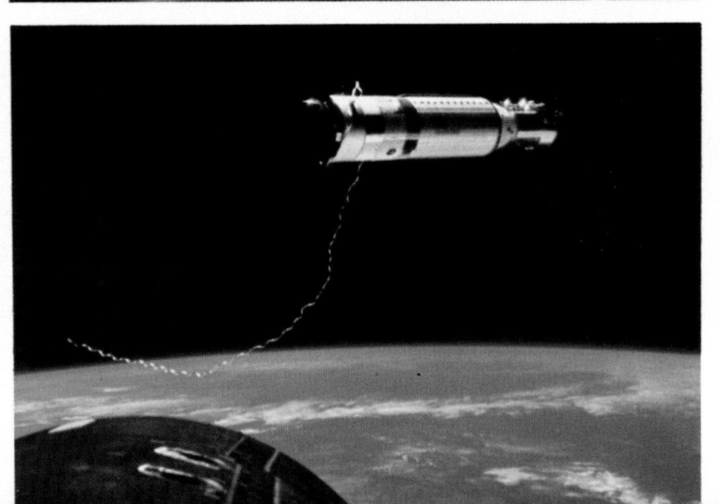

After successful artificial-gravity trial, Gemini 11 (partly visible in left foreground) casts off from Agena. For return to earth, retrorockets brake its speed in orbit and it drops from sky to a splashdown.

CHAPTER 10

Sounds and Pictures on the Air: the Story of Radio and TV

THE air is alive with radio waves—from tens of thousands of broadcast radio stations around the world, thousands of TV stations, hundreds of thousands of amateur radio transmitters, millions of "citizens band" radio transceivers and walkie-talkies, and dozens of space satellites. And that's not all—countless aircraft radios, ship-to-shore rigs, police-car two-way radios, scores of other special-purpose radio transmitters, and a great many radio-controlled garage-door openers.

Yet, just seven decades ago, there was agreement among many scientists that "wireless telegraphy" (ancestor of modern radio) was an interesting, but limited, development. In the February 1902 *Popular Science* a technical reporter warned that wireless telegraphy was "much slower than the standard Morse telegraph apparatus that uses a wire, and there is no assurance of secrecy." His conclusion was that "the proper field of wireless telegraphy appears to be the overspreading of limited areas of water with telegraphic facilities."

In other words, the only important use of radio would be ship-to-shore communications. Why this lack of foresight?

In 1902, "radio" transmitters and receivers were crude *electrical* devices. Fledgling electronics had not yet produced the vacuum tube that made possible radio as we know it. There was no radio voice communication—only Morse code, sent by transmitters that used huge high-voltage electric sparks to generate "Hertzian (radio) waves." The sparks were manufactured in machine-gun-like sequences: a short burst of a few sparks generated a Morse "dot" signal, a longer burst of more sparks produced a "dash."

At the receiver, an amazingly simple gadget called a "coherer"—a glass tube of metal filings—responded to the received waves, and a "telephone" (earphone) or an electric clicker made the code signals audible. "Skywires" (antennas) at the transmitter and the receivers were enormous structures—often hundreds of feet high—but of primitive design and inefficient.

Few early transmitters or receivers could be tuned to broadcast or receive on a specific channel or frequency, so operating two transmitters simultaneously within range of one receiver was impossible. And although the transmitters used thousands of watts to generate their output waves, communications were only marginally reliable: time of day, weather, and distance all affected radio communications in ways the experimenters didn't know how to control.

But in spite of its shortcomings, wireless communication fired the public's imagination. The ability of radio to communicate instantaneously over vast distances without connecting wires was an astounding phenomenon. In the four decades centered at the turn of the century, newspaper dispatches and news of radio in *Popular Science* bubbled over with the excitement induced today by trips to the moon. Wireless communication seemed sheer magic, and the great radio pioneers were wizards.

The discovery of radio waves was a result of nineteenth-century science's preoc-

cupation with the study of light. Years of optical experiments had shown that a beam of light behaves as if made up of a train of waves traveling through space. In 1865 James Clerk Maxwell, a Scottish physicist, produced a landmark series of equations that mathematically described the waves.

One by-product of Maxwell's theory was the concept that a rapidly vibrating electric charge will produce electromagnetic waves —waves carrying both electric and magnetic energy—that will travel at the speed of light. These are precisely what today we call radio waves; they are generated when electrons vibrate in an antenna, impelled to do so by a transmitter.

The man who turned Maxwell's prediction into reality was Heinrich Hertz, a German physicist. In 1887 he built a rudimentary transmitter that eerily resembles a modern car's ignition system. Its heart was an induction coil—2 coils of wire side by side. The primary coil had a few turns of thick wire; the secondary coil, many turns of fine wire. Nineteenth-century physicists

A radio pioneer, Heinrich Hertz built a rudimentary transmitter and ingenious receiver as early as 1887. Old engraving was frontispiece in July 1894 *Popular Science*.

Reproduction of apparatus used by Marconi to send messages long before his attempt at transatlantic wireless. At left, below, his first wireless transmitter and receiver.

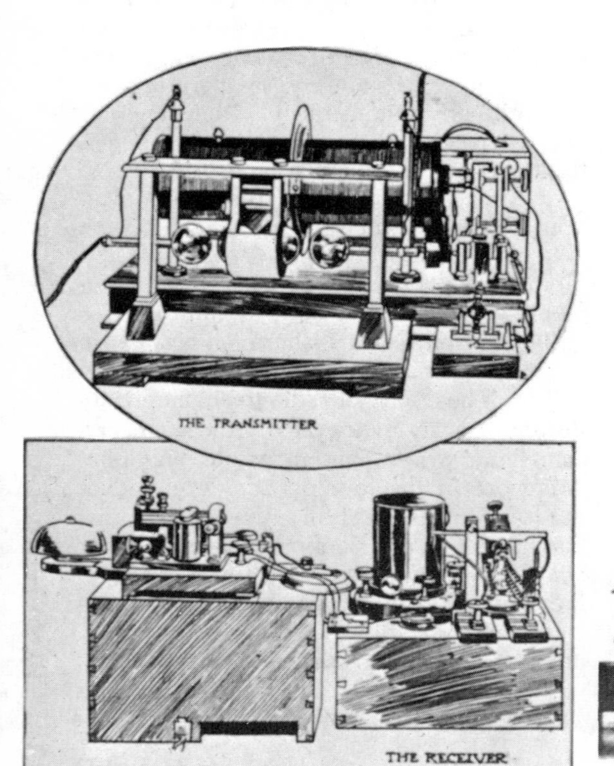

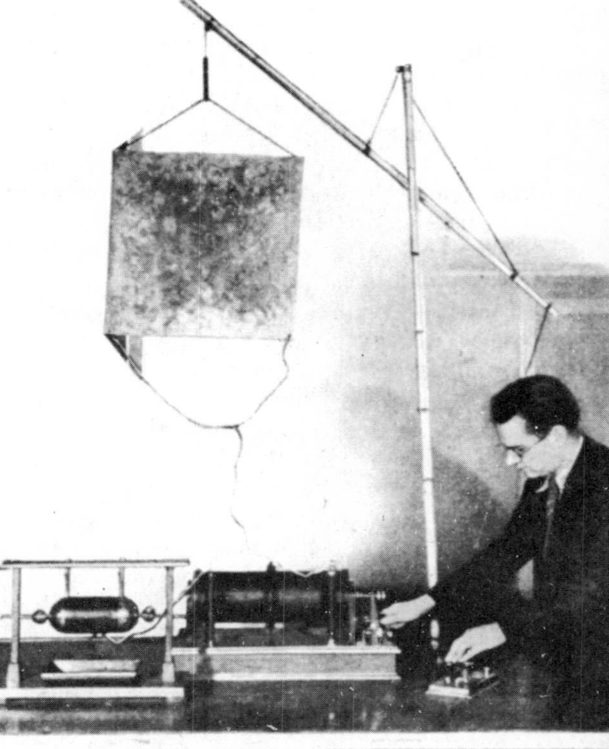

called this transformer a "Ruhmkorff coil."

Hertz connected the primary coil to a battery to establish a steady current flow, the secondary coil to a pair of metal balls separated a few inches to create a spark gap. A small paddle-like metal plate fastened to each ball acted as an antenna.

When Hertz interrupted the flow of current in the primary coil, a very high voltage was induced in the secondary by electromagnetic induction. Almost instantly, a bright spark snapped between the balls and, for a fraction of a second, very high-frequency oscillating currents raced through the paddles, radiating radio waves.

Hertz crafted an incredibly simple receiver to verify that his transmitter was generating electromagnetic waves. He joined two metal balls with an opened brass hoop, leaving a tiny (fraction of a millimeter) air gap between the balls at one point. Placing his receiver several feet from the transmitter, Hertz operated the induction coil. A weak spark jumped between the balls of the receiver, proving that waves were radiating from the transmitter and carrying energy to the receiver.

Hertz was not aiming to develop a communication system; he was trying to establish "the identity between electricity and light" as he explained in December 1890 in *Popular Science*. Curiously, in one regard the discoverer of radio waves had a less accurate understanding of their nature than has an elementary-school pupil today. Hertz, and most scientists of the period, believed that light and radio waves flow through an invisible medium permeating all space, which they called the "ether."

The concept that electromagnetic waves must be carried by *something* may sound logical, but it isn't true. Radio and light waves travel with ease through empty space. Continuous interchange between electric and magnetic forms of energy in the waves propels them forward, not the alternate "stretching" and "relaxing" of an ether.

Hertz died in 1894 and missed, by only months, seeing his apparatus transformed

Wooden towers at power station in England supported Marconi aerial for transatlantic wireless. Kites like that held by assistant G. S. Kemp, at right, suspended aerials at receiving end. Picture of Marconi at lower right was taken in Newfoundland shack on historic December 12, 1901, after he received first transatlantic wireless message from England.

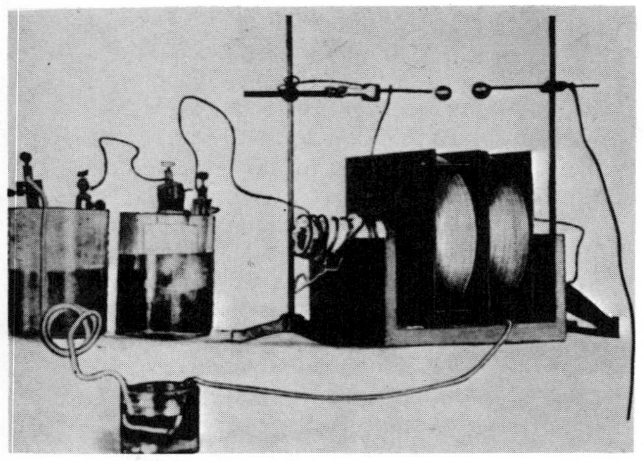

This was transmitter of Marconi-style outfit, at Harvard, pictured in 1899 *Popular Science*.

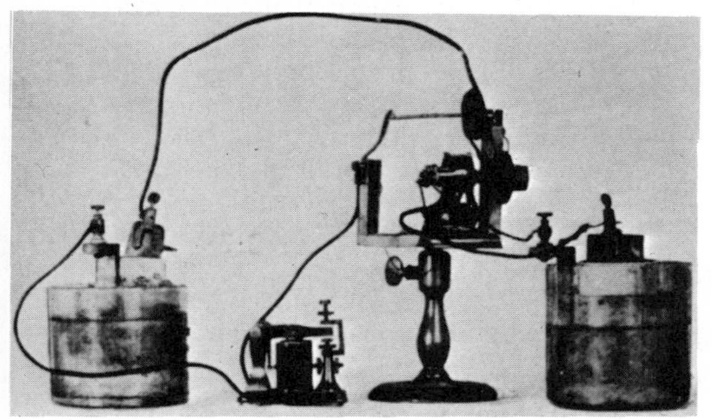

Photo of Harvard outfit's receiver showed batteries, sounder, motor to shake coherer.

Coherer, shown above 1 ⅓ times actual size, was this tiny tube containing iron filings.

into a practical communications device by the Italian, Guglielmo Marconi.

Marconi was but one of many experimenters who recognized the potential of Hertzian waves. In 1892 the famed English scientist Sir William Crookes wrote about "ethereal vibrations" in *Popular Science:* "Here is unfolded to us an astonishing new universe—one which it is hard to conceive should be powerless to transmit and impart intelligence." And, it is history that both David Hughes and Alexander Lodge in England and Alexander Popov in Russia all built working wireless telegraphs that preceded Marconi's first models in 1895.

Yet Marconi has been credited with ushering in the age of radio, and with good reason: he proved that *long-distance* radio communication was technically possible and commercially practical, when skeptics were numerous and persuasive. In 1901, when Thomas Edison was told of Marconi's successful transatlantic signaling, his comment was a blunt: "I don't believe it."

Marconi's first wireless patent was issued in Italy in 1896. Early messages spanned distances of only hundreds of feet, but in 1898 he sent a message 32 miles; and in 1899, across the English Channel to Boulogne, in France. By 1901, 12 ocean liners carried Marconi wireless equipment for ship-to-shore telegraphy to the few large wireless stations established on the coasts of Europe and the United States.

And, at 12:30 P.M. on December 12, 1901, in a shack atop Signal Hill in St. John's, Newfoundland, Marconi adjusted his receiver and heard three dots—Morse code for the letter "S"—repeated several times. The transmission had been sent by G. S. Kemp, his associate, from the coast of England. The Atlantic had been bridged by radio waves.

It was a minor technical miracle that Marconi's crude receiver picked up the feeble "S" from the wheezy transmitter 1,800 miles away. The stark simplicity of the equipment is astounding.

In 1903, Professor J. A. Fleming, of University College, London, wrote a definitive series of articles for *Popular Science* on "Hertzian Wave Wireless Telegraphy." (One year later, Fleming would become world famous as the inventor of the first vacuum tube.) The series was written because (in Fleming's words) "the periodicals

Special Price for 30 days on Complete

Receiving Set

$3.45

With One 1000 OHM Receiver

¶This fine Receiving Set, comprising three-slide Tuner, Variable Loading Coil, Fixed Condenser, Silicon Detector and Head Set is self-contained and well adapted for either portable or stationary use.

¶Detector is of radical design, making it possible to place point on any side of mineral.

¶Loading Coil is enclosed within tuner cylinder and its inductance is controlled by instrument switch on end of tuner.

¶Tuner inductance controlled by sliders of improved design.

THIS OFFER GOOD FOR ONLY THIRTY DAYS

¶We will send this complete receiving set with one 1,000 ohm receiver head band and cord on receipt of $3.45 for thirty days only. This is an unusual offer. Send us your check today. If not entirely satisfactory money refunded if goods are returned within 10 days.

Universal Wireless Co., 19 East 32d St., New York City

Anybody can own one, but rush in your order—you have only 30 days! Advertisement was in a 1915 *Popular Science*.

and daily journals, whilst eager to describe in a sensational manner these wonderful applications of electrical principles, have done little to convey an intelligible explanation of them." Fleming's descriptions of Marconi's equipment are fascinating.

He speaks of the "aerial wire" (soon to be shortened simply to aerial) as an "ether organ pipe" because, just as organ pipes set the air vibrating, the wireless aerial (in that day's view) caused ethereal vibration.

Marconi tried a variety of antennas. Some were single wires suspended from kites; others were elaborate geometric arrangements of wires strung between tall wooden masts.

The induction coil in a Marconi transmitter (almost exactly the arrangement developed by Hertz) was capable of generating a spark 10 inches long. According to Fleming: "The construction of the coil is a matter requiring great technical skill . . . the primary circuit [primary coil of wire] consists of 400 feet of thick copper wire . . . the secondary circuit of 10 miles of thin wire, making 50,000 turns around the core."

An important departure from the Hertz transmitter, though, was the use of a mechanical interrupter to repeatedly break the flow of current in the primary circuit. Each interruption generated a spark, so the automatic interrupter produced a chain of sparks when set in motion.

Fleming explained various devices favored by other experimenters: spinning wheels, platinum wires dipping into conducting solution, and streams of mercury hitting moving contacts. Marconi used a "hammer break" (a vibrating hammer repeatedly pushing open a metal contact) to interrupt current flow. The telegrapher controlled it to generate dots and dashes.

Marconi significantly improved the Hertz design in the vital area of receiver circuitry. He developed a very sensitive version of the metal-filings coherer—invented in 1891 by Édouard Branly of France.

Marconi's success was quickly followed by other triumphs. In 1905 messages were flashed from Washington, D.C., to Paris and —5,000 miles—to Honolulu. In 1907 the first transatlantic wireless-telegraphy company opened for business linking Ireland and Nova Scotia. By 1914 all passenger ships of the major maritime nations carried wireless equipment for emergency use. During

1917 amateur radio transmitter of low power (15 watts). Spark coil was capable of producing a spark 1 inch long; the interrupter mechanism was a kind of buzzer.

Make-it-yourself version of a "detector stand." It used a cat-whisker wire, made from a mandolin string, mounted on binding post. Article appeared in 1917.

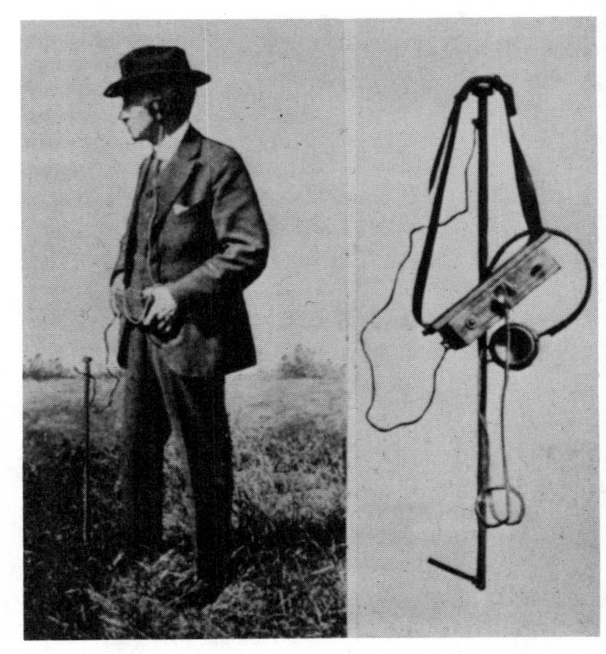

Ancestor of walkie-talkie, circa 1915. "With his pocket wireless attached to his body, H. B. Cox, the inventor, is a veritable radio station," said the magazine.

THE HEART OF THE WIRELESS

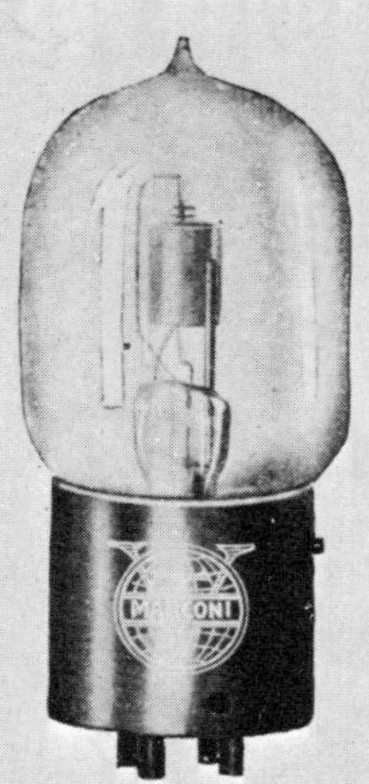

An Amateur Station Without a Vacuum Tube Is Years Behind the Times

MARCONI V.T.

\$7.00 each

Under agreements recently effected the Marconi V. T. is the only vacuum tube, or audion, which may be sold to amateurs, laboratories, schools of instruction and experimenters.

Fleming Pat. No. 803684
De Forest Pat. Nos. 841387-879532

The approximate operating life of the MARCONI V. T. is 1,500 hours.

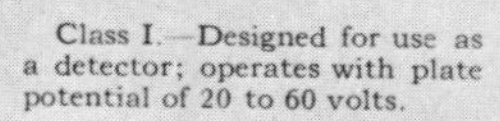

Class I.—Designed for use as a detector; operates with plate potential of 20 to 60 volts.

Class II.—Designed for use as an amplifier; plate potentials from 60 to 110 volts may be applied.

Tubes in either class may be used for detection or amplification, but those of Class I are best as detectors, and Class II tubes are superior as amplifiers.

Standardized Socket \$1.50 additional

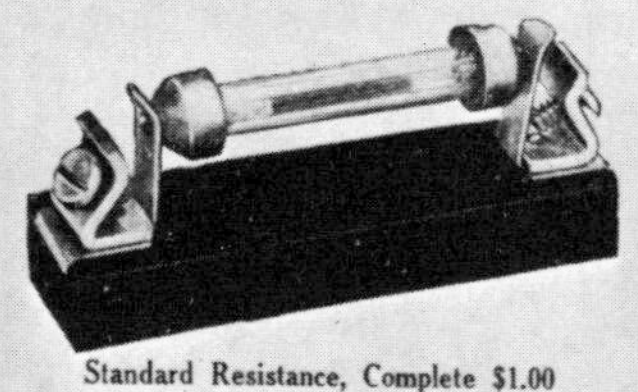

Standard Resistance, Complete \$1.00

The Marconi Resistance, connected in the circuit between the grid and the filament of the Marconi V. T., is made in the following standard sizes:

½ megohm, 1 megohm, 2 megohms, 4 megohms, 6 megohms.

Resistances of any special fractional values up to 6 megohms can be supplied.

Send all remittances with order to COMMERCIAL DEPARTMENT

MARCONI WIRELESS TELEGRAPH CO. OF AMERICA

Sole Distributors for De Forest Radio Telephone & Telegraph Co.

Room 1868, Woolworth Building - - 233 BROADWAY, NEW YORK

BRANCH OFFICES:

Schofield Bldg., Cleveland, Ohio
American Bldg., Baltimore, Md.
Insurance Exch. Bldg., San Francisco, Cal.
136 Federal St., Boston, Mass.
301 Commercial Bank Annex, New Orleans, La.
109 South 2nd St., Philadelphia, Pa.

Historic photo shows engineers and announcers at work on November 2, 1920, at KDKA in Pittsburgh, when the first commercial broadcast went out over the airwaves.

Ad for vacuum tubes (left). In lower photo: interesting tubes in collection of Dr. Lee De Forest, inventor of the Audion, or original 3-electrode tube. Largest tube shown is a ½-kilowatt oscillator.

the first years of commercial wireless telegraphy (and experimental radiophony) a key criterion of performance was the distance covered by a transmitter. This led to a race between nations to develop more-powerful transmitters and the longest transmitting antenna. In 1920 one of the largest radio transmitters ever designed was announced: RCA's Radio Center on Long Island, in Port Jefferson, New York. The antenna systems and buildings would occupy 10 square miles; the available transmitter power would be 2,000 kilowatts, enough electrical power to run a medium-size city.

Sparks and metal filings ushered in a new age, but the vacuum tube (and to some extent, the cat-whisker diode) changed the course of radio history. Without the development of the vacuum tube, which permitted sensitive receivers capable of receiving voice communication, wireless might have remained a mere adjunct to telegraph lines.

Fleming built the first vacuum tube—the vacuum valve—in 1904; Lee De Forest's invention of the amplifying vacuum tube—the Audion or triode—followed in 1907.

And in the next decade, World War I provided the impetus for rapid development of radio-communication equipment.

In receivers, vacuum tubes were soon

Get the Worlds News by Wireless

Every Home Can Listen-in with a Simple Radio Outfit ~ See Page 21

When this cover appeared in November 1921, *Popular Science* was so excited about getting the "world's news" over the air, it forgot to watch punctuation! Below, early editor's note.

Wake Up to Wireless!

Do you realize that the use of radio outfits for entertainments in the home is spreading through America like wildfire?

Do you know that there are nearly half a million wireless fans in the country today and that the thing is only started?

If you aren't awake yet to the recreation you can get from a wireless receiving set—the concerts, dance music, news, and public speeches it will bring you—this article by Armstrong Perry will prove unusually fascinating to you—The Editor.

March, 1922 83

The Radio Magnavox, Type R-2, with 18" horn. The original loud speaker, and now the world's standard. The most attractive and most efficient piece of apparatus of its kind manufactured. Beautiful black enamel finish, nickel plated trimmings, and solid mahogany base with piano finish. Indispensable to those who take pride in their set and the results achieved.

The Radio Magnavox, Type R-3, with 14" horn. Beautiful and durable as it is efficient. Reproduces wireless music and messages in volume without distortion. Anyone can operate it, every one can enjoy it. No adjustments or extras required. No radio set complete without one. *There is no substitute for the Radio Magnavox.*

Wireless music for home entertainments

ENTERTAIN your friends with radio concerts, enjoy the fascination of radio as a hobby, make wireless a profitable part of your business, get news and market reports before they are published, take

hear perfectly. This MAGNAVOX equipment enables everything received by radio to be swelled in volume to the full sound intensity required for any occasion, without losing even the most delicate tone modulations or a single bit of the original clearness and distinctness. It makes a radio set adaptable for office, store or factory use, and the use of radio music practical

Big-horn speaker, above, reminiscent of old phonograph, was beginning to square up as in lower advertisement.

Facts to Help You Choose Your Radio Set

HOW much must I pay for a radio receiving set? What will its range be? What will it look like when I install it? Of what will it consist?

These questions have been asked so frequently by our readers that we publish herewith photographs of receiving sets of various prices, with conservative estimates of their "normal" range.

No. 1

Crystal Detector Set

1—As a beginner's outfit and for amateurs living near some broadcasting station the crystal detector set, costing around $25, is simple to adjust. This set consists of a tuning coil, a fixed condenser and phones. The receiving range of these crystal detector sets, even with a good aerial, is usually limited to 25 miles.

No. 2

Sets for around $75

2—These outfits, popular with beginners, will tune more closely to the wave lengths you wish to hear than will the "crystal" sets. The cabinet contains the variocoupler, single variable condenser and vacuum tube with filament current control. Outside range, 50 to 100 miles. A loudspeaker cannot be used satisfactorily with this or set No. 3 unless very near a broadcasting station.

No. 3

A 200-Mile Range for $125

3—Receiving sets in the class of that shown at left cost about $125 and comprise a tuning device connected with a vacuum tube detector. A variocoupler and two variable air condensers are provided. If connected with an outside aerial of proper design, it should pick up radio-phone messages from stations 200 miles distant.

Outfits over $225 Use Loudspeakers

4—For $225 and up, the finest sets are obtainable, using a loudspeaking device as shown at the right. With a set of this type, containing an excellent tuning unit and two stages of audio frequency amplification, broadcasting stations 500 miles away and, under good conditions, two or even three times that distance, should be heard.

No. 4

employed as detectors (to strip the code signals—and later, voice signals—from the received radio waves) and as amplifiers (to strengthen the tiny signals produced by the detector circuitry).

The first vacuum tubes could not oscillate at high enough frequencies to actually generate radio waves, and for many years improved spark generators—some built like large electric dynamos—pumped Hertzian waves around the world. Vacuum tubes were eventually added to spark transmitters, though, to make voice or radiophone communications possible. Here, amplifying tubes boosted the audio signal from a microphone; the stronger signals were used to control a modulator that varied the current flow into the spark generator. Thus, the instantaneous amplitude of the output radio wave varied in step with the announcer's voice. This process is called amplitude modulation, or AM.

Paralleling the early development of vacuum tubes was invention of the "cat-

Herbert Hoover Joins One-Bulb Fans

Novelties of the Month in Radio

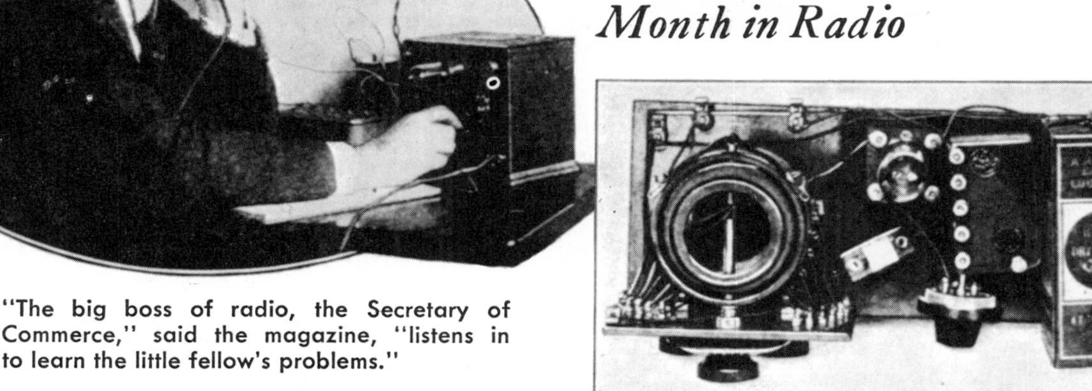

"The big boss of radio, the Secretary of Commerce," said the magazine, "listens in to learn the little fellow's problems."

Top and front views of a make-it-yourself set: tuning unit, or variocoupler, is at left; vacuum tube, grid condenser, and grid leak, in center; rheostat and batteries, at right. And all this you could put together for $15.30.

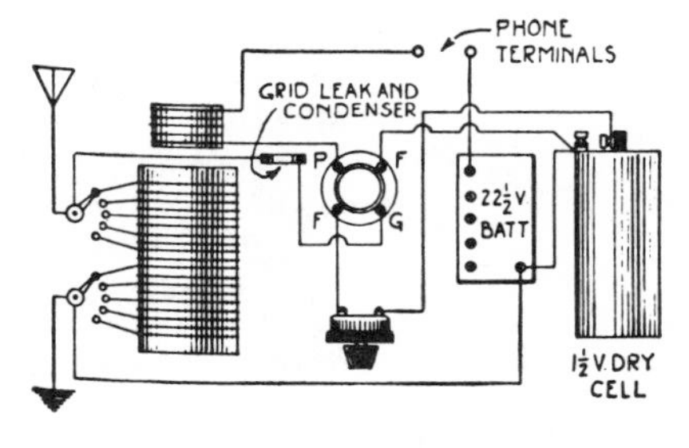

whisker" crystal diode. In 1902 G. W. Pickard, an American engineer, discovered that a sharp needle point resting against chips of certain crystalline materials would act as a detector of radio waves. By 1904 crystal detectors using Carborundum crystals were receiving ship-to-shore communications, and by the 1920s galena crystals (lead sulfide) were widely used by radio hobbyists (and by many professionals, too) to tune in the few regular radio broadcasts on the air.

The cat whisker was the pointed metal contact placed, by trial and error, on a sensitive area of the crystal. These primitive semiconductor devices were the heart of the popular crystal set.

The great transition from wireless telegraphy (electronic conqueror of the North Atlantic) to radio (nightly entertainer of millions of Americans) took place during a 10-year period beginning about 1915. The purveyors of change were an odd amalgam of engineers, hobbyists, businessmen—and,

Parts To Be Bought and Their Approximate Cost

Vacuum tube	$5.00	Binding posts (4)	$ .20
Socket	.75	Annunciator wire, 15 ft.	.10
Grid condenser	.25	Wood screws, 1 doz., ¾ in. No. 6	.10
Rheostat	.50	Aerial wire, 150 ft.	1.00
B battery	.75	Aerial insulators (2)	.50
1½-v. dry battery	.40	Lead-in insulator	.10
Variocoupler	2.00	Lightning arrester	1.25
Panel	1.00		
Inductance switches (2)	.40		
Instrument dial.	.30		
Contact points (14)	.70		

The total cost for these is $15.30, to which must be added the price of phones, which ranges from $4 upward.

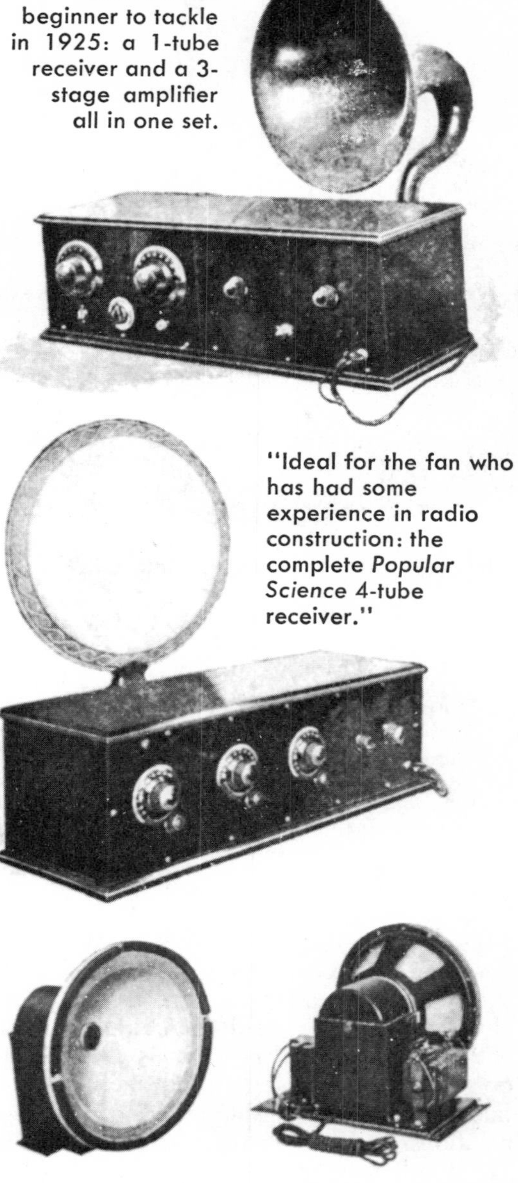

One for the beginner to tackle in 1925: a 1-tube receiver and a 3-stage amplifier all in one set.

"The Rolls-Royce of radio," *Popular Science* called Edwin H. Armstrong's first superheterodyne in 1924, and its circuit "the basis of most of the latest radio sets."

Broadcasting in 1924: boxing-match announcer speaks into microphone (1); voice is carried over telephone wire through nearest exchange (2) to broadcasting station's control room (3). Here operator observes broadcasting quality; corrects defects before they pass to transmitter (5). In transmitter, voice waves are made to modulate vibrations of "carrier wave," produced by an oscillating current supplied by high- and low-voltage generators (4). Carrier wave is put on ether by broadcasting aerial (6) and carries programs to listeners.

"Ideal for the fan who has had some experience in radio construction: the complete *Popular Science* 4-tube receiver."

A battery-powered dynamic speaker and (at right) same unit with built-in rectifier to "allow actuation from the light circuit."

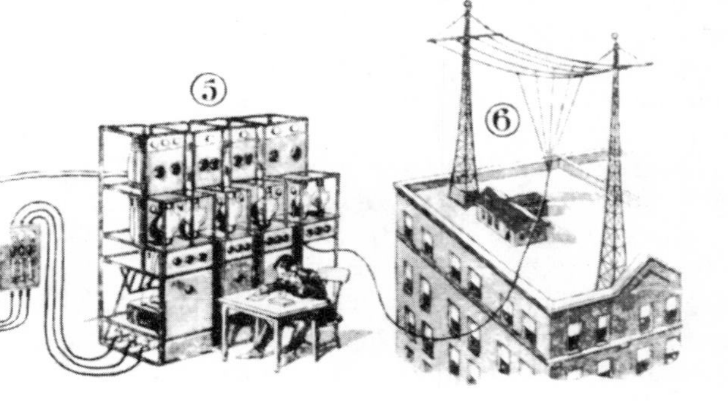

possibly, magazine editors. *Popular Science* claimed a part of the credit for starting the "national wireless craze" by publishing, in November 1921, "How I Listen In on the World by Radio," by Armstrong Perry.

Armstrong's first article triggered a flood of successors that taught America how to use their radio receivers. But equal credit must go to a happy fact of radio life:

From the earliest days of Hertz's experiment, it has been relatively easy and cheap to build a radio receiver. An effective receiver can be cobbled together out of wire, a few electronic components, and a catwhisker crystal. This aroused enormous interest in radio during the twenty years following Marconi's broadcast. In 1915, for example, $3.45 would buy a "complete receiving set" (as advertised in *Popular Science*).

What was on the air in 1915? Wireless telegraphy, of course, and a few time signals. By late 1921, though, the air waves had become considerably more interesting. As *Popular Science* reported that year: "In these times when the Bureau of Markets is broadcasting daily reports of market conditions from a chain of stations all across the country, naval stations are sending out weather and navigation reports, some commercial companies are operating radiophone stations (broadcasting music and special events), and thousands of amateurs are using both radiophone and radiograph, it behooves a good many of us to build receiving sets and receive some of this valuable and interesting free information."

And by the early 1920s, all the ingredients for radio growth were at hand: technical capability, audience demand, financial backing, and a slowly developing national prosperity after a recession. The pivotal

"Most powerful station" in 1923—WJAZ in Chicago—was on ground floor of hotel, walled by plate glass, so visitors could "see how broadcasting was carried on."

year seems to have been 1922. Herbert Hoover (then Secretary of Commerce and "the most important government official concerned with wireless") wrote that year in *Popular Science:* "The wildfire spread of radio has . . . been much more amazing than any other thing we have seen in our time. One is at a loss in trying to think of any phenomenon to which it can be compared."

Hoover also talked about programming: "While the broadcasting of music has been the first phase of the radio boom to attract attention, the transmission of the spoken word in addresses, sermons, and lectures will eventually be one of the most popular and best appreciated functions of the broadcasting stations."

The first scheduled commercial broadcast was aired on November 2, 1920, by station KDKA in Pittsburgh. By mid-1922 most people in the U.S. were in range of at least one station—if they fiddled with antennas and tuning mechanisms carefully.

By 1924 bland news-and-music had been replaced by "realistic dramas" and variety shows (only to return triumphantly—and blatantly—in the 1950s). And a new breed of radio receivers was under development —the brain child of a famed radio engineer, Edwin H. Armstrong.

The simplest home radios were crystal sets. They were cheap and reliable (almost nothing to fail electrically, although prone to mechanical breakdown). But they were not very sensitive and only one person could listen with headphones.

Next in line was the crystal set equipped with a vacuum-tube audio amplifier to boost audio signal strength, and possibly drive a loudspeaker, eliminating headphones. Then came sets that used vacuum-tube detectors instead of a crystal. And finally, sets in which tubes amplified the radio signals before they were detected.

But vacuum-tube amplifying circuits had the unfortunate propensity of self-oscillating: They would generate audible squeaks and, often, interfering signals that would disturb other nearby radios. So Louis A. Hazeltine developed the "Neutrodyne" circuit in which the tendency to oscillate was neutralized (hence the name). All these designs, though, required substantial outdoor antennas and were fairly finicky to tune.

Armstrong's first radio circuit was the so-called "super-regenerative" design—a scheme that placed an amplifying vacuum tube in an oscillator-like circuit. Because the tube was literally on the verge of oscillation at all times, it produced enormous amplification, and it would work with relatively small antennas.

But Armstrong's most important design was the "superheterodyne" circuit, the circuit found in virtually all modern radio receivers. Reportedly, he thought out the basic concept during World War I while a Signal Corps officer assigned to eavesdrop on German radio communications.

The superhet circuit uses a sensitive radio-signal amplifier that is sharply tuned to a single—intermediate—frequency (IF) which is much lower in frequency than the stations the set will pick up. The station frequencies are converted to the IF by a circuit called, logically enough, the frequency converter. When the radio's tuning knob is turned, the converter scans along the band of received frequencies—at each point on the dial a single frequency is converted to IF.

The superhet has several advantages. Because the IF circuitry only works at a single frequency, it can have enormous sensitivity. And the receiver can be tuned by one knob instead of the three or four of older receivers. Finally, the superhet arrangement is interference-free and very selective; the tuning can be sharp.

The first superheterodyne sets appeared for sale in the mid-1920s: huge, complicated, 8- or 10-tube, and costly units. But, two decades later, millions of low-cost, 5-tube "all-American"-circuit superhet radios would be cranked out each year. And today, the tiny imported transistor radios that cost a few dollars are superheterodynes too.

Armstrong—designer of the best amplitude-modulation radio was never satisfied with AM transmission. AM's chief handicap is vulnerability to static and noise. Any electrical phenomenon that produces radio waves, be it lightning or a sparking motor, can cause interference. For years, Armstrong searched for a static-free system of radio broadcasting. Finally, in 1935, he perfected frequency modulation (FM).

In AM, the instantaneous amplitude (strength) of the carrier radio wave is

78 POPULAR SCIENCE MONTHLY February, 1928

How *to* Build *an* Electric Set

Assembled from Standard Parts, This Sensitive and Selective Set Gives Superb Tone and Great Volume

By ALFRED P. LANE

HERE is a new and remarkable electric radio receiver especially designed for construction in the home workshop. It is sensitive. The selectivity is of a very high order, and the great volume, combined with true-to-life tone quality, will prove a revelation to anyone who has never heard a receiver using such tremendous power. Yet the alternating current hum has been reduced almost to the vanishing point, without sacrificing tone quality on the low notes in the slightest degree.

Full electric operation has brought up several important problems in design and construction. The question of how to control the volume, a relatively simple matter with battery operated receivers, is one of them. With battery operated sets you can turn down the filament current of the radio-frequency amplifier tubes, and the

This article, first in an unusual series, describes our new electric receiver, which has been tested and approved for home construction by the Popular Science Institute of Standards

want to have the usual trouble caused by the antenna coupling, which makes it impossible to get the first and second stages to tune alike on different antennas.

tened on the condenser shaft moves the primary farther and farther away from the secondary when you tune to the low waves. The result is more than normal volume on the high waves and extraordinary selectivity on the short waves.

The detector circuit is tuned by a separate drum, because it is extremely difficult to get the detector circuit to tune exactly like the radio-frequency stages.

Shielding is used on all tuned stages to increase the selectivity and because it makes the balancing much easier.

Satisfactory full electric operation means that the entire receiver and power supply unit must be designed as a complete system. This is particularly true if the power unit is to be constructed throughout with fixed resistances so that there will be no voltage adjustments or biasing adjustments.

Wall-socket radio arrives. This ambitious project was a reader favorite in 1928. The 5-tube receiver was unusual because it was powered by the AC power line (via a power supply) rather than by batteries. And it would play phonograph records, too. Blueprint to build it cost 25 cents.

Looking down on the receiver (with top plates of shields removed). Above is the power-supply unit —"one of the most powerful ever designed for home construction," said *Popular Science*.

NEW!
RCA Victor's AUTO RADIO
installed in 30 minutes

ONLY $34.95

Slightly higher in Canada and west of Rockies

DEMAND four things from an automobile radio. First, compactness. Second, easy installation. Third, tone and performance. And fourth, a fair price.

RCA Victor offers an automobile radio which has all this—and more. It is in a single unit . . . you can install it in 30 minutes, only two simple electrical connections, no soldering, and only one hole to drill . . . It has the tone and quality that RCA Victor alone can build into a radio . . . and the price is so low that everyone can enjoy a radio in his car—just $34.95.

The new RCA Victor Auto Radio also has Tone Control, a full electro-dynamic speaker and a tube-less "B" battery eliminator. There's no better buy at any price . . . It's top in quality, yet bottom in price. On sale at any RCA Victor dealers and many auto accessory shops.

"HIS MASTER'S VOICE" —On the Road

RCA Victor
Company, Inc.

A Radio Corporation of America Subsidiary

Camden, New Jersey

Mahogany and fretwork—vintage 1931. Getting all the bulky parts of the time into a clock-type radio was quite a trick.

First FM radio station used this 400-foot-high antenna tower overlooking Hudson River to broadcast daily test programs in 1939.

Early ad for radio "on the road" reproduced at the left stressed ease of 1933 install-it-yourself job. Tone and performance? Not up to later standards.

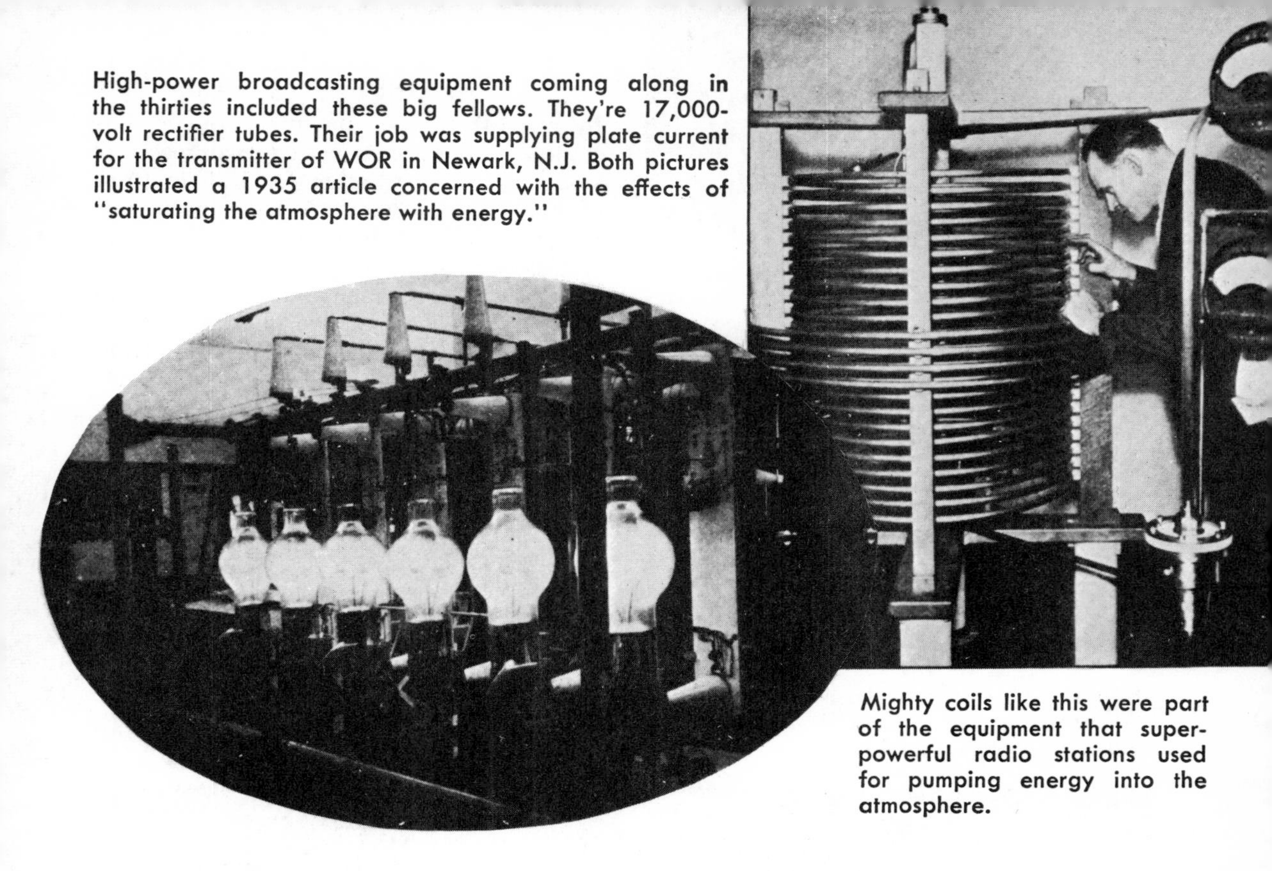

High-power broadcasting equipment coming along in the thirties included these big fellows. They're 17,000-volt rectifier tubes. Their job was supplying plate current for the transmitter of WOR in Newark, N.J. Both pictures illustrated a 1935 article concerned with the effects of "saturating the atmosphere with energy."

Mighty coils like this were part of the equipment that super-powerful radio stations used for pumping energy into the atmosphere.

varied in step with the transmitted music or voice. In FM, the *frequency* of the carrier wave wobbles back and forth in step with the music or voice. FM is static-free; AM isn't, for a simple reason: Static and noise radio waves are AM signals. An AM receiver reproduces them as snaps, crackles, and pops. But an FM receiver ignores them almost completely.

FM broadcasts went on the air in 1940, but the cost and complexity of FM receivers limited the audience until the fifties brought mushrooming interest in high-fidelity sound.

During the twenties and early thirties, radios were expensive. Considering the average income in 1925, a high-performance console radio was more than twice as expensive as a 1972 big-screen color TV. And yet everyone wanted a radio. Hence the sudden spurt in "build-it-yourselfers."

Popular Science ran a steady flow of "how-to-make" articles during the period and also prepared blueprints ("send 25¢ in stamps or coin"). A man who built his own radio could save almost 70 per cent of the cost of a purchased unit. And, performance was excellent!

The hobbyist/craftsman made many of the components himself ("anyone handy with tools can make a variocoupler for about 50¢"). When he was finished, $30 or $40 for parts and a few evenings of loving labor had produced a $120 radio that would "bring in stations within 100 miles with unfailing clearness and fine volume of tone."

Like the phoenix, radio broadcasting seemed to rise from its own ashes in mid-century. Television had nudged radio out of its role as the entertainer America listened to and converted it into the ubiquitous purveyor of background music, weather, and news that it is today. And for a few shaky years, it looked as if radio might be dead. A bright spot during this dark period was the development of stereophonic broadcasting, commonly called FM stereo multiplex.

Early attempts at stereo in 1953 smacked of a Rube Goldberg invention. One of the two stereo sound channels was transmitted

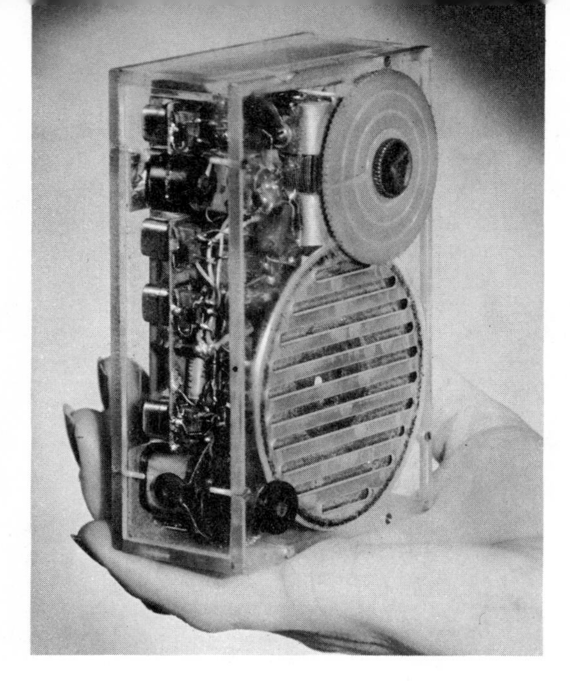

World's first transistor radio came from Texas Instruments in 1953. It used 4 germanium junction transistors, cost $40.

by an AM station; the other by an FM station. The listener placed two receivers —one for AM, the other for FM—in front of him, spaced about 6 feet apart, and heard fairly realistic sound.

This scheme was doomed to failure, though, for a fundamental reason: each station's broadcast carried only part of the music—the left *or* the right channel. Thus, if you tuned in only the co-operating AM station or only the FM station, you'd hear distorted music. And so the 1-radio owner would tune in neither station. Since the sensible men running radio stations would never tailor the programming for the few listeners who owned 2 radios, the idea fizzled away.

Another problem lay in the simple fact that the AM channel has significantly lower fidelity than the FM half.

But by 1960 a practical and workable

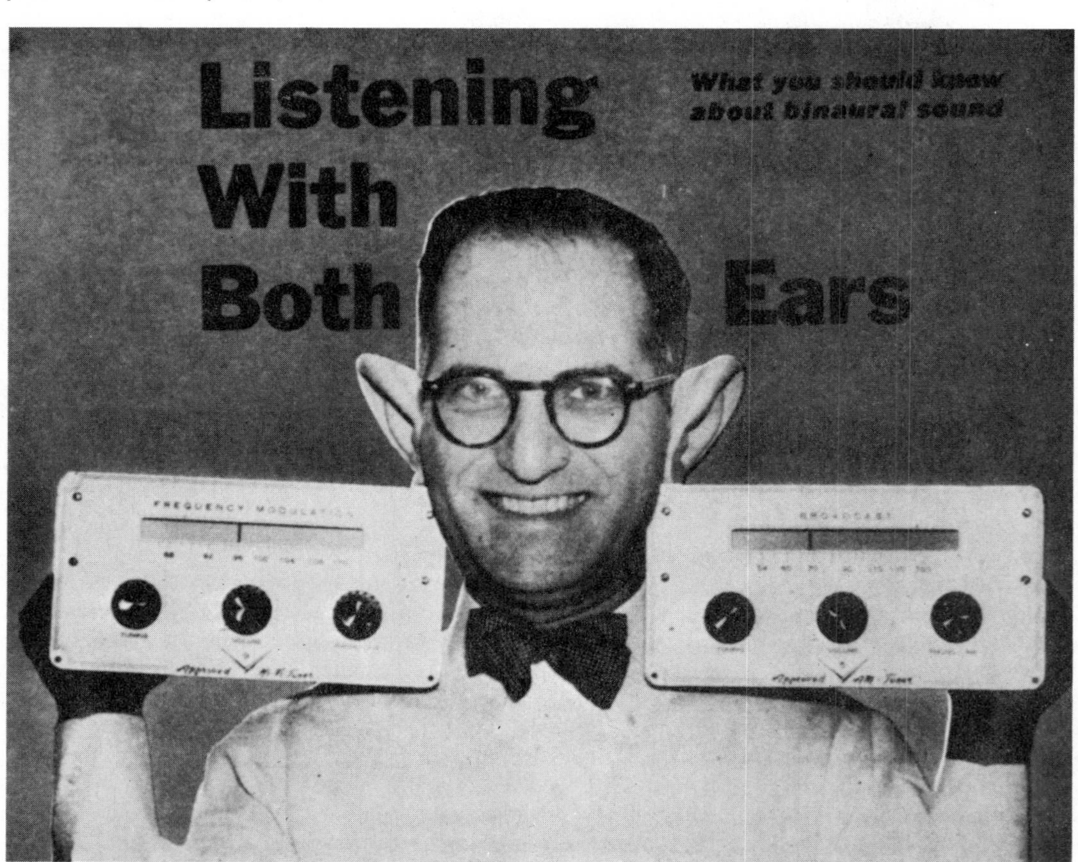

When stereo was new (1955): "Some prefer to call dual-speaker sound stereophonic, rather than binaural," said this article, "but whatever you name it, it's the most."

stereo broadcast system was in operation. It worked solely on FM frequencies; it required only a single station to operate; and it was compatible. A listener with a stereo-equipped receiver heard 2-channel stereo sound; a listener with a conventional FM receiver heard monaural sound.

The secret is a technique of electronic legerdemain called multiplexing. Two different electronic signals are squeezed onto a single radio wave. Here is how it's done:

Let's say a radio station is broadcasting a stereo phono record. Left channel (L) and right channel (R) are electronically added together to create L+R. This is the equivalent of a monaural phono record.

The L+R signal is one of the two signals multiplexed onto the carrier wave. The other is a *difference* signal made by electronically subtracting R from L: L—R.

These twin signals are placed on the carrier in such a way that an ordinary FM receiver ignores L—R. It takes L+R and cranks out full-fidelity monophonic music.

But a stereo FM receiver looks for both signals and decodes them by combining them electrically two ways:

$$\begin{array}{r} L+R \\ +\ L-R \\ \hline 2L \end{array} \qquad \begin{array}{r} L+R \\ -\ L-R \\ \hline 2R \end{array}$$

Thus, it produces separate L and R signals that are fed to separate audio amplifiers and speakers.

* * *

The history of television is a tale of scientists and engineers tilting at a windmill. For years, hopeful inventors tried to create a *mechanical* TV system, although simple arithmetic proved it impractical for producing anything but a small, fuzzy image.

The word "television" first appeared in *Popular Science* early in 1921 in an article about the French experimenter Édouard Belin. Belin had developed a workable "telephotography" system—an ancestor of the facsimile machines used today to transmit photographs over telephone lines and radio links. Belin was optimistic that television apparatus would soon be developed that would permit "seeing over the telephone wires." It was probably the success of the mechanical telephotograph that lured technicians onto the reefs of mechanical TV and induced them to resurrect a device invented in 1884 by Paul Nipkow, a German physicist.

Shortly after the telegraph was developed, there was a flurry of activity to transmit pictures by wire. One fact was obvious to the early experimenter. The image would have to be scanned and broken up into many individual elements at the sending end. Then each element would have to be converted into an electrical signal. Finally the elements would have to be reconstructed into a picture at the other end of the line.

Nipkow's spinning perforated disk made it possible to dissect an image into a chain of horizontal strips. Near the circumference of the disk was a series of holes positioned along a spiral curve. As the disk spun, each hole, in turn, "saw" a slightly lower strip within a narrow field of view.

During the 1920s 4 spinning-disk TV systems were proposed:

John L. Baird, a Scottish inventor, developed his first prototype in 1922. To build his camera, he placed a glass lens in each hole on the rotating disk, as well as a light-sensitive photocell a short distance behind. The subject sat in front of a picture frame that defined the field of view. As the disk whirled, the light-and-dark pattern of each horizontal strip was transformed into a varying electric signal, which was broadcast by the transmitter.

At the receiver, the varying electric signal controlled the brightness of a neon lamp. The varying light of the lamp was scanned across a ground glass by another rotating disk to reconstruct the image.

In the U.S. in 1923, C. Francis Jenkins revised the camera scheme. Glass prisms in 2 rotating disks deflected a beam of light to and fro across the stationary subject. A photocell responded to the differing intensity of light reflected by different regions of the subject, and produced the varying video signal. Jenkins first proposed his system to broadcast movies "over radio," rather than to photograph live events.

General Electric's Dr. E. F. W. Alexanderson, used a similar "flying-spot" arrangement. His 1928 camera used a spiral-pattern rotating disk to scan a subject with a moving arc-lamp beam. A bank of photocells captured the reflected light and transformed variations into a video signal.

The receiver was similar to Baird's ex-

Rotating-disk TV scanner of C. Francis Jenkins was so new in 1924, *Popular Science* called it "contrivance" for "radio-vision . . . instantly viewing far-away events."

Crude TV image seen at left, below, was achieved with transmitter (above) and receiver (below). System was developed by J. L. Baird, one of the first to succeed in transmitting actual "pictures."

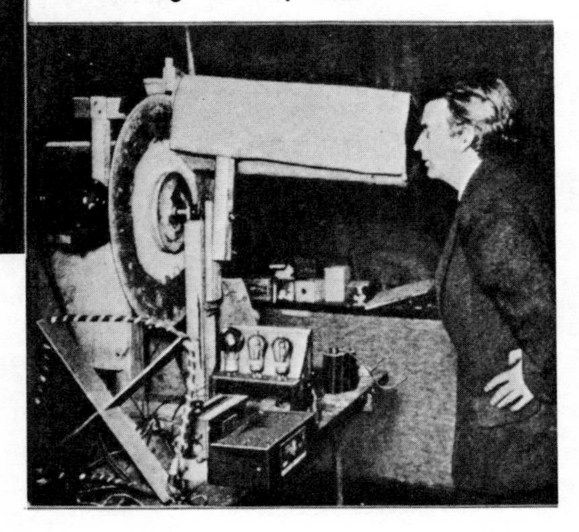

cept that a separate magnifying lens enlarged the image on the ground glass.

The Bell Telephone Laboratories system, first demonstrated in 1927, was perhaps the ultimate in complexity. The Bell camera, too, used a rotating disk to move a light beam across the subject, but the receiver was markedly different:

It had a kind of picture tube. This was a specially made neon bulb with 2,500 internal sections—a 50-by-50 grid—each of which could glow independently. The receiver circuitry accepted the video signal and controlled each section individually (the tube had 2,500 wires!) making each glow to reproduce the brightness of a single dissected element of the subject to be viewed. It worked, but it took a roomful of equipment.

All of these systems had common shortcomings. Since the neon tubes used in the receivers glow red, the flickering images were black and pink. It was impossible to build a system that would dissect the image into elements fine enough to create a high-resolution image. And it was close to impossible to make the mechanical components move fast enough to produce flicker-free images.

Synchronizing the motion of the spinning disks in the camera and receiver to get a stable picture was difficult. Alexanderson's receiver had a hand controller that the viewer worked to correct bad synchronization. And none of the systems were

Alexanderson's TV system was first that was small and simple enough to be a candidate for home use. There were lots of bugs: note the hand controls that had to be operated by the viewer to maintain spinning-disk synchronization. Drawings show transmitter and receiver of 1928. Size and position of some parts were altered for clarity.

any good at photographing large or outdoor subjects.

In spite of these problems, by 1928 several radio stations were using spinning-disk systems, telecasting experimental TV programs on a steady basis.

Amazingly, the spinning disk was knocked off its undeserved perch by 2 inventions that are kissing cousins to one of the earliest electronic devices. The Kinescope (TV picture tube) and the Iconoscope (TV camera tube), invented by Dr. Vladimir Zworykin of RCA, both utilize the concept that beams of electrons can be deflected by a magnetic field, a phenomenon demonstrated by Sir William Crookes in the 1880s. In fact, both of Zworykin's tubes bear close physical resemblance to a Crookes tube.

Inside a Kinescope (cathode-ray tube, or CRT), a beam of electrons from an electron gun scans across, up, and down a phosphor coating behind the faceplate. Varying magnetic fields produced by coils outside the neck propel the beam. The phosphors glow white when struck by electrons, the brightness depending on the current level of the beam.

TV-receiver circuitry transforms the received video signal into a varying beam current. Thus, as the electron beam sweeps across the phosphor screen, it "paints" the picture that you see. The first cathode-ray tubes had only a few square inches of picture surface, yet they operated on the same principle as today's big-screen picture tubes.

In the Iconoscope is a mosaic of thousands of tiny photosensitive elements deposited on an insulating mica wafer. The image to be broadcast is focused on it, and a moving electron beam scans the mosaic to generate the video signal. More recent camera tubes—including the Vidicon and image Orthicon—use a similar beam-scanning principle.

By 1938 black-and-white television employing these tubes was under test in New York City. Television as we know it today had arrived . . . except for color.

In 1927 Dr. H. E. Ives, of the Bell Telephone Laboratories, demonstrated a spinning-disk color-TV scheme that, in many ways, resembles our modern system. In the camera, red, green, and blue filters (in front of phototubes) broke down the colors of the subject into their primary color components—one red, one blue, one green image. At the receiver, a mirror system reblended the three separate images to create a full-color image.

Dr. Peter Goldmark, of CBS Laboratories, also used a three-filter approach in 1940, with one significant difference: the viewer's eye blended them to create the picture. The camera was basically a black-and-white unit equipped with a rotating color drum to place red, green, and blue filters successively in front of the pickup tube. The receiving set was also a black-and-

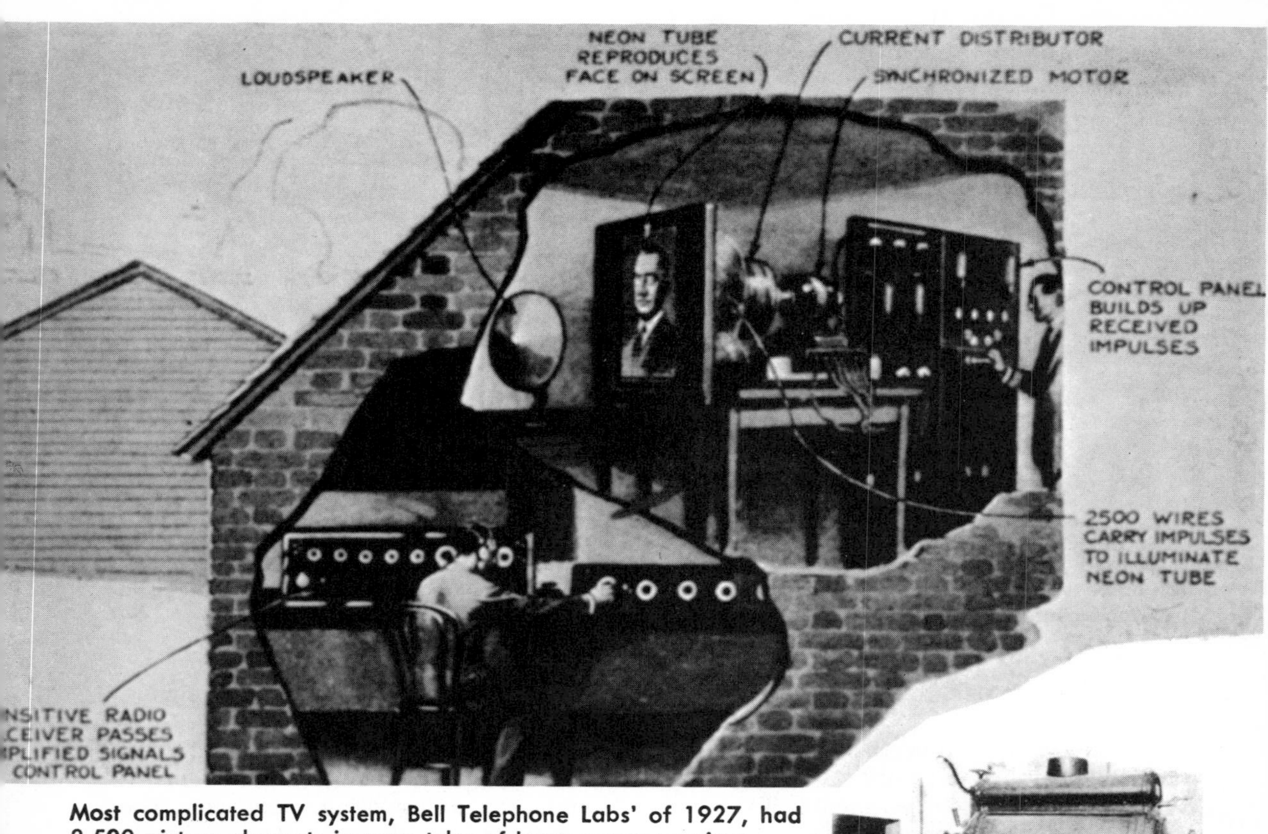

Most complicated TV system, Bell Telephone Labs' of 1927, had 2,500 picture elements in neon tube of large-screen receiver.

New York's first TV broadcast station put boxer Primo Carnera on air, above, in 1931. Viewers watched, below, on home receiver.

For broadcast, spinning-disk device above scanned faces with spot of light. Photocell bank below served as camera.

white model, with a rotating color wheel synchronized to the camera's drum. In step with the camera's drum it spun red, green, and blue filters in front of the picture tube.

The system produced a superb color picture and, for a time in the early fifties, it was the official American color-TV system. It had two defects, though: it was not compatible; owners of black-and-white TVs would see a jumbled picture during a colorcast, and the size of the color wheel would get prohibitively large when scaled-up for a big-screen picture tube.

The RCA-designed compatible color system soon nudged it out. But the Apollo moon project TV cameras employed a Goldmark spinning filter disk before the camera lens to obtain a three-color sequence. Earth-based electronic converters then showed you a normal broadcast signal.

Today's RCA-system camera has 3 pickup tubes, each mounted behind one color filter: red, blue, or green. (Some newer cameras have a fourth tube that captures a black-and-white picture.) The outputs of the multiple tubes are combined into a composite TV signal that is sent to the transmitter.

Basically, this composite signal is a black-and-white television signal carrying an extra complement of color information tucked into empty nooks and crannies. A black-and-white set ignores the color information; a color TV uses it to operate its fantastic three-beam picture tube.

The faceplate of a color CRT contains hundreds of thousands of phosphor color dots arranged in triads (groups of three), one red, one blue, and one green dot. The dots are made to glow by beams from 3 electron guns. A perforated-steel shadow mask, just behind the dot array, shields the dots so that only one beam strikes dots of any one color. Thus, the "red" beam activates the red dots; the "blue" beam, the blue dots, and the "green" beam, the green dots.

The receiver circuitry decodes the composite signal into three signals that control the strength of the three beams. Thus, as the beams sweep across the screen, they

How to make this home TV set, with a scanning disk rotating in front of a neon lamp, was told in *Popular Science* in 1928–29.

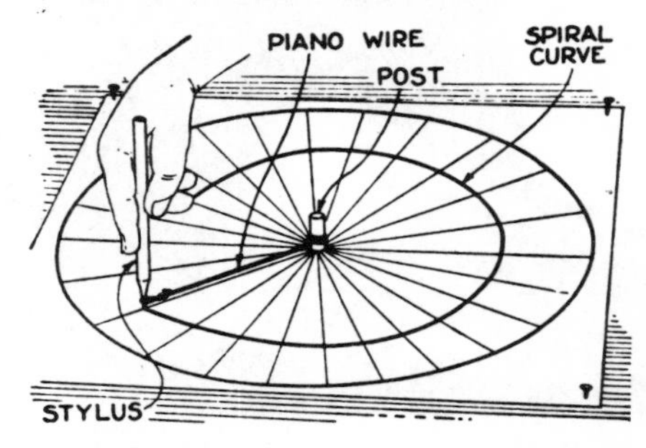

Divide a circle on a metal disk into 24 equal sectors and scribe a spiral with a stylus and piano wire as shown above.

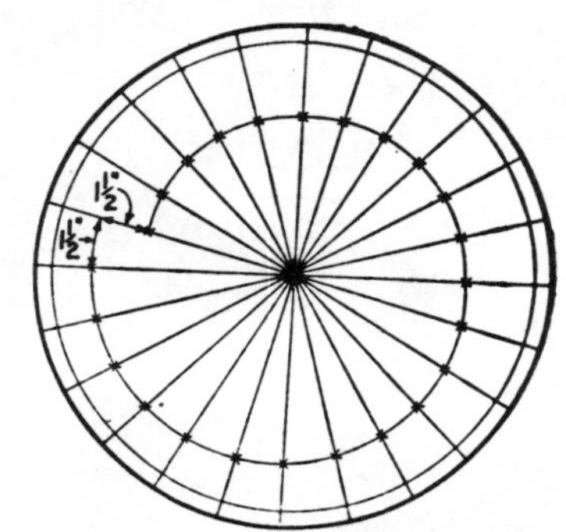

Drill holes at intersections (x's). Assemble apparatus, said magazine, "and receive 'visions' as they are being called."

Another do-it-yourself disk: a spiral of lenses projected a postcard-size picture on a ground-glass screen.

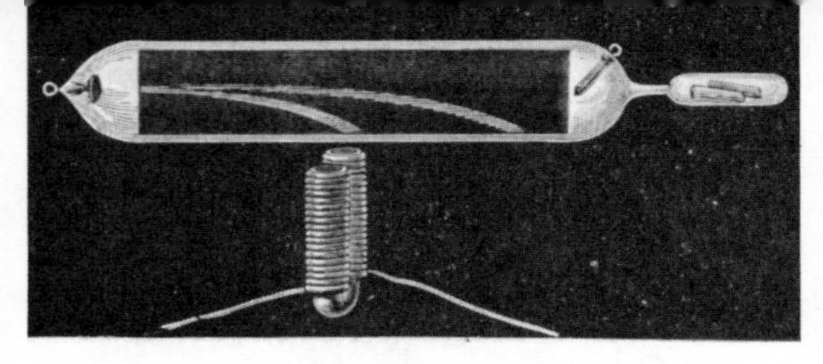

Modern TV sets use same principle as this ancient Crookes tube (1879 *Popular Science*).

Dr. V. K. Zworykin holds his Kinescope tube which uses the beam-deflection principle as did the Crookes tube above.

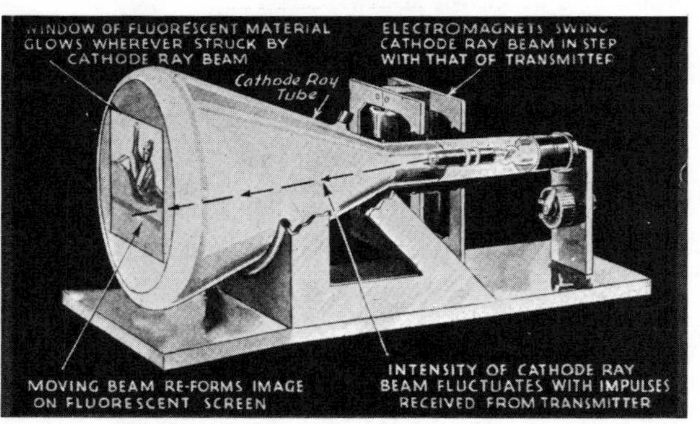
WINDOW OF FLUORESCENT MATERIAL GLOWS WHEREVER STRUCK BY CATHODE RAY BEAM
ELECTROMAGNETS SWING CATHODE RAY BEAM IN STEP WITH THAT OF TRANSMITTER
Cathode Ray Tube
MOVING BEAM RE-FORMS IMAGE ON FLUORESCENT SCREEN
INTENSITY OF CATHODE RAY BEAM FLUCTUATES WITH IMPULSES RECEIVED FROM TRANSMITTER

How a Kinescope worked in a home receiver is shown above. Drawing below shows how Zworykin's Iconoscope in a TV camera picked up the picture. The new tubes got rid of spinning disks and humming motors, made outdoor television practical by eliminating need for glaring studio lights, and brought compactness to TV equipment.

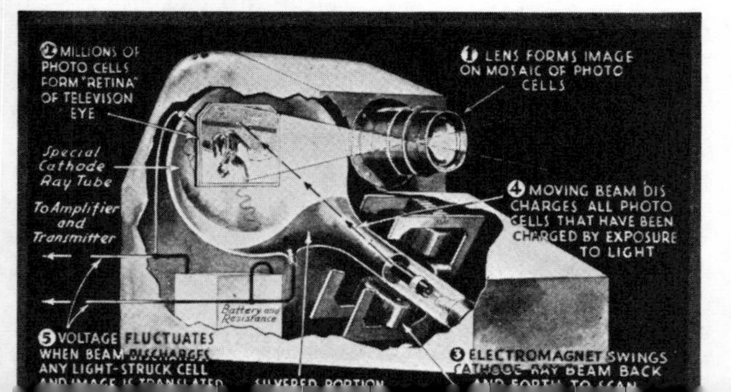
2 MILLIONS OF PHOTO CELLS FORM "RETINA" OF TELEVISON EYE
1 LENS FORMS IMAGE ON MOSAIC OF PHOTO CELLS
Special Cathode Ray Tube
To Amplifier and Transmitter
4 MOVING BEAM DIS CHARGES ALL PHOTO CELLS THAT HAVE BEEN CHARGED BY EXPOSURE TO LIGHT
Battery and Resistance
5 VOLTAGE FLUCTUATES WHEN BEAM DISCHARGES ANY LIGHT-STRUCK CELL
3 ELECTROMAGNET SWINGS CATHODE RAY BEAM BACK

make individual color dots glow in response to the three color signals produced by the camera's pickup tubes. Your eye merges the color-dot pattern into a full-color picture.

In a single decade, color broadcasting has taken front stage in television entertainment. Even TV from outer space is in "living color."

In nine decades, Hertz's spark generator has spawned an incredible assortment of inventions (and mis-inventions), concepts, and techniques. Radio and TV have been steadily trudging toward one goal—the universal communications media, possibly even supplanting the printed word in importance. The next ten years should all but complete the journey:

UHF channels will increase (partially spurred on by easy-to-use UHF tuners on TV sets).

Pay TV country-wide will tie TV sets into major theatrical and cultural enterprises.

Cable TV will cease to be a mere conveyer of a good signal. Already cable companies initiate programs. They will offer their facilities as a public service to local groups, eventually tie in the cable to data banks to provide instant information (stock listings, market prices, etc.) at the turn of a knob.

Four-channel stereo will provide total concert-hall realism for FM music.

Three-dimensional TV, an old dream, may become a reality when TV technology teams up with laser holography, the new technique of producing 3-D images.

Live TV around the world—and the universe—via satellite may soon bring about truly international programming, and advance co-operation between nations.

Closeup view of Zworykin's Iconoscope. Truly a television eye, it derived its name from Greek for "image observer."

Smallest, and cheapest, TV in 1949 was this 3-inch set: it cost only $100.

GE's 12-incher in same year gave you 18 tubes and aluminum-backed screen, all for $390.

Most popular were 10-inchers, $325 up. This Westinghouse was $400.

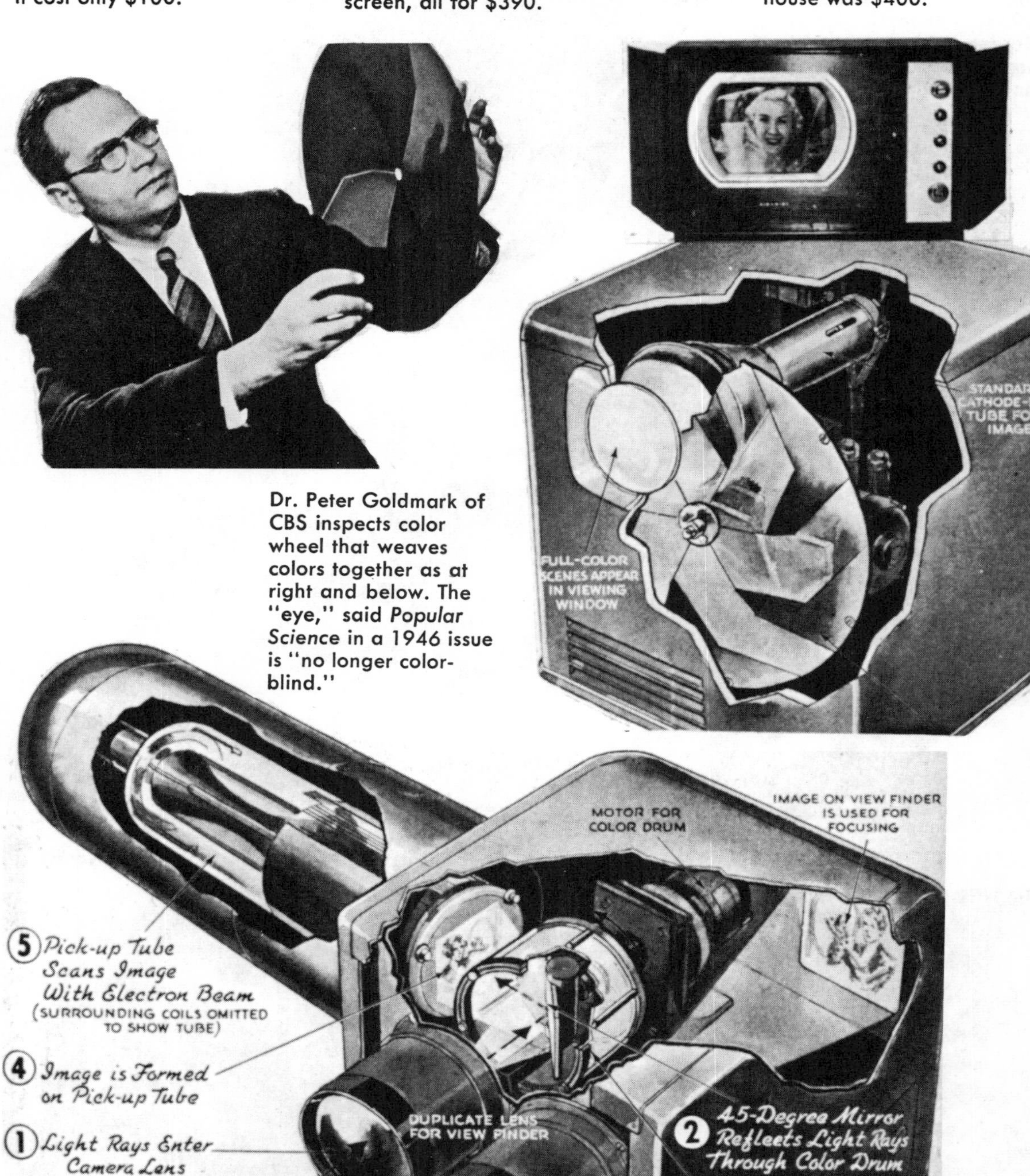

Dr. Peter Goldmark of CBS inspects color wheel that weaves colors together as at right and below. The "eye," said *Popular Science* in a 1946 issue is "no longer color-blind."

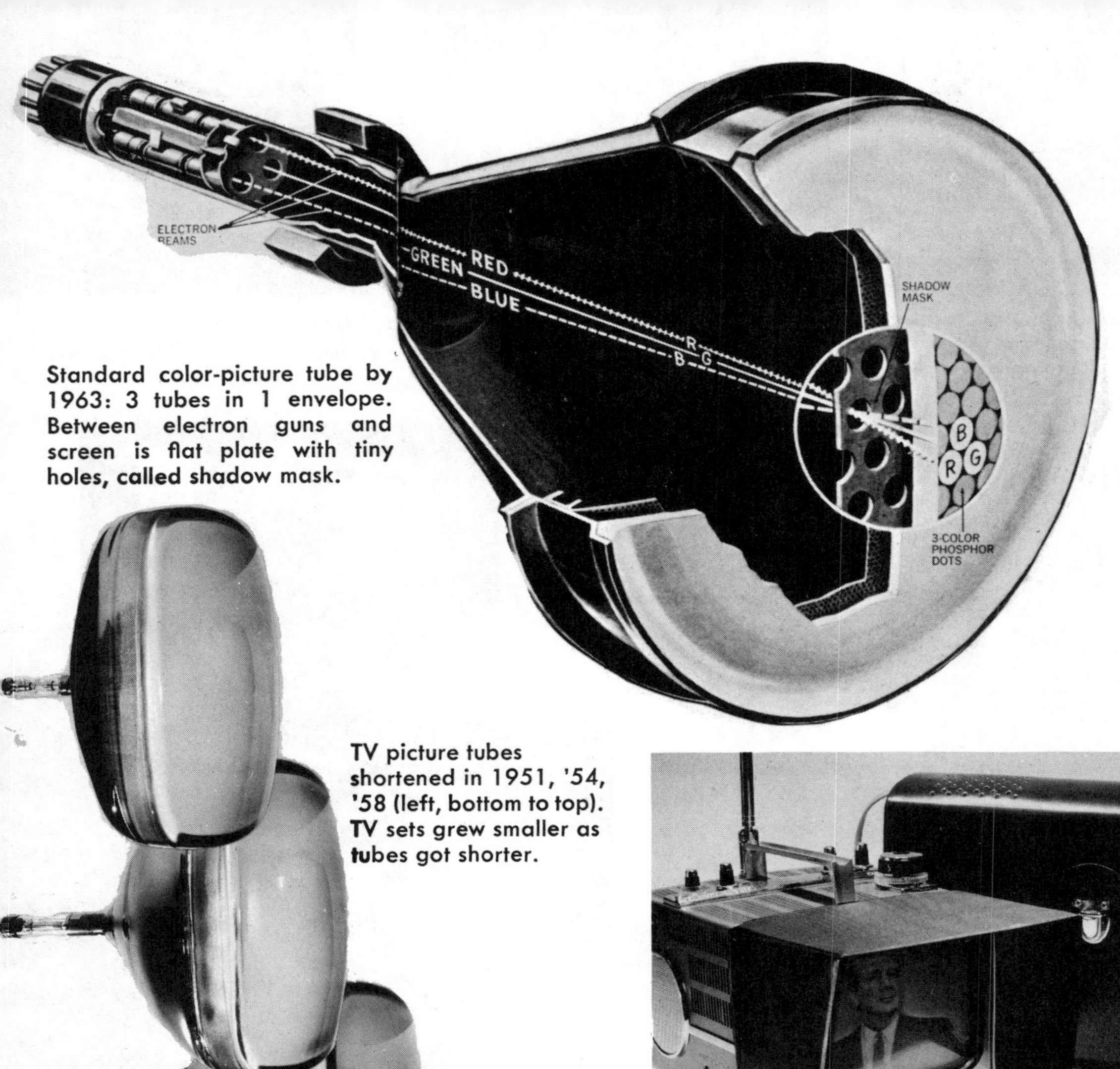

Standard color-picture tube by 1963: 3 tubes in 1 envelope. Between electron guns and screen is flat plate with tiny holes, called shadow mask.

TV picture tubes shortened in 1951, '54, '58 (left, bottom to top). TV sets grew smaller as tubes got shorter.

Portables arrived, with rechargeable battery packs to free you from the power line. Note sunshade for outdoor viewing.

"All the world is jumping on the color bandwagon," said *Popular Science* in 1967, showing this drawing of how color TV works.

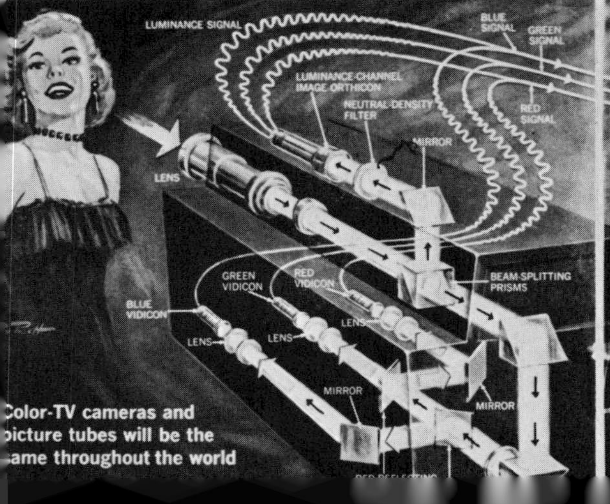

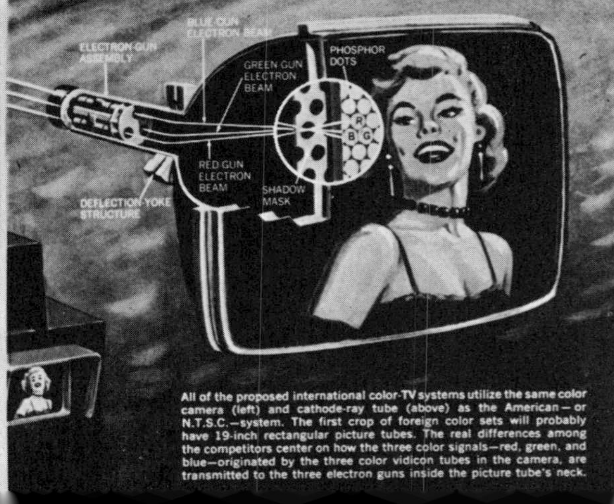

All of the proposed international color-TV systems utilize the same color camera (left) and cathode-ray tube (above) as the American—or N.T.S.C.—system. The first crop of foreign color sets will probably have 19-inch rectangular picture tubes. The real differences among the competitors center on how the three color signals—red, green, and blue—originated by the three color vidicon tubes in the camera, are transmitted to the three electron guns inside the picture tube's neck.

CHAPTER 11

Look What Happened to Photography

WHEN the first issue of *Popular Science* appeared just one hundred years ago, the art and science of photography were already well advanced—indeed, it was fifty years old. It had a full armory of sophisticated techniques, a record of artistic triumphs, and a pantheon of great masters, such as Mathew B. Brady, who had indelibly captured the essence of our Civil War in thousands of photographs.

Yet the photographic methods of those first fifty years bear little resemblance to those of today. A mere decade after *Popular Science*'s appearance, they were revealed to be hopelessly clumsy, an intolerable brake on progress.

The great transition came when wet plates gave way to dry. For wet plates were an abomination. They required that the photographer coat, expose, and develop them while they were still wet. Thus Brady made his famous photos under conditions that would leave most modern photographers in tears. He and his teams of combat photographers each had to take a wagon-load of equipment, materials, and chemicals into the field (the soldiers dubbed them "What-Is-It?" wagons), including a tent so the work could be performed in the dark. Those memorable portraits of Union soldiers and the scenes of battle carnage were usually made by a two-man team, working with their glass, collodion, silver nitrate, developer, and so on and so forth.

Like most great discoveries and inventions, photography is the culmination of efforts over a long time by numerous pioneers, some working together; most, independently. And so, like most, photography's beginnings are clouded in claim and counterclaim for priority. But it is probably safe to say that the first photograph was made by the Frenchman Joseph Nicéphore Niepce in 1822; it was a copy of an engraving (a portrait of Pope Pius VII). The photograph was made on glass coated with a light-sensitive material, a resin called bitumen of Judea.

In 1826 Louis Jacques Mandé Daguerre, a Parisian painter and operator of the Diorama, a theater of illusions featuring gigantic paintings and phenomenal lighting effects, came to Niepce and offered to work in partnership.

After years of experimentation, a practical, complete system of photography emerged, based on the daguerreotype. It was announced in the Paris newspaper, *Gazette de France*, on January 6, 1839, and has assumed the status of the "birth of photography" with Daguerre its father (although it was really the fruit of union between Daguerre and Niepce, who had died in 1833). The process was presented formally to the scientific world by François Arago, the eminent astronomer, in a joint meeting of the Academy of Sciences and the Academy of Fine Arts.

The newspaper account said, in part: "This discovery seems like a prodigy. It disconcerts all the theories of science in light and optics, and, if borne out, promises to make a revolution in the arts of design.

"M. Daguerre has discovered a method

to fix the images which are represented at the back of the camera obscura [originally a darkened room for viewing images of outdoor scenery projected by a lens upon a white screen], so that these images are not the temporary reflections of objects, but their fixed and durable impress, which may be removed from the presence of those objects like a picture or an engraving."

Here is how the daguerreotype was made: A sheet of copper was plated with silver on one side, which was then cleaned, buffed, and polished to a mirror finish. The sheet was placed on top of a small box, with the silvered side exposed to the box's contents—iodine crystals. The vapors of iodine given off reacted with the silver to form silver iodide, a light-sensitive compound. Daguerre halted the process, which took from 5 to 30 minutes, when the film over the copper sheet turned a golden yellow color.

The prepared plate was put into a lightproof box, which could then be attached to the back of a camera and exposed by admitting light to the lens. The exposures ordinarily took from 15 to 30 minutes. Even though the silver iodide surface was not visibly affected, it had changed enough chemically (some being reduced to dark particles of metallic silver) to incorporate a *latent image* of the scene, which could be developed into a visible image subsequently.

After the exposure, the daguerreotype plate was taken from the camera in semidarkness and—suspended in a closed box—bathed in the fumes of liquid mercury heated to 167 degrees F. The mercury vapor "developed" the latent image by depositing a layer of a whitish amalgam in direct proportion to the amount of light that had fallen on the plate originally.

In other words, where the original scene was very bright, the most mercury was left; where there were deep shadows, no mercury was deposited. To remove the sensitive silver iodide from these areas, the whole plate was washed with "hyposulfite of soda" or the "hypo" known to modern darkroom workers (sodium thiosulfate). Then the picture was rinsed with distilled water, dried, and mounted under glass.

The daguerreotype was capable of remarkable brilliance and detail, so much so

First photographic portrait ever taken was made by John William Draper of his sister in 1839, using improved daguerreotype process. Exposure: 6 minutes, sunlight.

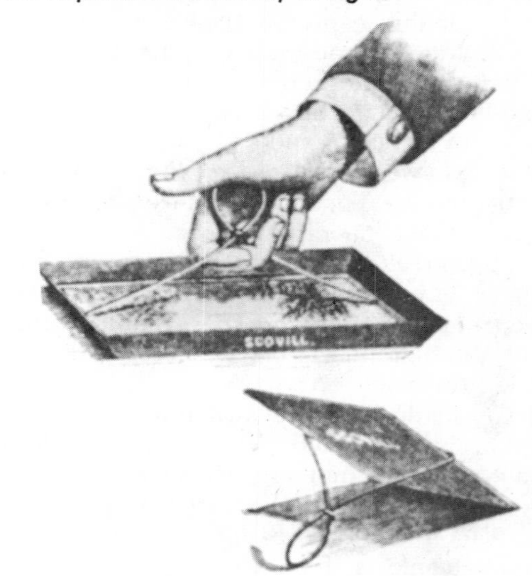

Dry-plate gadget of 1880s made it easy to lift a slippery plate from a developing tray, or to stand it up safely for draining and drying. Dry plates liberated photography from wet-plate procedures.

Wet plates had to be coated, exposed, and developed while still wet. Cumbersome system required lots of equipment; here is an amateur of the 1870s with a portable outfit.

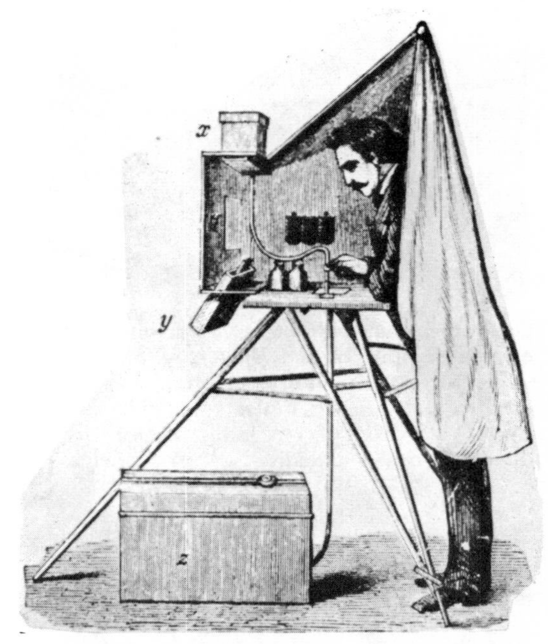

Wet-plate photography demanded that the photographer have a "dark tent" in the field so he could process the collodion negatives. This is a compact 1870s' model.

This 1880s' darkroom lantern has a kerosene lamp inside. The front glass is ruby red for safety, but the side door can be opened to provide full light for inspecting a developed plate, once it has reached the hypo.

Simplicity of this view camera is typical of 1888 camera equipment: a lens, a bellows, and a plate holder. No shutter—you just remove the lens cap and put it back to make a "time exposure."

that Paul Delaroche, the French painter, said, "From this day painting is dead." Lewis Gaylord Clark, editor of the New York *Knickerbocker Magazine,* wrote in 1839: "Their exquisite perfection almost transcends the bounds of sober belief."

Yet the daguerreotype had many and obvious shortcomings. For one thing, it had to be held at one specific angle to the viewer for him to see the image correctly. And although exposure times had been considerably reduced from the 8 hours required for Niepce's first efforts, they were still so long that portraiture was almost impossible.

Two Americans vied for the honor of having produced the world's first real photographic portrait in 1839. Professor John

Eadweard Muybridge's camera house, built on the campus of the University of Pennsylvania to study animal locomotion, housed 24 camera batteries to make successive motion-stopping "still" photos that, viewed in sequence, gave a primitive sort of "movies."

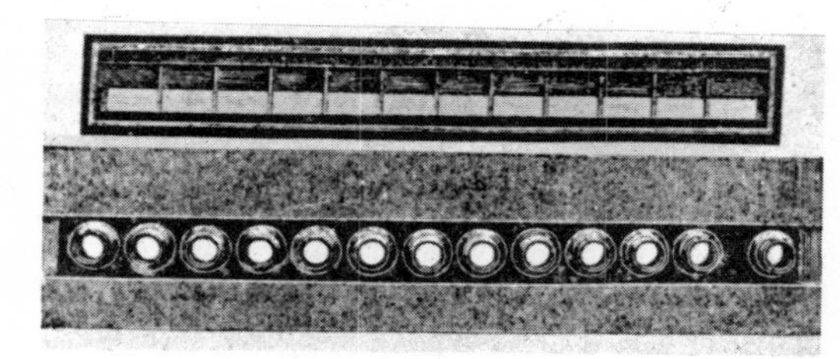

Muybridge's camera batteries were divided into compartments, each having a similar lens, and were operated electrically by a motor clock so that 12 successive pictures could be made in just a fifth of a second.

Arguments about just how horses moved their legs in different gaits were finally settled by famous Muybridge photo series, made in 1870s. In 1880 Muybridge projected sequence photos of a running horse onto a screen with a special "magic lantern."

W. Draper took a picture of his sister in just 6 minutes; the famed inventor Samuel F. B. Morse photographed his wife and daughters by having them sit outdoors on the roof of a building from 10 to 20 minutes "in the full sunlight and with the eyes closed."

In 1840, the world's first photographic portrait studio was opened in New York, with large mirrors to intensify sunlight from outside. Still, the exposures were so long that, to cite one photographic historian, the subjects ran the risk of being sunburned. As the mania for daguerreotype portraits swept the world, photographers discovered that sheets of blue-tinted glass placed between the subject and the sun would reduce the discomfort without materially lengthening the exposure.

Very soon, improvements in the chemistry of the daguerreotype process succeeded in reducing exposure times to more reasonable levels. Treating the mercury with bromine and chlorine vapors (the combination was known in the trade as "quick-stuff"), for example, brought the time down to 20–40 seconds. Faster lenses, such as the 1840 Petzval, also helped. And the original hefty equipment was reduced in weight from 110 pounds to a mere 10.

With these improvements, the daguerreotype process boomed for 15 years. But beginning in 1834, the English scientist William Henry Fox Talbot had been working on a process to register a photographic image on paper—using silver salts—in negative form, from which any number of positive copies could be made. (In fact, when the announcement of Daguerre's achievement was made, in 1839, Talbot wrote to the French Academy of Sciences to make a prior claim.)

Talbot's calotype process was far simpler and less expensive than the daguerreotype process and free of noxious fumes. It used any good-quality drawing paper instead of handcrafted silver-plated copper sheets. Any number of positive copies could be made from a single negative, and these paper copies were easy to view and handle. But paper made a poor support for the negative (because its texture came through

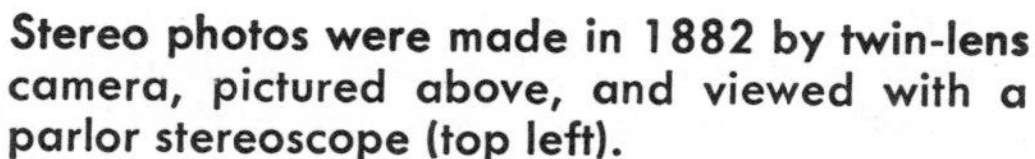

Stereo photos were made in 1882 by twin-lens camera, pictured above, and viewed with a parlor stereoscope (top left).

Cyclist's camera of 1886 was attached to handlebars for photos on the move.

along with the image) so the positives never had the striking quality of the daguerreotypes.

Glass would obviously make an ideal medium, but efforts to find a way to bind the light-sensitive silver salts to a glass support proved unsuccessful. Then, in 1851, an English sculptor, Frederick Scott Archer, published his method of coating a glass plate with collodion—a syrupy liquid made of cellulose nitrate in ether and alcohol—to which potassium iodide or another salt of iodine was added. While the plate was still wet and in a dark room, he dipped it into a solution of silver nitrate. This sensitized the plate by forming silver iodide in suspension. Still wet, the plate was exposed in the camera, developed on the spot with gallic or pyrogallic acid, then fixed with hypo, washed, and dried.

By the mid-1850s, the new "wet-plate" process had swept aside both the calotype and the daguerreotype. As one contemporary account put it: "Daguerreotypists everywhere threw down the instruments of a laborious and complicated business, the buff and the buffwheel, the iodine,

This Eastman Kodak camera of 1888 is the progenitor of the famous "Brownie" line of inexpensive box cameras for amateurs.

Kodak's first folding camera dates from 1890. It could make sizable quantity of 48 4-by-5 pictures at one loading.

First real movies appeared on film strips like this 1897 one of moving train. *Popular Science* called earliest movies "animated pictures."

This early example of a motion-picture camera was made in 1895 and was in the Los Angeles Museum when *Popular Science* described it in 1942.

bromine, and mercury pots, the developing boxes and gilding stands, for the simplicity of the collodion process."

The wet-plate process was the fastest and simplest known to photographers. But its limitations were, as we have seen, still formidable.

The breakthrough came when *Popular Science* was a gleam in its first editor's eye. It was in 1871, when Dr. Richard Leach Maddox, an English physician and amateur photographer, tried using gelatin in place of collodion, with silver bromide rather than silver iodide, the whole forming an emulsion that was allowed to dry. At first slower than wet plates, the new gelatin-bromide dry plates were rapidly improved by other inventors until they were faster than any other process. And since they could be bought ready-made, kept until needed, and exposed and developed whenever the photographer chose, amateur and professional photographers gleefully abandoned the messy wet plates in their favor by 1880.

An 1881 *Popular Science* review called "Progress in Photography" said, with considerable understatement, that the new negative process offered "decided advantages." The same article prophesied the roll-film camera, saying that it only waited for the introduction of a flexible support for the gelatin emulsion.

There was not long to wait. George Eastman, by day a bank clerk in Rochester, New York, and by night an amateur photographer, had begun experimenting with the dry plates in 1878, and in 1880 he went into business selling a perfected version. In that year a small 4-by-

Edison, one of the pioneers of motion pictures, invented this kinetoscope in 1889. It was a peep show with a moving film of still photos.

The phantascope, one of first movie projectors, was invented about 1897 by C. Francis Jenkins. Hooked up to an "electric lantern," it projected motion pictures onto a screen.

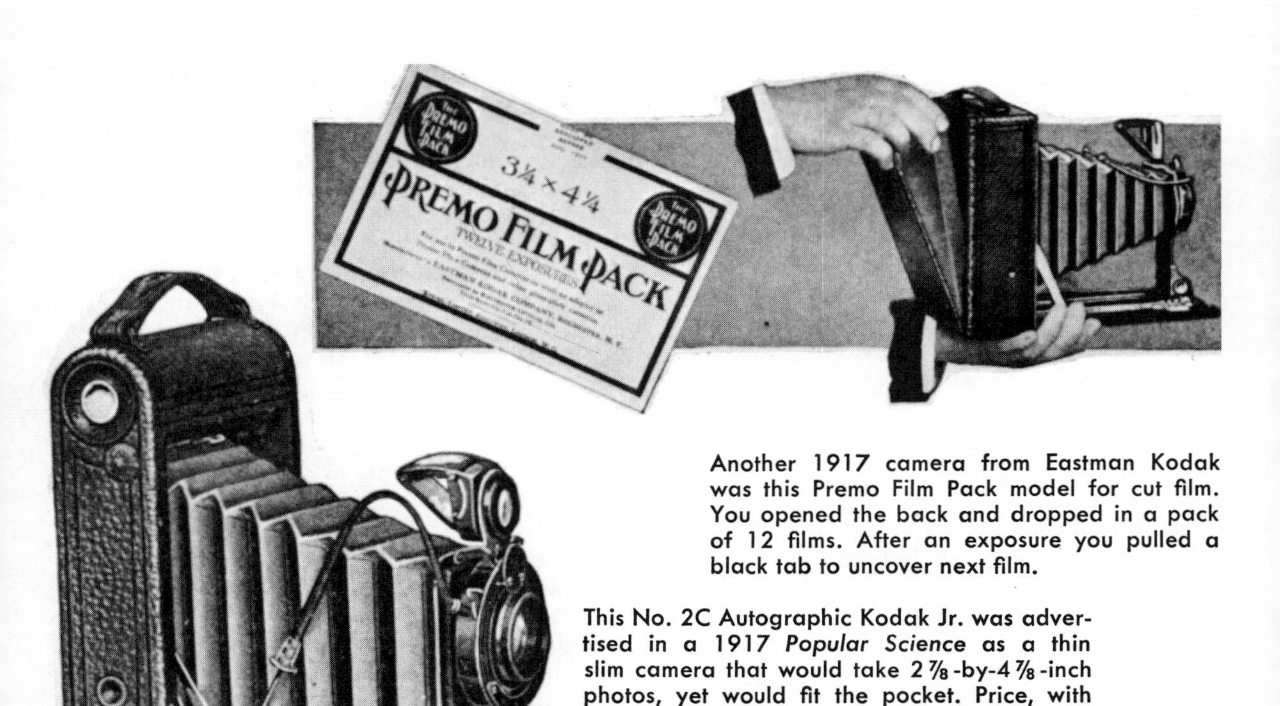

Another 1917 camera from Eastman Kodak was this Premo Film Pack model for cut film. You opened the back and dropped in a pack of 12 films. After an exposure you pulled a black tab to uncover next film.

This No. 2C Autographic Kodak Jr. was advertised in a 1917 *Popular Science* as a thin slim camera that would take 2⅞-by-4⅞-inch photos, yet would fit the pocket. Price, with a ball-bearing shutter with speeds up to 1/100 of a second and an f/7.7 lens—$19.

Autographic cameras and films were introduced by Eastman Kodak in 1914. They enabled a photographer to scribe the date on the film itself, through a slot in the back of the camera.

Graflex camera, right, was a true single-lens reflex. The image was reflected by a mirror onto a ground glass, where it could be viewed. At moment of exposure, mirror swung out of way, allowing image to register on film. Date of this ad was 1917.

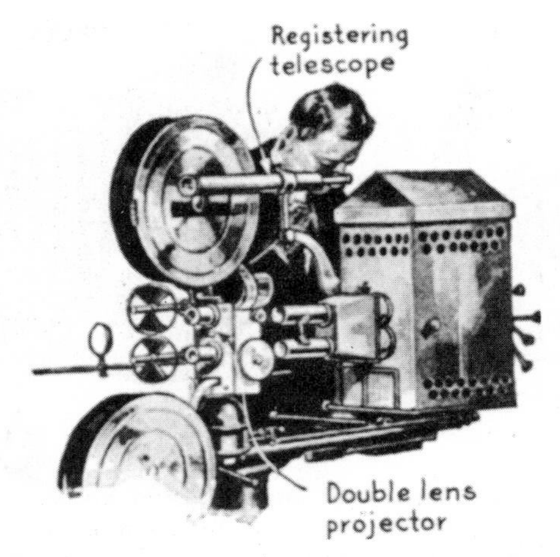

This 1918 Technicolor projector showed movies in full color from black-and-white films. Secret was a process that superimposed 2 shots made through different color filters.

5 camera with lens, tripod, and an initial supply of a dozen Eastman dry plates cost $12.25. Eastman's business prospered, but he was not satisfied. After much research, in 1885 he developed a paper-based emulsion, which he called "American film." The paper was stripped off after development, leaving a thin film negative that could then be mounted on thick gelatin or glass for printmaking.

The great virtue of the new film was that it could be rolled on a spool, and such spools could be adapted to existing cameras. In the words of that 1881 article, "the operator is able to do away with glass and its weight, and may store rolls of sensitive material in the camera itself. Then, by turning a screw, he may place fresh portions of the band in a condition for exposure."

Eastman followed up his improvements with a camera that would take advantage of them—and revolutionize photography for good, making of it the folk art we know today. In 1888 he introduced the Number One Kodak camera. Sold to eager customers for $25 complete, it was loaded with enough Eastman American film for 100 pictures.

A French firm made its remarkable Pathex home-movie camera and projector available to Americans in 1926. The combined weight of both machines was 5 pounds; close-ups of moving objects could be made at 5 feet.

George Eastman, age seventy, poses for *Popular Science* in the garden of his Rochester home with Ciné-Kodak 16mm movie camera, first introduced in 1923. Eastman said he was "just an amateur photographer."

Popular Science

MONTHLY

SEPTEMBER, 1928

SUMNER BLOSSOM *Editor*

VOL. 113, NO. 3

Making Over the Movies

The Movietone film presenting to America for the first time George Bernard Shaw speaking. On left edge of the film is the photographic record of his voice.

ENTIRE Film Industry Being Revolutionized As 400 Theaters Show Pictures That Speak and First Great Full-Length Talking Drama Is a Reality—Complete Musical Comedy Is Arranged For and in a Few Months It May Be Produced in a Thousand Auditoriums—How Remarkable Inventions Give Voice and Music to Silent Screen

By ALDEN P. ARMAGNAC

WILL talking movies, newest competitor of the silent drama, eventually usurp its place? That is the question on the lips of everyone who ...atched its phenomenal spread ... Today the

Movietone newsreels. And the Warners' Theater presents "*The Lion and the Mouse*," a Warner Vitaphone picture with Lionel Barrymore and May McAvoy—and other Vitaphone features."

The smaller cities and towns have talking movies, too. Aberdeen, S. D.,

of sound in movies: "The field of its artistic possibilities is not yet furrowed."

Despite popular impression, all ... dramas are a new thing ... New York saw the first ... *New York*, only a ... viously talking ...

This is how *Popular Science* heralded advent of the talkies, a phenomenon that was to "make over" the movies indeed. On left of the strip of film of George Bernard Shaw is optical sound track of his voice.

Early talkies were plagued with problems of keeping extraneous sounds from being recorded along with the actor's voices. This 1928 sound camera had to be used in the soundproof booth so it wouldn't pick up the noise of its own clicking.

When all had been exposed, the user returned the camera with film to Rochester to have the film processed, mounted, prints made, and the camera reloaded. The customer got back his camera, ready to shoot, and his 100 mounted prints with negatives—at a cost of $10. The camera itself was only 6 inches long, 3½ inches wide, and 4 inches high, approximately. To operate it, said the Kodak Manual, all that you needed to do was "(1) Point the camera. (2) Press the button. (3) Turn the key. (4) Pull the cord."

For his new "complete system of photography," Eastman coined one of the world's most successful advertising slogans: "You push the button, we do the rest." Eastman, incidentally, also coined the word "Kodak" because he wanted a short trade name that could be easily spelled and pronounced in any language—and the letter "K" happened to be a special favorite of his.

Amateurs who tried this new painless photography were bowled over, and the age of the box-camera snapshooter was ushered in. Introduction of the first folding Kodak followed soon, in 1890.

Because of the comparative portability and convenience of these cameras, candid photography became possible. To capitalize on their new freedoms, photographers clamored for spy or "detective" cameras that would allow them surreptitiously to capture every mood and nuance of behavior of unsuspecting subjects.

Manufacturers were soon turning out camera models that could be hidden in derby hats, stickpins, canes, and even pistols. Many of the earliest "detective" cameras, which were simply small box cameras, were made by the Scovill Manufacturing Company and by the firm of E. and H. T. Anthony, the largest photographic dealers in the country. The two later merged, forming Ansco.

In the late 1880s Eastman perfected the first commercial transparent roll film, with a flexible base of cellulose nitrate. It was this advance that opened the way for Thomas Edison to develop his motion-picture camera and projector soon after.

The scientific foundations of what was to become "the movies" date from almost the same year as the beginnings of scientific photography. In 1824 Peter Mark Roget presented a paper before the Royal Society in London entitled "The Persistence of Vision with Regard to Moving Objects." The effect he described was eventually incorporated into devices that passed a series of pictures, representing different stages of motion, before the eyes of the viewer, with the resulting illusion that a single picture was "animated."

One device is the familiar zoetrope, a simple parlor toy in which successive pictures are seen through slits in a rotating cylinder. A more complex version, the praxinoscope, used mirrors instead of slits. An American mechanical engineer named Coleman Sellers was the first to use photographs, rather than drawings, in a zoetrope-like device that he called a kinematoscope. A yet more complicated device capable of projecting the photographic images—the phasmatrope—was unveiled by Henry R. Heyl to a rapt audience of 1,600 at the Philadelphia Academy of Music in 1870.

Then came one of those piquantly oddball moments in history, which, seen in flashback, is momentously important. In 1872 ex-Governor Leland Stanford of California, a railroad magnate and owner of a large and prosperous stable of race horses, bet a friend the trifling sum of $25,000 that race horses *did* have all 4 legs off the ground at some point during the gallop.

To prove his contention, Stanford asked one of California's best-known photographers, Eadweard Muybridge (né Edward Muggeridge in England, his name he Saxonized from some obscure impulse), to give him photographic evidence. Stanford won his bet, and subsidized seven more years of research, at a cost of 40 or 50 thousand dollars. And in the process, man's notions of how animals move were turned topsy-turvy, and the basis for modern motion-picture photography was laid.

In an article in 1882, *Popular Science* said: "Mr. Muybridge has been exhibiting some remarkable rapid-process photographs in Paris, one of which is said to have been taken in one hundredth of a second. . . ."

Actually, by this time Muybridge had succeeded in photographing at the then incredible speed of 1/500 of a second, using a brilliant white background, full sunshine, a very fast lens, the new fast collodion plates, and a complex battery of

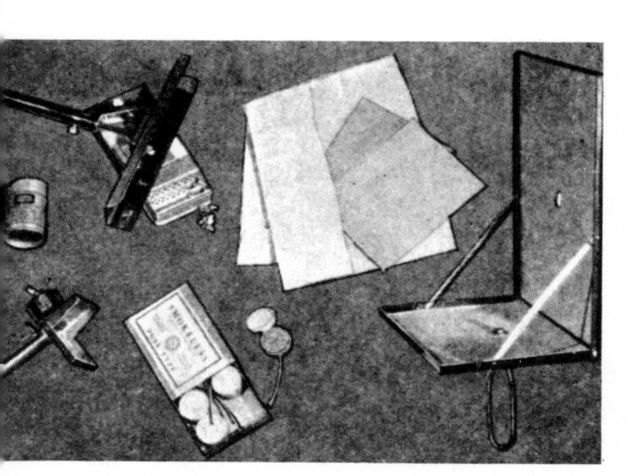

A 1932 photo shows flashlight guns, one fired by small cartridge-like primers, the other by a sparking mechanism; a box of flash cartridges; and a flash-sheet holder with sheets.

After 1932 introduction of the flashbulb, life for the photographer became simpler. Here a "photoflash bulb" is inserted in a combination reflector and dry-cell holder.

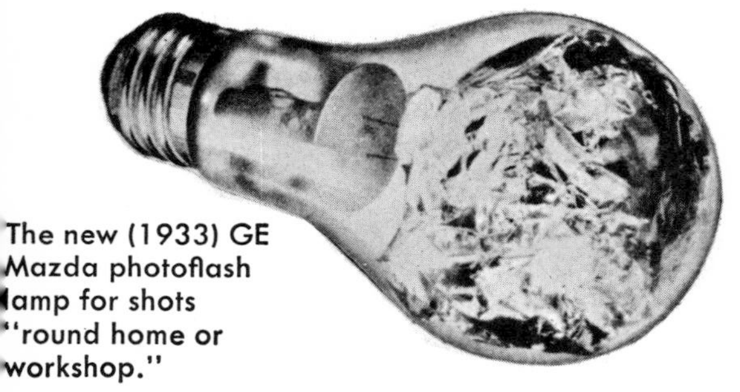

The new (1933) GE Mazda photoflash lamp for shots "round home or workshop."

For Snapshots
at NIGHT
USE
THIS
LAMP
...with the new
SUPER SPEED FILM
List
Price
35c

Photoflood lamps arrived around 1934. This GE ad advised using a fast lens—at least f/6.3, together with "super speed film." The lamps lasted for about 2 hours.

The world's first drive-in movie, for 400 cars, opened in 1933 in Camden, N.J. Patrons watched the latest flicks on a 60-foot screen, heard talkies via a special sound system.

The introduction of Kodak's Kodachrome film for transparency color in 1935–36 stimulated amateur photography tremendously. Inventors were 2 musicians, Leopold Mannes and Leopold Godowsky.

The 35mm camera transformed the art. This 1935 model's ultra-fast lens and superb portability allowed it to get "impossible" shots.

In 1939 the Rolleiflex was the pre-eminent example of twin-lens reflex design (a viewing and a taking lens). It still is today.

electrically operated cameras that were triggered by threads broken by the speeding horses. His photographs astounded the world of painters and natural scientists alike—no human being had ever seen a horse's legs in such "strange" positions before.

In 1879 Muybridge projected series of his photographs on a device he called the zoopraxiscope, which had a maximum capacity of about 200 still photographs. Something better was required, and for the next two decades, dozens of inventors all over the world tried their hands at a practical system for motion-picture photography. The first commercial success was scored by Thomas Edison.

Pouncing on Eastman's introduction of transparent, flexible roll film, Edison designed a workable movie camera—the kinetograph—and a peep-show viewer—the kinetoscope—around it. The kinetoscope contained long belts of film that ran continuously between a light and a magnifying lens, with a revolving shutter making it possible for the viewer to see each individual picture for an instant. On April 14, 1894, the first of several kinetoscope "parlors" opened in New York.

In 1896, *Popular Science* wrote that "in Mr. Edison's kinetoscope, photographs made at each forty-sixth of a second follow one another so quickly under an eyepiece as to fuse with the effect of life and action." A profusion of inventors then competed to perfect a projector, rather than a viewer, several succeeded, and the great movie boom was on in earnest.

Who could have foreseen what would follow? In a long 1897 article on the new "animated pictures," as the writer called them, *Popular Science* mourned the lack of color and predicted: "Instantaneous photography in color is not yet possible, nor is it likely to be achieved in the near future." The writer also complained about the lack of "the effect of solidity, due to our binocular vision," and the lack of "sounds appropriate to certain scenes."

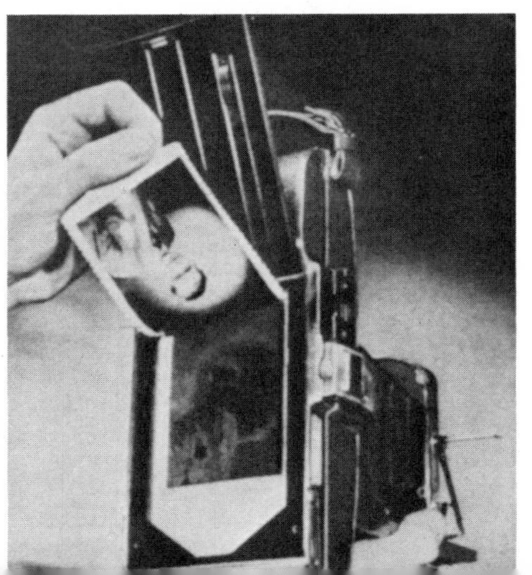

In 1947 Edwin H. Land, then the thirty-eight-year-old president of Polaroid Corporation, announced a startling breakthrough in photography—pictures in a minute, processed inside the camera. In upper view of 2 at left, he holds a 1-minute picture of himself. Below it is the first (1949) production model of Polaroid camera. It sold for under $100.

Well, by 1908 Charles Urban was exhibiting a 2-color (green and red) film process called Kinemacolor, which, though not as realistic as later 3-color systems, was full-color cinematography. (The first film made in a 3-color process—Technicolor—was a 1932 Walt Disney cartoon, *Flowers and Trees.*) And stereoscopic movies were shown commercially now and again, starting in the early 1920s, usually with red and green filters at the projector and in the viewers' spectacles to provide separation of the stereo pairs. Movies in 3-D made a brief comeback in 1947, using polarizing filters, developed by Edwin Land, instead of colored gels. (Stereo still photography stretches back to the 1840s.)

As for the "appropriate" sounds, the vacuum was filled during the 1920s, with a variety of experimental attempts to synchronize sound and picture, either from separate sources or combined on the film. A *Popular Science* article of 1921, headlined "First the Movies; Now the 'Talkies'," discussed a means of synchronizing a phonograph with the film.

Ironically, the movie that overturned an industry and forced the sound revolution was one that used an already obsolete system—sound on synchronized disc. It was, of course, Al Jolson's *The Jazz Singer* of 1927, a silent movie with 4 sound interludes. In a few years, modern systems of sound on film, such as Movietone, had swept the field.

One development the author of that 1897 *Popular Science* article on "animated pictures" did not foretell—that amateurs would be able to film and show their own movies, in color and sound, almost as easily as they could make snapshots.

In 1923 Eastman brought out 16mm reversal film on a cellulose acetate base—non-combustible safety film—along with the first Ciné-Kodak 16mm motion picture camera and Kodascope projector. Together, they made "home movies" practical for the first time. In 1926 the French firm Pathé marketed a home movie camera and pro-

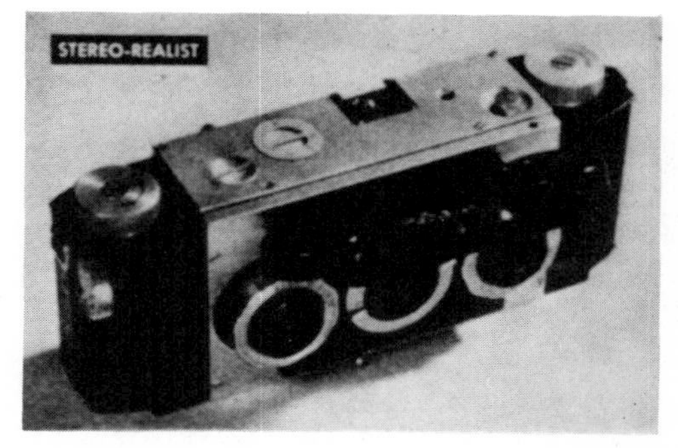

After years of neglect, stereo photography became popular when World War II ended and such cameras as this Stereo-Realist bowed in. Each shot exposed 2 slightly different images which merged in viewing.

Stereo photographer got back his transparency pairs in a mount that slipped into a viewer like this Stereo-Realist. Interocular distance—crucial to stereo—was adjustable.

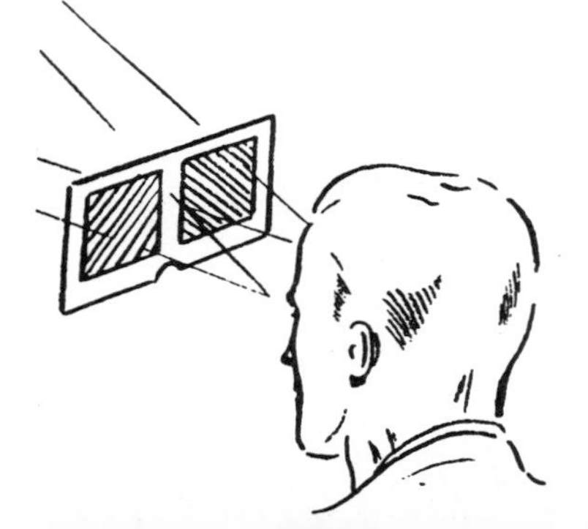

Three ways to 3-D theater movies: a 1923 rotating-shutter viewer, above, that had to be installed at each seat; Lumière's colored glasses, right, and Land's polarizing glasses, each half polarized differently, left.

"You can't take snapshots in here," said this trainman, "there isn't enough light in Grand Central Station." But Kodak's Tri-X film had just been introduced (1955)—fastest ever.

Box cameras in the mid-1950s got more sophisticated by incorporating built-in flash, double-exposure prevention. Flash enabled amateurs to get "natural-light" shots.

Cameras above used this GE peanut M-2 flashbulb (shown here actual size) which gave almost as much light as older, bigger bulbs that cost more.

Snapshooting became even simpler when Kodak introduced its Instamatic system in 1963. Cartridges dropped into camera (right, above); film never had to be rewound.

jector with a combined weight of just 5 pounds. In 1928 the amateur could film his family in glowing color, with the new Kodacolor film, a curious ribbed film used with a banded color filter. In 1932 the introduction of 8mm Kodak film and equipment made amateur moviemaking less expensive and simpler.

More recently, the face of home moviemaking has been altered by several significant developments. Sound became available to the movie amateur via magnetic striping on the film itself. Electric drive, now provided by battery power, freed the operator of even a light, portable camera to run as much film in a take as his camera holds, instead of having to stop the action to rewind a spring mechanism.

And in 1965 a new format, introduced by Kodak, gave the amateur 50 per cent greater image area from 8mm film. The seemingly impossible job was done by reducing the size of the sprocket holes on the film. The new film format, called Super 8, resulted in a picture much closer in quality to the far more expensive 16mm film. Furthermore, it was packaged in cartridge form, so that it could be instantly loaded and unloaded in the new cameras that were designed for the Super 8 format. The cartridge-loading concept was soon extended to projectors. (And this "Instamatic" system became available for still cameras too.)

Still photography, of course, went through an equally remarkable evolution after roll film gave it extreme mobility.

Color came with the Lumière Autochrome process in 1907, then others—too

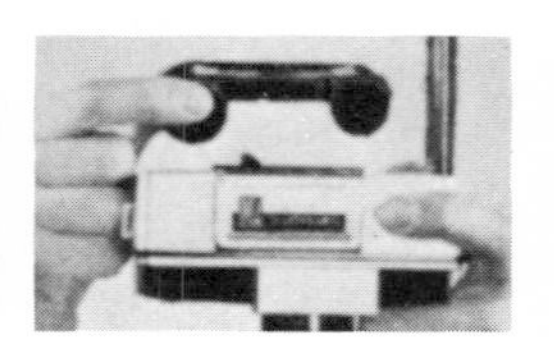

complicated or costly for amateurs at first, but magazine ads' color photos bespoke their commercial success.

The 1930s saw a spate of changes in still photography. The flashbulb and the photoflood lamp liberated the photographer from messy, unreliable, and often dangerous "flash" materials. The miniature camera, with its ultra-fast lenses, brought new dimensions to candid photography and permitted the grabbing of hitherto "impossible" shots by natural light—giving rise to a whole school of "available-light" photography.

Most important, perhaps, was the development in 1935 of a superb new color film—Kodachrome. The new film was literally the creation of a pair of gifted young musicians for whom photography was an abiding passion—Leopold Godowsky and Leopold Mannes. As boys, in 1916, they saw a poorly colored motion picture and resolved to find a better color system for photography. Eventually they decided to "invent" a multilayered color film—seeking a chemical rather than an optical solution to the color problem. In the *World Almanac* they are credited with a 1928 invention titled, "Photo, color, controlled penetration." They took their research to Kodak, and in 1935 Kodachrome, the first commercially successful amateur color film, was introduced in 16mm movie film. Kodachrome for 8mm movies and for 35mm slides followed in 1936. It is still the standard against which all other color films are measured.

Perhaps the most dramatic photographic innovation of the post-World War II period was that announced in 1947 by Edwin H. Land, president of the Polaroid

Electric-drive 8mm Minolta of 1964 used 4 penlight batteries in its handle (top photo), had zoom lens, and could be run by cable release (bottom photo) or remote control for nature photography. Electric drive was important development for home movie equipment, eliminating the need to hand-wind the traditional spring motor and permitting much longer "takes."

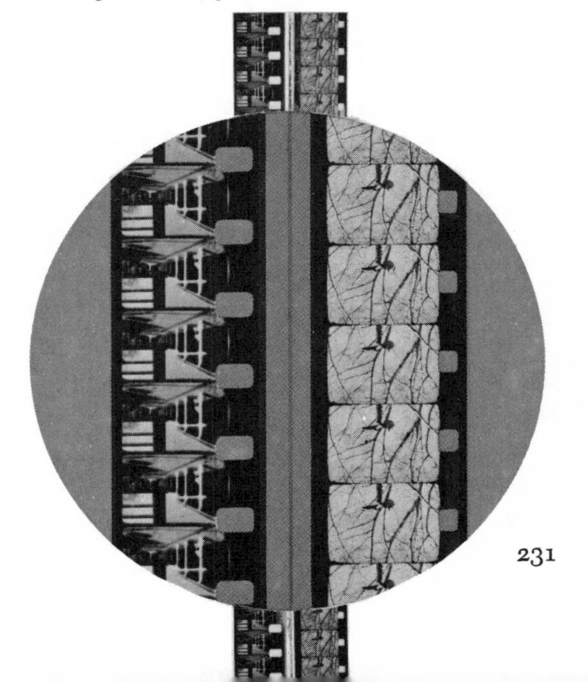

Greatest improvement in home movies over the last decade was introduction of Kodak's Super 8 system. The photo contrasts standard 8mm (left) with Super 8 (right). Note that perforations for Super 8 are smaller. Result is usable picture area 50 per cent larger than with standard 8, for sharper, clearer, and grain-free pictures; also, it was easier to get sound stripe on film. Instamatic packaging made for easy loading.

Corporation. He demonstrated a camera that produced a picture in just one minute after exposure—processing the print *inside* the camera with self-contained chemicals. When the Polaroid camera appeared in 1949, it cost less than $100. Today, black-and-white Polaroid prints take only seconds to make, color prints—introduced in 1963—a minute, and the cameras cost as little as $20. Polaroid has been a tremendous success, and Dr. Land is now one of the wealthiest men in America.

In 1972 miniaturized light sources, such as Flashcubes, Magicubes, and compact electronic flash, combined with fast films and faster lenses, have enabled even the rankest amateur to capture—in color—images that would have been beyond the wildest dreams of advanced professionals only a few years ago. And just think of poor Joseph Nicéphore Niepce and his 8-hour exposures!

What's coming up in photography? In 1971 Kodak demonstrated what they called a "concept model" of a videocartridge player. It would accept your Super 8 movies packaged in cartridges (the way you have them now if you use a cartridge-loading projector) and play them over your color TV set, Devices of this kind should be coming along in the early seventies.

Farther along, look for something to perk in that most fascinating new branch of the photographic science—holography, or lensless photography. This is the only real way to take—and view—a true 3-D picture. But right now it needs lasers at both ends of the process, and is therefore expensive. In ten years . . . who knows?

The Instamatic concept was extended to home-movie projectors in 1969 with a new breed of machines that automatically played and rewound film in cartridges—fumble-free.

Four flashbulbs on 1 base inside a 1 ⅛-inch plastic cube—that was Sylvania's 1965 Flashcube (in use above).

Flashcubes needed batteries to fire, but not the Magicube at right, a 1970 wonder flashed by its own ignition system.

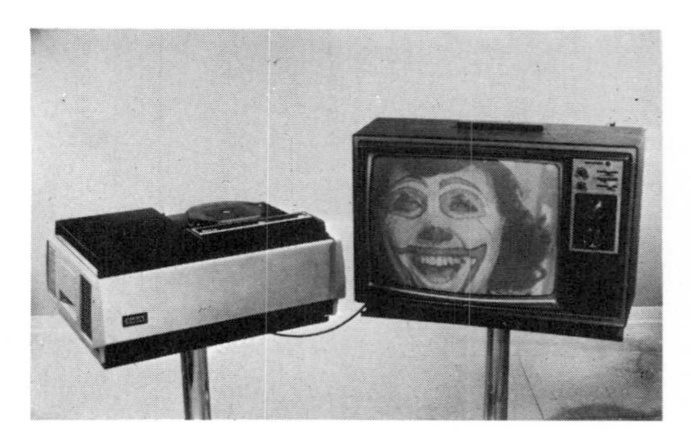

Kodak's "concept model" of Super 8 video-cartridge player would accept both home movies and prerecorded features on Super 8 film, replay them directly on color TV set.

CHAPTER 12

Adventure Under the Sea

RESULTS of the highest scientific importance are to be expected."

That prediction appeared in the February 1873 *Popular Science* and it applied to the voyage of a ship, the *Challenger*. The ship had left Great Britain to explore the depths of the sea—on the first round-the-world voyage for that purpose.

For her voyage to probe beneath the surface of the sea, the 2,306-ton Royal Navy corvette was prepared as carefully, for her time, as is an Apollo spaceship today. She was given a steam engine in addition to her sails and a laboratory was built in. It was equipped to the last detail: "spirits of wine" (alcohol) for preserving specimens.

On December 30, 1872, the *Challenger*, George S. Nares, captain, made her first station (stop at sea) to collect data. She was off the Spain-Portugal border, at latitude 41° 52′ North, longitude 90° 42′ West. The line she lowered had a dredge at its end. The dredge brought up bottom mud, of no special significance. But it was cold; it was used to cool champagne.

Nevertheless, with this first station, modern oceanography began.

Not long before, the basin of an ocean was thought to be smooth and shaped like the inside of a bowl. Then, just prior to the *Challenger*'s voyage, a few ships preparing the way for laying telegraph cables across the seas had found signs of canyons and elevations. Now the *Challenger*'s soundings indicated that there was a 10,000-mile-long range of mountains from Iceland to the Antarctic—today's Mid-Atlantic Ridge.

The maximum depth of the ocean was believed to be 3 or 4 miles. North of the Virgin Islands, the *Challenger* made a sounding of 3,875 fathoms (almost 4½ miles). She was over what today is called the Puerto Rico Trench—not the deepest point known in the Atlantic.

In the Pacific, the *Challenger*'s hemp rope with a weight on its end was lowered 200-odd miles south of Guam. It went down and down 5 miles deep. It was the deepest spot discovered, up to that time, in all the oceans. The ship was over the Marianas Trench. The Marianas and Puerto Rico trenches are great gashes in the bottom of the sea, by far larger than any Grand Canyons on land.

Before sailing as a scientist on the *Challenger*, Charles Wyville Thomson had collected sea-floor sediment that had seemed to contain a filmy net resembling egg white. This white stuff seemed to make slow movements and was thought to be the simplest known living thing. The English scientist Thomas Huxley called it *Bathybius haeckelli*. A U.S. scientist called it "the most interesting organism . . . known except man." It was pictured as the original beginning of life on earth—in the sea.

The *Challenger* did not seem to collect any. Then Thomson and J. Y. Buchanan noticed something. When someone poured "spirits of wine" into a bottle of water from the depths, bathybius appeared. It was a non-living, chemical precipitate. It was, in fact, sulfate of lime (gypsum) formed from sea water when alcohol was added to preserve specimens. That and nothing more.

Little was known of minerals on the deep bottom before the *Challenger*'s crew hauled up some rocklike lumps. These proved to be mostly manganese (today's manganese nodules, discussed later).

A professional scientist interested in strange animals being found in the sea, C. Wyville Thomson became a pioneer in oceanography.

In this dredge aboard the Royal Navy ship *Porcupine,* in 1869, Thomson caught the first animals taken from 2 ½-mile depth.

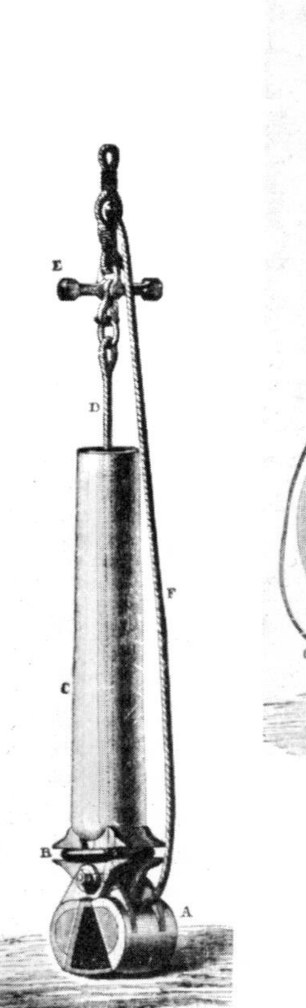

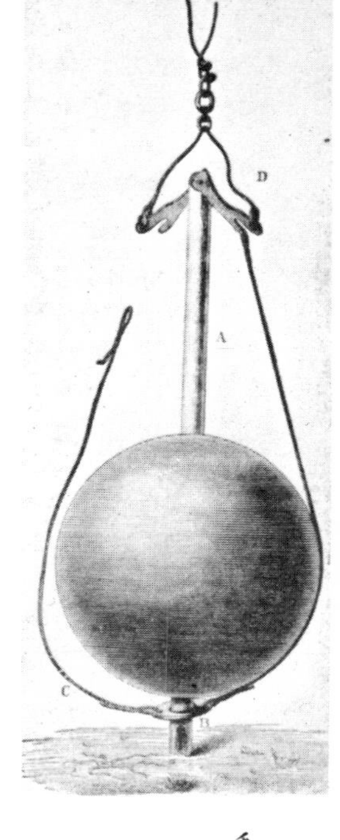

Deep-sea sounding device (above) grabbed sample of bottom. Apparatus above at right, invented by J. M. Brooke, U. S. Navy, also took bite of the bottom. At right, a 225-pound dredge—a bag of light canvas inside a bag of twine netting.

At a time when few weather readings were taken, the *Challenger* obtained 50,000 a year. Her scientists studied currents, including deep currents, and analyzed the water. She took temperatures all the way down, and found cool water in the equatorial Atlantic, indicating that Antarctic water creeps along the bottom. She took samples of the bottom and found most of it was ooze composed of microscopic animals. It held billions of their fossils.

In 1872, 30 deep-dwelling fish were known. The *Challenger* increased the known number to 370; by 1971 around a

26 *THE POPULAR SCIENCE MONTHLY.*

ON SOME OF THE RESULTS OF THE EXPEDITION OF H. M. S. CHALLENGER.

By Prof. THOMAS H. HUXLEY, F. R. S.

IN May, 1873, I drew attention, in the pages of this *Review*, to the important problems connected with the physics and natural history of the sea, to the solution of which there was every reason to hope the cruise of H. M. S. Challenger would furnish important contributions. The expectation then expressed has not been disappointed. Reports to the Admiralty, papers communicated to the Royal Society, and large collections which have already been sent home, have shown that the Challenger's staff have made admirable use of their great op-

From 1872 to 1876 H.M.S. *Challenger* sailed around the world exploring the oceans, Thomson aboard her. Thomas H. Huxley summarized results in May 1875 *Popular Science. Challenger* found 4,417 new animal species, deep currents, great depths.

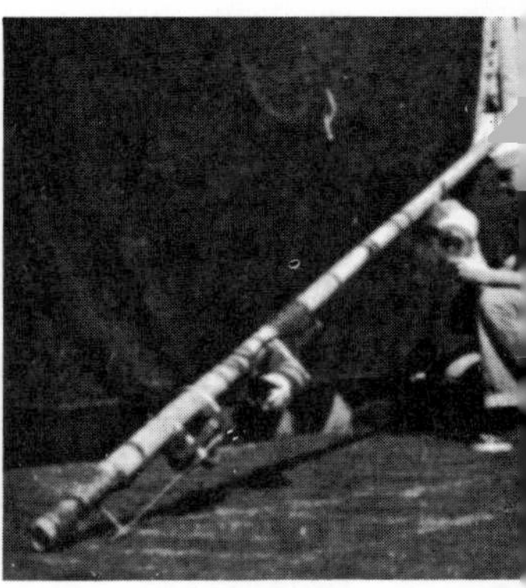

1952 tools to explore depths of sea included improved versions of Thomson's. Left to right: Nansen bottles to collect water samples; snapper to bite into sea floor; midwater trawl to capture fish between surface and bottom; underwater camera.

thousand were known. Her crew caught fish in all sizes, from tiniest infusoria to the great whale. She showed what many deep-sea fish—brotulids, lantern fish, hatchetfish—are like. Small, a few inches long. Big-headed. Big-eyed. Stringy-tailed. Dark-colored or luminescent. Some big-mouthed with jagged teeth.

Besides fish, they hauled up "myriads of curious creatures"—some with numberless eyes, others without any; starfish, "growing on long and slender stalks." They found new lacelike and basket-like sponges, sea spiders (in Arctic and Antarctic waters), and large sausage-shaped crawling sea cucumbers. Today, some scientists believe these cucumbers, usually under a foot long, are the most numerous large animal on the sea bottom in the abyssal zone, 20,000 feet down and deeper.

The *Challenger* found a uniform cool temperature and constant salinity in deep regions of all oceans. Two of her scientists, Thomson and John Murray, therefore concluded that deep-sea animals would be worldwide in distribution. They are not.

The ship carried 4 improved models of a device invented almost 200 years before

A WONDER FROM THE DEEP-SEA.*

BY M. L. VAILLANT.

DURING the last voyage of the French deep-sea dredging-ship Travailleur, a fish was found, off the coast of Morocco, at the depth of about 7,500 feet, which may certainly be regarded as one of the most singular beings yet brought to light in any of these investigations. It is about eighteen inches long, and three quarters of an inch thick at the thickest place, and is deep black. Its body, the form of which is marked in front by an enormous mouth, somewhat resembles that of a macrouran, and tapers regularly from near the anterior quarter, where the external branchial orifice may be seen, till it terminates in a point at the caudal extremity.

After success of Thomson and the ships *Porcupine* and *Challenger,* the French decided to go deep-sea fishing and sent out the *Travailleur. Popular Science* for May 1883 (above) reported what they caught (right).

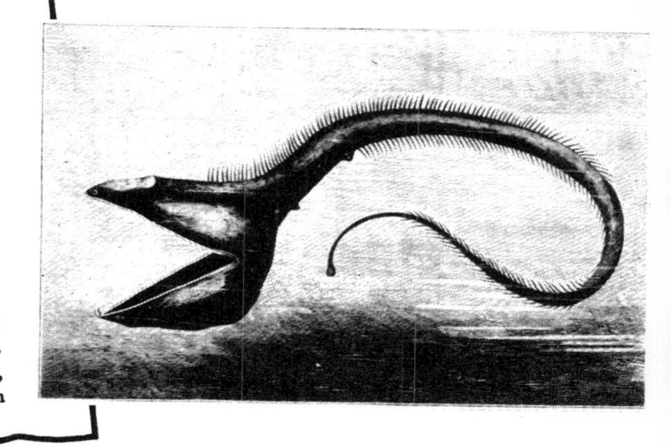

Great gulper—one of deep-sea fish that can swallow fish larger than they are—was taken by *Travailleur.* This ugly specimen, 18 ½ inches in length, most of it mouth and stringy tail, was captured 7,500 feet down, off Morocco. It was new to science when it was hauled up.

Sea monsters do exist: illustration from an 1889 *Popular Science* shows men hauling aboard a giant 10-armed squid. Such monsters—weighing up to 2 tons—were even washed ashore in the 1880s, but not since.

Flower-like animals with long bare stems buried in the bottom were hauled up by the *Challenger.* In 1968 first photo of a living Umbellula was made by Walter Jahn. Camera found it 3 miles down.

The dark sea a few hundred to a few thousand feet down is full of luminous fish. Deep-sea squid in the Sea of Japan, much smaller than the giant squid, sparkle with bluish light from 400 light organs. This picture appeared in a 1922 *Popular Science*.

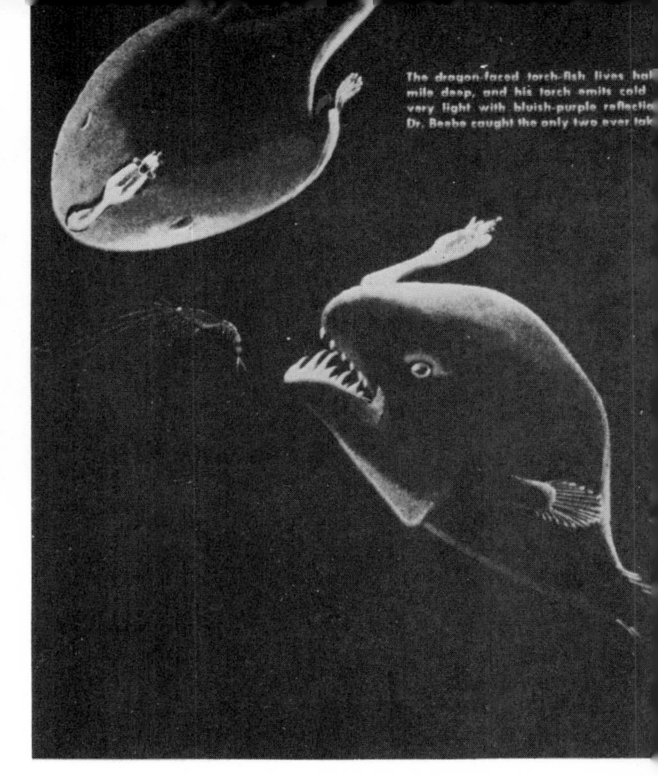

Dragon-faced torch-fish lives half a mile down. Its torch emits cold silver light with bluish-purple reflections. Bronx Zoo's William Beebe took first 2 captured.

her voyage: the microscope. Through them, Thomson and Murray peered at tiny animals and plants in sea-bottom mud. With Buchanan, H. N. Moseley, and Dr. von Willemoes-Suhm, they stared at scum collected in tow nets. They saw thousands of new species of tiny creatures and proved that the microscopic animals are the most common on the globe. Today, counting the infinitely small life and other animals, plants, and fish (the most common backboned animals, perhaps 45,000 species), the oceans are estimated to hold four-fifths of all earth's living things.

The commonest animal? Scientists of the 1970s, including Sir Alister Hardy, vote for the copepod—a jointed-leg crustacean not unlike shrimp, weighing as little as 1,140,000 of an ounce, they scull along on 5 or 6 pairs of legs. They may be red, black, or sparkling, and as much as an inch long. Some have a single median eye.

The *Challenger* found 4,417 new species of living things. Her trip lasted over three years. She traveled 68,890 nautical miles and took up 362 stations. She sailed every ocean except the frozen Arctic.

In 1877 Professor Viktor Hensen of Kiel University, Germany, coined a name for the tiny animals and plants that the *Challenger* and its microscopes had made known. Hensen wanted to distinguish them from animals that can swim against a current (the nekton) and from animals that

"Ten small light sources scattered over tiny body," said *Popular Science* in 1922, "turn deep-sea shrimp into swimming lighting plants." Light has steady glow.

Submarine was invented at least 200 times in the 1800s. Some of them almost worked. Then in 1898 John Philip Holland launched one that did. U. S. Navy bought it in 1900.

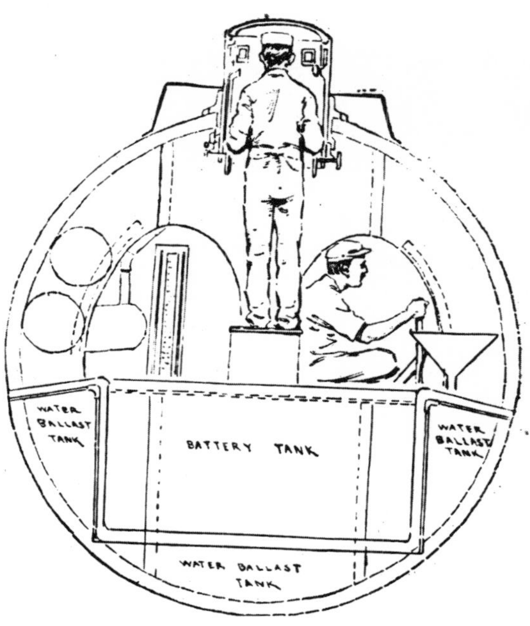

Standing in conning tower, crewman peered into the murky water in the hope of seeing where Holland sub was going. *Popular Science* published these drawings in 1900.

live on or in the bottom (the benthos). The word he came up with, and the name this teeming microscopic life has been known by ever since: plankton.

The first practical submarine was the invention of John Philip Holland—who, in the 1890s, built a series of undersea craft. His ninth one, named after him the *Holland*, became the United States Navy's first submarine, in 1900.

The 54-foot-long *Holland* was propelled by electric storage batteries and motors when submerged, and by gasoline engines on the surface. She carried 3 Whitehead torpedoes and surfaced to launch them from her torpedo tube.

Admiral George Dewey, hero of the Battle of Manila Bay, declared after viewing a 1900 trial that his attack on the Spanish fleet would have been impossible had submarines been in Manila Harbor. The Navy soon had 6 more Holland-type submarines, and later also bought submarines from Simon Lake. A Lake submarine was the first to sail the open sea, from Norfolk to New York City.

Both Holland and Lake thought submarines would help men explore the depths. They planned small windows, and Lake designed a sub that let a diver step out onto the bottom, as do some submarines today.

In 1913 John Ernest Williamson decided he'd be the first to take photos under the sea. A reporter on a Norfolk paper, he planned to make the attempt in Chesapeake Bay. Captain Charles Williamson, a deep-sea salvager, provided just what his son needed: a flexible metal tube designed so that it could be lowered from the surface 30 feet to the bottom. There was an observation chamber with glass windows on the tube's end, and electric lights to illuminate the water.

Williamson wriggled down the tube and stayed in the observation chamber all afternoon. He got his first underwater photos: 2 fish swimming, and a growth on the sea bottom.

In February 1914 he went to the Bahamas, with a 4-ton steel diving bell with large glass windows and mercury-vapor lighting. Through the windows he took the first undersea movies. In 1915 he made a second film, a version of Jules Verne's *Twenty Thousand Leagues Under the Sea.*

In 1930 two men prepared for a voyage no man ever had made: into the depths of the sea. "1,300 feet is divers' goal," said *Popular Science* for August 1930. "A quarter of a mile beneath the sea!"

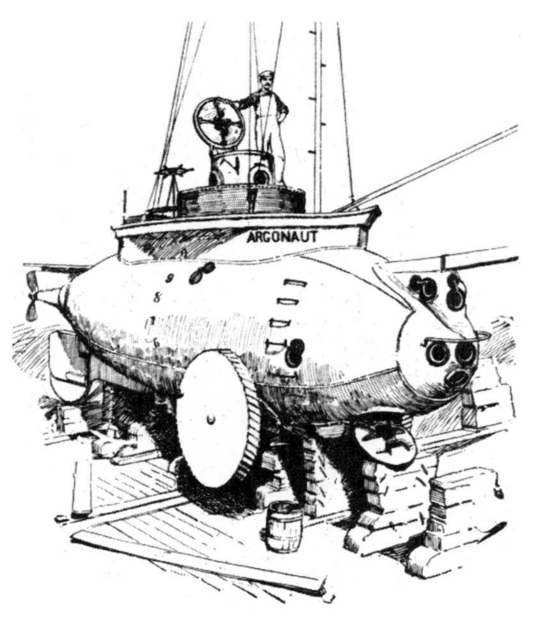

Early sub *Argonaut* was invented by Simon Lake. Wheels rolled over the sea floor. A hatch in bottom (at lower front) let diver out of pressurized compartment into sea.

After the *Argonaut* came a series of Lake submarines of varied size and design. Here, 2 men spear fish through open hatch of one of them as it rests on bottom.

In a steel ball dangling on the end of a cable, William Beebe and Otis Barton planned to be lowered into the ocean to eyewitness the depths. Barton and Captain John H. J. Butler made the blue-steel ball. Will Beebe called it a "bathysphere." It had three small quartz windows. Tanks on walls provided oxygen to breathe. There were no seats; the men inside squatted the whole trip. Once Beebe sat on a monkey wrench so long that it imprinted its profile on his rear. There was no air conditioning; Beebe waved a palm-leaf fan.

When Beebe and Barton began their series of dives, which spanned from 1930 to 1934, the deepest Navy divers had descended 306 feet. The deepest submarine had reached 383 feet. The deepest diver of all had reached 525 feet in a lake in Bavaria.

Then in 1930 Beebe made a telephone call to their barge on the surface. Said Beebe: "Six hundred feet down—only dead men have sunk below this!"

At that depth, he saw a pale, large, light body with telescopic eyes—an unknown fish. At 700 feet, a jellyfish had a stomach filled with glowing luminous food. At 1,000 feet, there was "a loop of black, sea-serpenty hose." At 1,750 feet, luminous fish—

How Lake's *Explorer* dived to bottom of Long Island Sound was told in 1933 *Popular Science*. From left: William Beebe, Lake, diver Frank Crilley who exited in Sound.

For rescue work, Lake's *Defender* (pictured in 1928 *Popular Science*) was designed to pick up men with Momsen lungs leaving a sunken sub.

How first underwater movies were made was reported in a 1916 *Popular Science*. Cameraman descended tube to steel diving bell lowered in the Bahamas. Behind scheme: John Ernest Williamson of Norfolk, Virginia.

Beyond heavy glass window, school of black-and-gold-striped fish is observed by Mrs. Williamson, as shown in 1929 *Popular Science*. At right: version of tube then in use.

with double rows of lights. At 1,850 feet, a 10-armed squid with big, almost human eyes. At 2,450 feet, a fish at least 20 feet long—maybe, Beebe thought, a whale shark—and a lantern fish "ablaze with their full armor of iridescence."

In 1934, en route to their deepest point, over a half-mile down (3,028 feet), Barton saw a fish with eyes on the ends of long periscope stalks. At over a half-mile down, a shrimp shot a cloud of luminous fluid at Beebe's porthole.

At any moment the cable from the surface barge to the bathysphere could have parted, and Beebe and Barton would have plummeted to their deaths in the sea.

In 1931, as Beebe and Barton went deeper and deeper, *Popular Science* reported on another voyage no man had ever made: Sir Hubert Wilkins would travel in a submarine beneath the ice of the Arctic. He had a diesel battery-powered hand-me-down submarine from the U. S. Navy. It was named the *Nautilus,* after one that Robert Fulton had worked on in the early 1800s.

Atop his *Nautilus,* Sir Hubert had runners to glide along under the ice. He planned to travel this way from lagoon to lagoon in the Arctic ice, surfacing in each and recharging batteries for the next hop.

At the edge of the ice pack between Spitzbergen and Greenland he tried to skid beneath the ice. He lost the stern plates, and controlling or diving the *Nautilus* was difficult. Frost formed inside; the crew grew pessimistic. Wilkins gave it up.

Nevertheless, in the years that followed, Simon Lake advocated a submarine cargo ship to travel under the ice. In World War II some German submarines managed to hide under the edges of Arctic ice.

In April 1938 *Popular Science* reported a major breakthrough. For years, as divers moved deeper into the sea, they had difficulty breathing air or oxygen. Now, they would have on tap a new breathing gas—a mixture of helium and oxygen "with which the U. S. Navy and the Bureau of Mines have been experimenting for over 12 years." At the Washington, D.C., Navy Yard, two divers—William Badders and J. H. McDonald—tested the gas. In a diving tank, under the simulated pressure of 500 feet of water, they thrived on the helium-oxygen.

In 1943 Jacques-Yves Cousteau and Emil Gagnan of France invented the aqualung. From a tank of air on his back, a valve system feeds air to a swimmer at the same pressure as the water at the depth he has reached. Used with a face mask and swim fins, the aqualung opened to men and women the shallower ocean depths. Now dives to 90 feet and of 20-minute duration are common for YMCA-trained divers.

In 1945 the Navy's *Cape Johnson*, Captain Harry H. Hess, located a number of seamounts in the Pacific. These flat-top submerged mountains may once have been islands that supported coral reefs and then sank.

The story of oceanography during the early years of *Popular Science* was the story of the voyage of one ship. The story of oceanography since World War II has been the story of many ships and submarines, many men, and dozens of voyages.

In 1948 Commander Skip Palmer took the *Carp*, a diesel-electric Navy submarine, on an Arctic voyage. He sailed the surface until about 50 miles inside the ice pack. There he found a lagoon a mile in diameter. He dived, then returned to the lagoon and surfaced. He had shown that men could travel by submarine beneath the ice—and resurface in a patch of open water.

William Beebe's and Otis Barton's world record descent of 3,028 feet, over a half-mile, stood for 15 years. During those years, 1934–49, there was no observation of the depths of the sea by the eyes of men. The man who finally broke the record was Barton himself. In 1949, diving alone in a new steel ball called a Benthoscope, he reached 4,050 feet.

In 1951 an oceanographic ship out of Denmark, the *Galathea*, with scientist Anton Bruun aboard, lowered her trawl almost 7 miles into what Bruun supposed was the deepest place in the sea—the Philippine Trench. The question: Was there life almost 7 miles down? Bruun brought the answer: In the trawl were "small whitish growths—sea anemones." This was proof enough. But there was more: 5 bivalves, 1 amphipod, 1 bristle worm, and 75 of the sausage-shaped sea cucumbers, now proving to be among the most common deep animals. There were also bacteria in the cold bottom mud. "An unex-

First men to go deeper than any diver or submarine were William Beebe and Otis Barton in 1930. In 1934 bathysphere (top left), they went more than half-mile down.

Alone in a Benthoscope (here raised by cable), Barton in 1949 went deeper—4,050 feet—where he saw light flashes from sea animals.

DEC. 1947
25 CENTS
POPULAR SCIENCE
MONTHLY
PICCARD
Diving Balloon
p. 82

Auguste Piccard of land-locked Switzerland, far from sea, next had an idea. He proposed to hang a steel ball, like that of Beebe and Barton, beneath a float to explore depths.

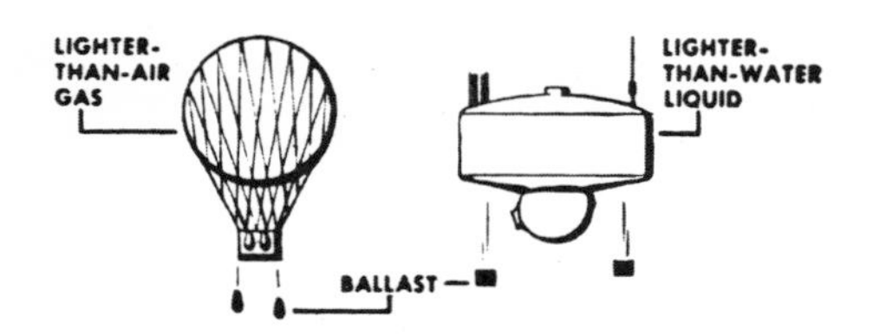

Principle of balloon is used by bathyscaphe. It valves off fluid (gasoline) lighter than surrounding medium to descend, drops ballast to rise. Piccard hoped to dive 2½ miles.

pectedly rich variety of bottom-dwelling animals," said Bruun.

Anton Bruun had reason to believe that his collection of animals had come from the greatest depth in the sea. Ever since the first *Challenger* had found a big hole in the Marianas Trench, other men and ships had been locating deeper places.

The American ship *Tuscarora* located a 27,900-foot depth (4,655 fathoms) off Japan in 1890. *Popular Science* in 1891 reported it as "exceeding any similar depression yet found in any other region of the great oceans." Britain's *Penguin*, Andrew Balfour, captain (he had been aboard the *Challenger*), got 5,155 fathoms in the Kermadec Trench—the first depth over 30,000 feet.

The Dutch ship *Willebrord Snellius* then found 5,539 fathoms in the Philippine Trench. In 1932 a U. S. Navy oiler, the *Ramapo*, measured 5,673 fathoms (34,038 feet) in the Izu Trench (south of the Japan Trench). Then Germany's *Emden* sounded 5,686 fathoms in the Philippine Trench.

The U. S. Navy's *Cape Johnson*, Captain Harry H. Hess, in 1945 found 5,740 fathoms (34,440 feet) in the Philippine Trench—the deepest place yet in the sea, and the sounding by which Anton Bruun was going. But the very year Bruun trawled his sea-floor animals from the Philippine Trench—1951—a new "greatest depth" was found in the sea.

That year a second British oceanographic ship named *Challenger* sailed the Pacific. Aboard her, her captain, Commander George Stephen Ritchie, Royal Navy, and oceanographers Thomas Frohock Gaskell and John C. Swallow bounced hydrophone echoes off the bottom. It took 10 seconds for their echo to return; they had 5 miles of sea under them. Now 11 seconds; 12 seconds; 12½ seconds. They had found a new world's-deepest spot.

They estimated it at more than 35,000 feet and named it the "Challenger Deep" of the Marianas Trench. Later estimates by Russians (36,198 feet) and Americans (35,-800 feet) have made it a few hundred feet deeper; it is still the deepest known place in the seas.

In 1952, with Navy sponsorship, the Scripps Institution sent two former ocean-going tugs to the South Pacific. On the cruise, Bob Fisher and Scripps director Roger Revelle measured the Pacific's Tonga Trench—now the second-deepest-known place in the seas.

Harris Stewart, aboard a tug, at one time radioed to the other a new find: "Ran into a pretty interesting seamount." What had been found was a giant, isolated, submerged mountain—26,300 feet from the bottom—towering up from the Tonga Trench.

Piccard's bathyscaphe *Trieste* (right) first dived 2 miles in Mediterranean. Lights blazed (as above) when in Pacific she made man's deepest dive: almost 7 miles.

Biggest bathyscaphe is the 70-foot *Archimède,* owned by the French. In 1962 she reached about 6 miles down in Kuril-Kamchatka Trench near Japan. In 1964 she made 10 dives into deepest hole in Atlantic: 5 miles into the Puerto Rico Trench.

Another type of undersea craft came along in 1959: the deep-diving submarine, or submersible. Jacques-Yves Cousteau's diving saucer, above, went down 1,000 feet.

Six thousand feet! In July 1965 the American submersible *Alvin* (named for Allyn Vine who helped plan her) became first craft of her kind to exceed depth of a mile.

It's the tallest isolated mountain on the globe: Capricorn Seamount.

The expedition brought back thousands of measurements of magnetism, gravity, temperature, currents, tides. Their scuba divers, Willard Bascom, Bob Livingston, and Walter Munk, found rugged black pinnacles of volcanic rock at sunken Falcon Island.

Then came the upside-down fathometer, invented by a Navy physicist, Waldo Lyon. A kind of sonar, it profiles the ice above a submarine. Now the atomic submarine *Nautilus*, with a power plant designed by Rear Admiral Hyman G. Rickover, could head for the North Pole. En route, the *Nautilus* made thousands of depth measurements of the Arctic Ocean. Her inertial navigation took her precisely to the Pole. She reached it at 11:14 P.M. on Sunday, August 3, 1958. Commander William R. Anderson, her captain, did not bring her up through the ice.

Nine days later, Commander James F. Calvert took the *Skate* to the Pole. En route, he zigzagged the submarine over the underwater mountains known as the Lomonosov Range. Watching the depths over closed-circuit TV, he saw a school of fish miles and miles long. Almost every sailor came to look at the TV screen. With Lyon's sonar, Calvert found polynyas (lagoons) in the ice overhead.

Four *Skate* crewmen—Dave Boyd, Dick Arnest, Dick Brown, and Sam Hall—were among the first scuba divers to enter the Arctic Ocean. Their heavy sponge-rubber

A 4-hour trip down on a night in 1965 put the American *Aluminaut* 250 feet deeper than the *Alvin*. Going down, said Jim Cooney, "I was concerned with strain-gauge readings" (checking stress on the hull). Bigger than any deep-sea craft before, *Aluminaut* had 2 mechanical hands (at right, below) and, like others of type, windows and undersea lights.

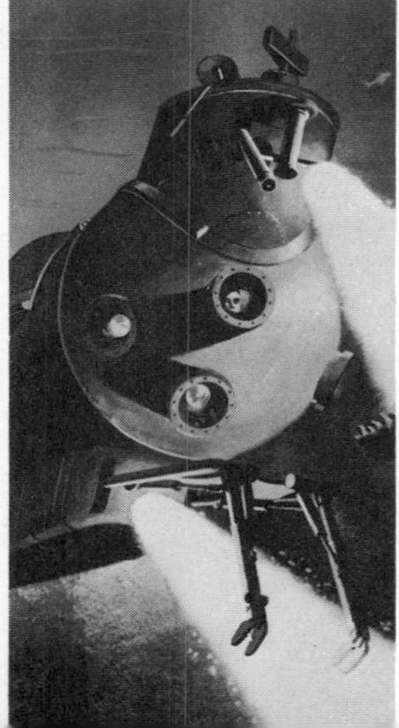

suits froze and the ice on them tinkled as the divers re-entered the submarine. Such diving in 1971 is becoming frequent.

Our biggest nuclear submarine, the *Triton* (447 feet long, 8,000 tons) made a submerged trip around the world in 1960. Captain Edward L. Beach tried to locate the deep bottom of the Philippine Trench. "The limits of the huge trench were well beyond the range of sonar," he reported, "the bottom so far away as to be like a void." In the Lombok Strait the *Triton* behaved like an airplane in an air pocket. She sailed from a layer of dense water into a layer less dense and dropped 125 feet in 40 seconds.

U.S. nuclear submarines have many times crossed the frozen Arctic Ocean, the one sea on earth that surface ships do not regularly navigate.

Still another means of conquering the deep had been developed—the bathyscaphe. Auguste Piccard of Switzerland and Max Cosyns of Belgium planned to hang a steel ball—like the Beebe-Barton sphere—beneath a float filled with lighter-than-water gasoline. The bathyscaphe was not built until the fifties, but *Popular Science* showed how it would look on its cover in December 1947 and described how it would work. It would be made to sink by loading it with shotgun pellets (or, when under way, by valving off gasoline). Dump the pellets on the bottom, and the gasoline in the float would make it rise.

Five bathyscaphes were built in all. They progressively reached ever-greater depths. France's Georges Houot and Pierre-Henri Willm broke Otis Barton's 4,050-foot record when they reached 5,115 feet, almost a mile. Auguste Piccard and his son Jacques reached almost 2 miles. On February 15, 1954, off Africa, Willm and Houot reached 2½ miles—the average depth of the ocean.

The U. S. Navy bought the bathyscaphe *Trieste*. Jacques Piccard came along with her as pilot. In 1960 he and Lieutenant Don Walsh reached 35,800 feet (close to 7 miles) in the Challenger Deep of the Marianas Trench. At that level Walsh and Piccard saw an ivory-colored flatfish about a foot long and a shrimp. There was life at the greatest depth.

Bathyscaphes couldn't travel far over the floor of the sea; they were elevators, to take men down and up. So men built newer vehicles to prowl the depths: deep-diving submarines or submersibles.

The first was Jacques-Yves Cousteau's diving saucer, built to go 1,000 feet down (deeper than scuba divers). Two men lie prone on foam-rubber mattresses, staring through viewports. Saucer divers have seen rare fish, undersea canyons, and great boulders being moved out to sea by no one knows what force. Dr. Francis Parker Shepard, a Scripps geologist, cruised between high canyon walls that almost closed over the diving saucer.

In the Bahamas, Woods Hole's *Alvin* in 1965 made the first deep submarine dive—6,000 feet, over a mile, 20 times as deep as World War II submarines usually went. Only 3 men knew how to pilot her: William Ogg Rainnie, Jr., Martin J. McCamis, and

Two-man *Deep Jeep*, a Navy deep-diving craft, was designed and piloted by Will Forman. An outstanding dive: 2,000 feet down near San Clemente Island.

***Deep Quest* in 1968 dived 8,310 feet—her 2-man crew and 2 scientist-observers protected in the pressurized spheres here being lowered into the outer shark-shaped hull.**

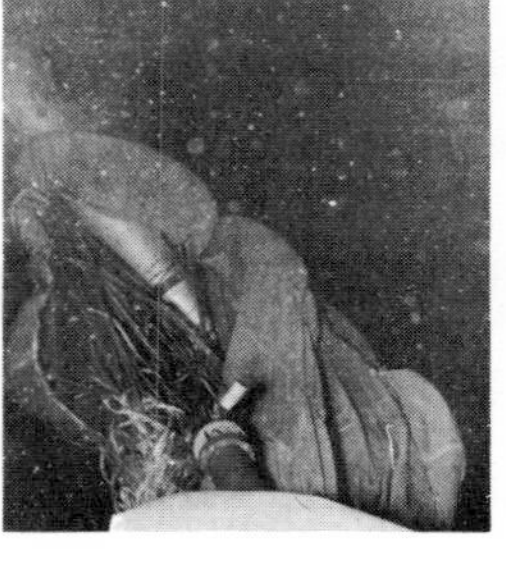

Unmanned devices are at work on sea floor. Navy's *CURV*, above, built to pick up practice torpedoes, recovered H-bomb that plane accidentally dropped off Spain in 1966. Right: bomb and its parachute on bottom.

H-bomb lay on steep continental shelf a half-mile down, off Palomares (left). U.S.S. *Petrel* lowered *CURV* and guided her by TV cable. *CURV* fastened lines to hoist bomb.

Valentine Wilson. On the bottom, Rainnie and McCamis saw a dim, rocky landscape.

J. Louis Reynolds' *Aluminaut* later dived deeper—6,250 feet, where she hung about 3,000 feet above bottom. Those aboard her saw plankton glowing in the sea.

Lockheed's *Deep Quest* in 1968 was driven by Larry Shumaker more than 8,000 feet deep in the Pacific. He found lilylike animals 1½ feet tall covering the bottom. That dive gained the *Deep Quest* the world's record for the deepest submarine dive—a record she still holds.

Other subs have been busy: General Dynamics' *Star III* cruised over an undersea cliff in the Bahamas. Aboard the Navy's *Deep Jeep*, pilot Will Forman saw basket sponges like those the first *Challenger* hauled up. In Westinghouse's *Deepstar 4000*, oceanographer Eugene LaFond, pilot Bob Bradley, and electronics engineer Dale Good were the first men to see a lifeless sea floor. Off California, 4,000 feet down, it was one of the dozen or so oxygenless ocean areas where nothing lives. Dead fish and squid lay there. "Eerie," said LaFond.

In another deep-diving submarine, Grumman's *Ben Franklin*, Jacques Piccard and five crewmen in 1969 stayed down 30 days and traveled 1,444 miles from Palm Beach,

Florida, to a point at sea well east of Cape Cod.

The men in the *Ben Franklin* cut off their battery-powered engines and were carried by the drift of the Gulf Stream. Through 29 viewports, the most of any deep-diving sub, they saw a big jellyfish with tentacles 30 feet long. On the bottom were hills 100 feet high and gigantic coral heads. They climbed to avoid both.

Like some other deep-sea subs, the *Ben Franklin* had a mechanical arm, with which it picked up sea-floor specimens for viewing. Operated from inside by a man at a viewport, the steel arm swivels around almost as if it had muscles, and a claw on its end grasps and lifts anything from a pebble to a boulder weighing hundreds of pounds.

The *Ben Franklin's* trip was almost flawless. It showed how far along by the 1970s is man's ability to explore the depths. Said Piccard: "Never had a group of oceanographers been able to live so intimately with the inner sea."

Divers today, too, go deeper. And they stay longer and get more work done. Many breathe a gas made of helium, oxygen, and nitrogen; or helium and oxygen alone. After breathing the gas awhile, the diver's body becomes saturated with it, and he can remain down—theoretically—indefinitely.

The Jacques Cousteau-Emil Gagnan scuba gear (*s*elf-*c*ontained *u*nderwater *b*reathing *a*pparatus) frees a diver of the old cumbersome suit, awkward heavy helmet, and long unwieldy air hose to the surface. With scuba, a man can bend over, turn somersaults, and even work on the bottom with head and hands down and feet up—some diving archaeologists do. Using advanced versions of scuba gear, breathing gas mixtures, descending to the bottom in steel diving bells, and swimming out of them, divers are setting records.

In 1964 divers Jon Lindbergh and Robert Stenuit spent 49 unbroken hours 432 feet down near Nassau, the Bahamas. On the bottom they slept in an inflatable 6-by-9-foot rubber tent, invented by Link Aviation's Ed Link. That was the longest deep dive men ever had made although Lindbergh was already looking ahead to a 600-foot dive, as he told *Popular Science* in 1965.

Two years later, with Jon Lindbergh as adviser, Art Pachett and Glen Taylor went down 636 feet off Morgan City, Louisiana, divers' capital of the world, and stayed under pressure of that depth 48 hours. From a barge on the surface, Dr. Joe MacInnis listened to their heartbeats and watched them by TV.

Taylor once slipped up to his shoulders in muck on the bottom. But the divers got 6 good hours of work done on a mock-up of an oil wellhead. They set two records: deepest dive and longest working time.

By the fall of 1970 their depth record was broken. Three French divers—Christian Cadiou, Michel Liogier, and Patrick Cornillux—went 820 feet deep in the Mediterranean. They breathed helium, with 2 per cent oxygen. A new combination, hydrogen and oxygen (hydrox), is now being tested.

Tragedy has struck record-setting divers. In 1961 a Swiss, Hannes Keller, and a British writer, Peter Small, went down 1,000 feet near Santa Catalina Island, California. They made the descent in a boiler-shaped diving bell. Keller swam out and planted U.S. and Swiss flags on the bottom. In less than 2 minutes, he returned. The diving bell started up, but the hatch wouldn't close. The chamber lost pressure; the men had to breathe air instead of their gas. Chris Whittaker went down 200 feet to fix the hatch and was lost; Small was dead of the bends when he reached the surface.

Today, inside of pressure chambers on dry land—not in the sea—divers have successfully weathered pressure equivalent to that of 1,000 feet (U.S.), 1,500 feet (Britain), and 1,700 feet (France). On their 636-foot dive, Pachett and Taylor rode down in an Ocean Systems' *ADS* (*A*dvanced *D*iving *S*ystem) *III* that provided them with breathing gas under pressure.

The Link-Perry submarine *Deep Diver* has a rear compartment that provides gas to divers, exactly as *ADS III* does. In 1968, off the Bahamas, Denny Breeze and Roger W. Cook stepped out on the bottom 700 feet down—the deepest any submarine has let out divers.

These days, for long sojourns beneath the sea, air or gas under pressure fills the inside rooms of underwater shelters. Men can live in one, making it a base from which to swim out into the cold depths. In 1962 France's Jacques-Yves Cousteau put the first shelter 36 feet down in the Red Sea. Divers lived in it for a month. He also had a deeper shelter, 90 feet, for scuba divers. They stayed at that depth for a week. In 1964 Cousteau divers stayed 3 weeks in a shelter at a 325-foot depth.

In 1964 the U. S. Navy put 4 men—Lester E. Anderson, Robert A. Barth, Sanders W. Manning, and Robert E. Thompson—190 feet down, off Bermuda. They remained 11 days in *Sealab I.* In *Sealab II* Scott Carpenter, first astronaut to turn aquanaut, stayed 30 days in 1965. He was 205 feet beneath the Pacific off California. Three 10-man teams of *Sealab II* stayed down for 15 days each.

Sealab II had a particularly welcome new aquanaut: Tuffy—a 270-pounder, 7 feet long. Tuffy was a porpoise. He visited under the guidance of veterinarian Sam H. Ridgway. To men on the bottom Tuffy carried letters, lifelines, tools. He answered sound signals. Now the Navy has gone further: it has trained porpoises so well they can be allowed to swim out into the open sea. Given a sound signal, they voluntarily come home.

Sealab III, to house 50 aquanauts at around 435 feet down, was postponed indefinitely after the death of aquanaut Berry L. Cannon at its start. The Navy said he probably died of carbon monoxide poisoning due to a malfunctioning of his breathing apparatus.

Undersea shelters, "habitats," are springing up all over—or under. One of the best known is General Electric's *Tektite.* In it, 50 feet down, Richard Waller, Conrad Mahnken, H. Edward Clifton, and John Van Derwalker remained undersea for 60 days, beating Carpenter's 30-day record. Five women aquanauts worked out of *Tektite*—a first.

Using a new GE rebreathing apparatus, divers lengthened "outside" work periods to 6 hours (its maximum is 10 to 12 hours).

In Hecate Strait, Alaska, in 1968, men aboard the oil rig *Sedco 135* added a footnote to oceanography. In 1933, during a week-long Pacific typhoon, the crew of the Navy's *Ramapo* had carefully estimated the height of a wave at 112 feet—the highest wave ever seen. The men of *Sedco 135* sighted a wave in the Pacific estimated at 95 feet, thus proving the *Ramapo*'s wave may have been no exaggeration.

In 1970 University of Miami scientists caught a brotulid 26,132 feet deep in the

Depth record of 300 feet, the limit for compressed air and hard-helmet gear, was set in 1915 by Frank Crilley, above, during salvage of sunken submarine.

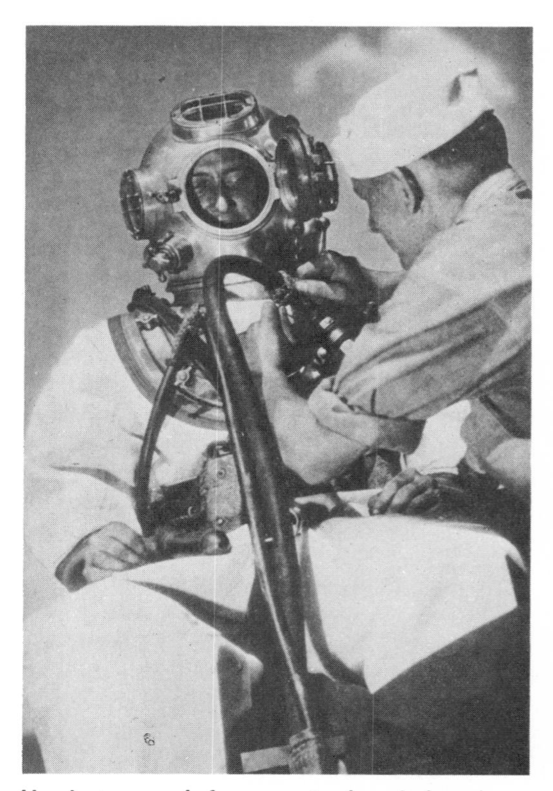

Hardest record for men to break has been that of distance dived into the sea. Divers in heavy bulky gear like standard Navy outfit above have struggled for years to gain inches more of depth.

Scuba gear, above, brought divers freedom of movement, enjoyed here by subsea cameraman reading waterproof exposure meter.

Armored diving suits were tried in 1920s to exceed depth limits of conventional gear.

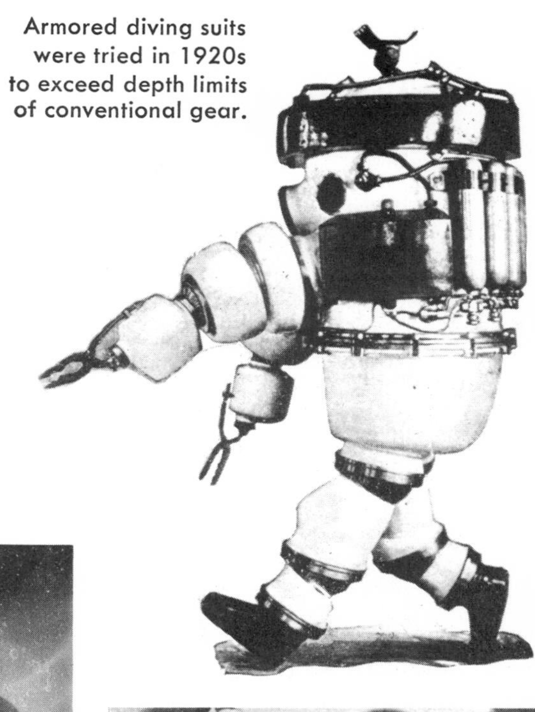

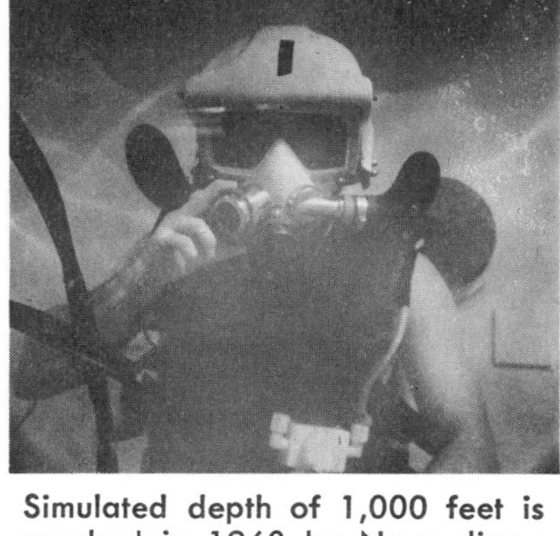

Simulated depth of 1,000 feet is reached in 1968 by Navy divers wearing advanced scuba gear (above) and breathing helium-oxygen mixture, in "wet pot" of dry-land test tank above.

Unarmored divers now use modern deep-sea scuba gear like Navy's Mark VIII at left. Diver breathes helium-oxygen from tanks on back or hosed from a sea-floor shelter.

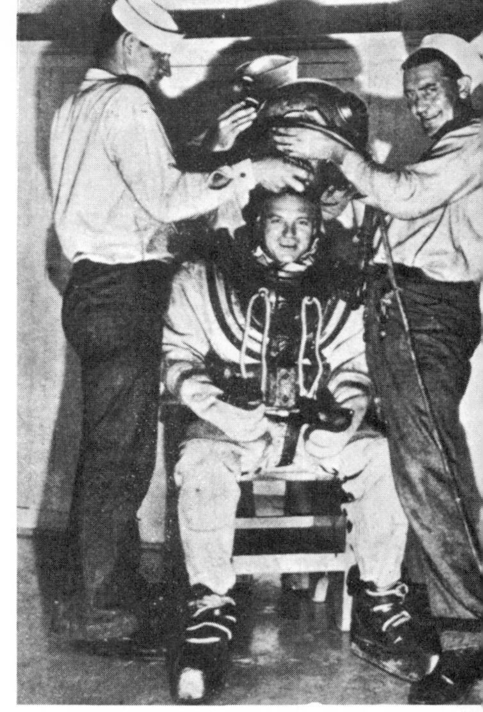

Breakthrough in diving came in 1938 when Navy master diver William Badders (above) and a companion proved divers could breathe helium-oxygen at pressure of 500-foot sea depth.

Navy's success with helium-oxygen mixture for divers climaxed experiments begun in 1920s in tank at left by U. S. Bureau of Mines.

New scuba gear, 1971-model Electrolung, serves to extreme depth of 1,000 feet. Diver rebreathes, purified by chemical.

Jon Lindbergh, noted diver, rides a Perry midget submarine. In a 1965 *Popular Science* story he predicted the dives of more than 600-foot depth that have since come true.

Puerto Rico Trench, the deepest fish ever captured. Beneath its chin, a brotulid has whisker-like barbels to feel for food on the bottom. Miami's 6-incher may have been doing so, as the trawl closed upon it.

A third *Challenger* is at sea today and will be till 1973. She is the *Glomar Challenger* operated by *Glo*bal *Mar*ine, supported by the National Science Foundation and run by a consortium: the Scripps Institution, Woods Hole, Columbia University, and the Universities of Miami and of Washington.

The *Glomar Challenger* drills oil-well-type holes. She has drilled through as much as 20,146 feet—almost 4 miles—of heaving ocean water, and her bits have penetrated as far as 3,320 feet into the sea floor. Her scientists want to know such things as how the earth and the universe were made.

Studying tiny fossils of plankton found in the samples of sediment they bring up, they have learned that the Gulf of Mexico has been the same for 100 million years; the Mediterranean has been sometimes dry, sometimes shallow, sometimes open at Gibraltar and Suez, sometimes not; oldest fossils in the Atlantic, according to John Ewing of Columbia, are 150 to 160 million years old. They suggest that the continents of North America, Europe, and Africa split apart about 180 million years ago and have been moving apart ever since.

In the Atlantic the *Glomar Challenger* has discovered that sediment piling up on the bottom can form undersea mountain ranges; she located an Appalachian-size deposit of sediment. In the Pacific the *Glomar Challenger* has found the oldest sediment farthest west, indicating that the Pacific floor is moving westward. It appears to be oozing into the great trenches and squeezing under the continent of Asia. This may be one cause of the earthquakes and volcano eruptions associated with the trenches, or Ring of Fire, as they are called—the world's earthquake zone.

Though the ship hunts no oil, she has found evidence of it on the floor of the deep ocean. What she learns about how our earth was made is also certain to show something about how minerals are formed and where to look for them.

In 1966 *Popular Science* asked John Steinbeck, the Nobel and Pulitzer prize-winning novelist, to write on the future of

Modern 2-chamber diving bell aids deep-sea operations. Upper chamber holds observers; divers work from lower one.

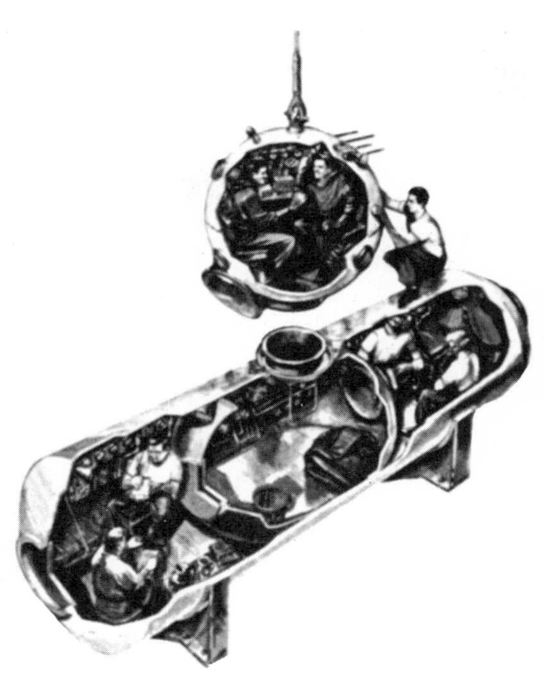

Between dives, tank on ship's deck keeps divers under pressure. They can repeat dives, work on bottom many hours.

A look at the future: men reach the bottom inside diving bells (left) or deep-diving submarines. They can enter undersea shelter as scuba divers or via submarines that attach to hatches of sea-floor hotels or laboratories. The first dry transfer, from a submarine to an underwater shelter beneath Gulf Stream off Florida, has been made.

Another way to keep divers at work: house them in undersea shelters. Picture from 1966 *Popular Science* shows Navy's *Sealab II* being readied for descent. Thirty men lived 15 days in *Sealab* at 205-foot depth.

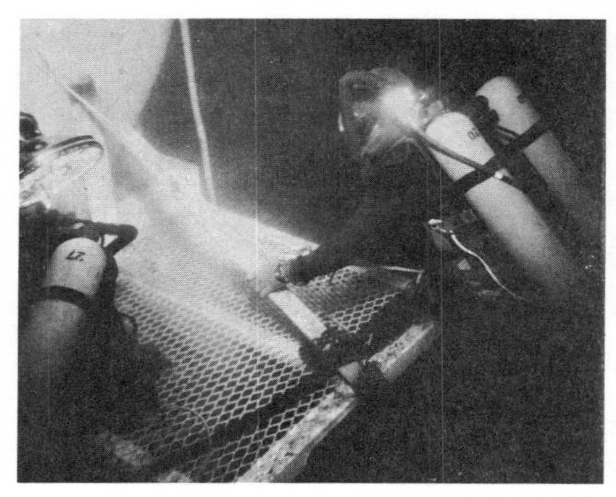

Astronaut turned aquanaut: Scott Carpenter in diving gear outside *Sealab II*. In 1965 he stayed at a depth of 205 feet for 30 days, the world's record for the longest continuous stay undersea.

man and the sea. He did. In the September 1966 issue Steinbeck said: "Three-fifths of the earth's surface is under the seas—but, with the washing down from the continents of minerals and chemicals, it is probable that four-fifths of the world's wealth is there . . . it is unknown, undiscovered, and unclaimed . . ."

The manganese nodules the first *Challenger* found on the deep-sea floor may at last be mined. John Flipse of Deepsea Ventures (Tenneco) has located them 3 miles down (by towed TV camera) in the Pacific. Their manganese (and copper, nickel, silver, iron) assay at high value. Flipse plans to vacuum them up in a giant suction hose and extract the minerals. If he succeeds, he'll be first to mine the sea deeper than a few hundred feet.

"More important in the near future," said Steinbeck, "the plankton, the basic reservoir of the world's food, live in the sea."

The Russians already are collecting for food the krill—small, shrimplike, planktonic animals that feed blue whales in Antarctic seas. Scientists are learning how to raise plankton plants and animals under controlled conditions.

Getting more fresh water from the ocean says Paul Fye, president of the Woods Hole Oceanographic Institution, may be one of the ocean's next great contributions to men. Medications may also be gained from the sea: scientists at the Osborn Laboratories of the New York City Aquarium say several thousand ocean animals may yield helpful drugs.

Wrote Steinbeck: "The planning, computing minds which so gently laid that crazy-looking scarecrow [the unmanned spacecraft *Surveyor*] on the moon could easily design the means, not only for exploring our watery world, but for placing whole producing cities on the sea bottom."

Robots are helping to explore the sea. A

In 1969, a new record: in GE's habitat *Tektite* (pictured before being submerged in Lameshur Bay in Virgin Islands), 4 scientists lived 50 feet beneath the surface for 60 days—and *Popular Science* told "what it feels like to live under water 2 months."

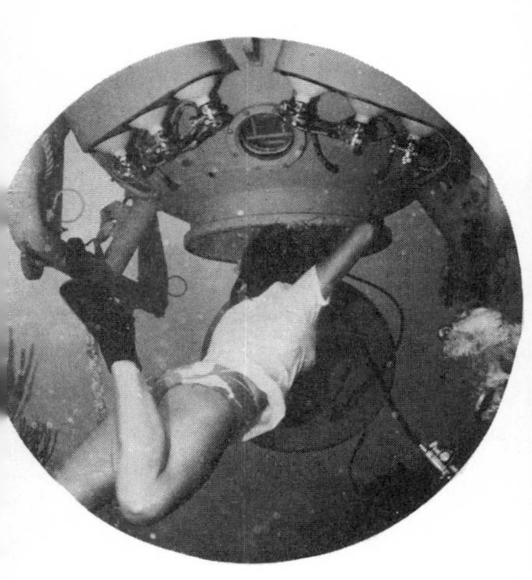

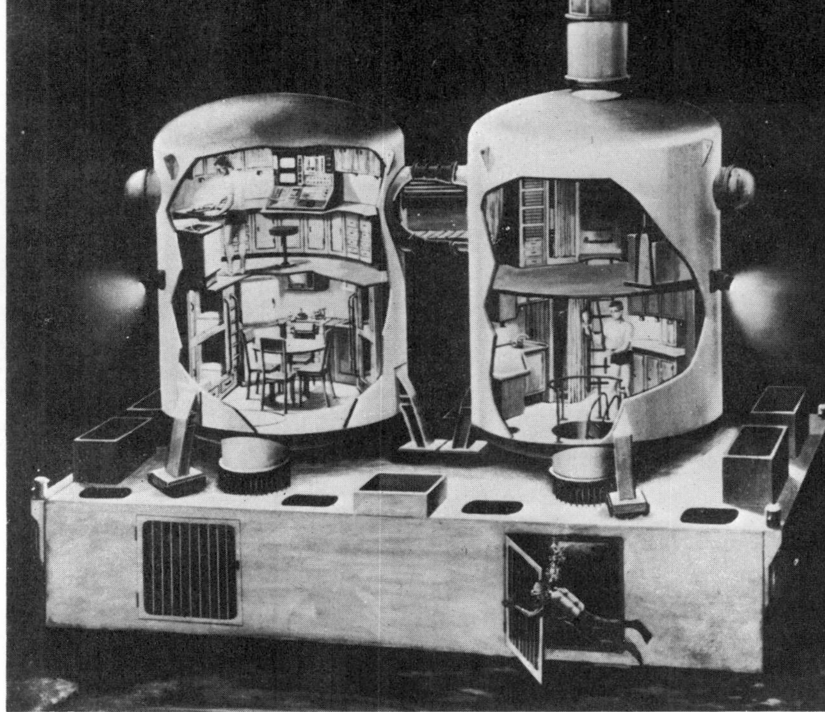

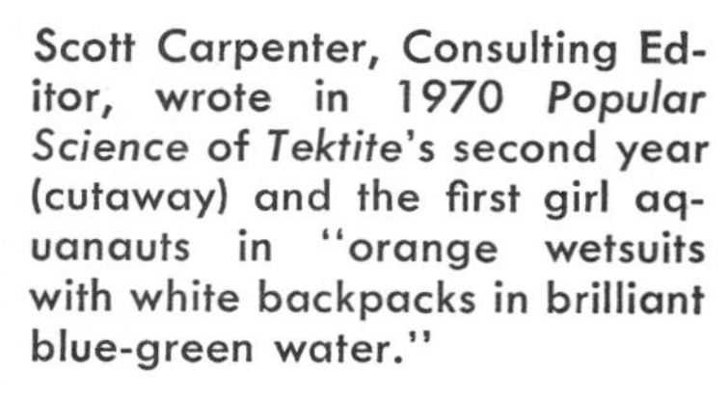

Scott Carpenter, Consulting Editor, wrote in 1970 *Popular Science* of *Tektite*'s second year (cutaway) and the first girl aquanauts in "orange wetsuits with white backpacks in brilliant blue-green water."

camera you heave overboard sinks to the bottom, takes photos, drops weights, rises to the surface, and signals you to come pick it up. In "the untold story" of the fabulous machines that recovered an H-bomb dropped off Spain, *Popular Science* in 1966 described the *CURV:* "a weird unmanned spacecraft of the deep." Remotely controlled, "it could maneuver in any direction, using two horizontal propellers and a vertical one." In 1969 *Popular Science* showed another unmanned jumble of pipes that found the sunken submarine *Scorpion.*

Robots on the sea floor today are watched from shore or surface ship on closed-circuit TV. They include dredges to dig ditches, bulldozers to move undersea earth, and a trencher to dig into solid coral. Manned undersea mechanical equipment includes the *Crawlcutter,* a dredge that moves sand to shore through a 12-inch pipeline; and an underwater pickup truck (its scuba-diver drivers are not enclosed, but are in the sea).

Says Steinbeck: "There is something for everyone in the sea . . . the excitement and danger of exploration for the brave and restless . . ."

About a dozen new deep-diving submarines to provide exploration and excitement are in the sea or about to enter it now. The U.S. has General Dynamics' new *Sea Cliff* and *Turtle;* Westinghouse's *Deepstar 2000.* Already at sea is a nuclear exploring submarine, the NR-1, that can cruise just above the continental shelf for months at a time. Many scientific projects require long periods of watching—something not possible until now on the bottom of the sea.

Built by Lockheed and already in the water are 2 Navy DSRVs—submarines that can dive 3,500 feet and rescue 24 men per trip from a sunken submarine. Scott Carpenter wrote about them in the September 1969 *Popular Science* in "Escape from the

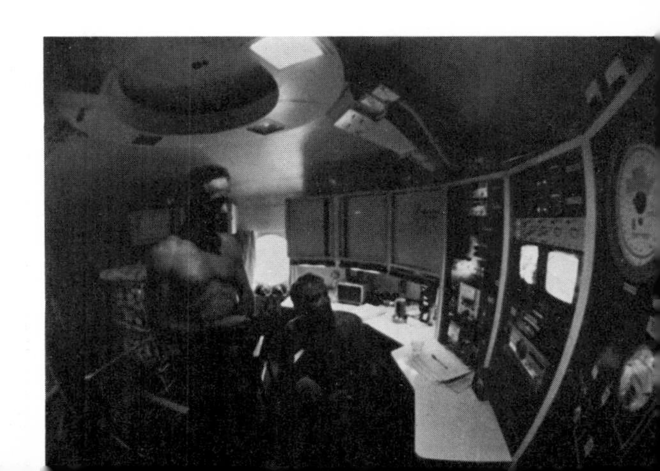

Inside habitat, Carpenter visits John Van Derwalker, director of *Tektite*'s scientific programs, at control and communications center. (No visitors were permitted after the first day of occupation by a crew.)

The oceans hold much promise as new sources of minerals. And off South-West Africa, diamonds up to 4.9 carats (below) have been brought up by the dredge *Seventy-Seven* owned by Sammy Collins of Texas. (Collins lays undersea pipelines.) A flexible hose lowered to the sea floor sucks up a mixture of air-and-water bubbles—and the diamonds.

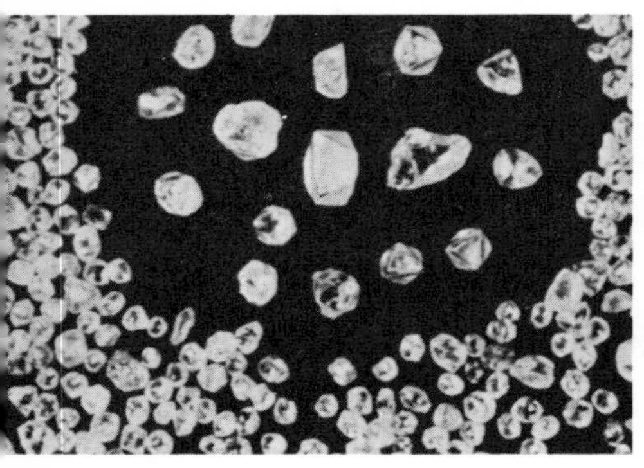

Diver emerges on sea bottom from little yellow submarine *Deep Diver* (left), designed by Edwin A. Link in collaboration with Perry Submarine Builders, and successfully tried out in 1967. Four-man, 22-foot craft was first of modern deep-sea type to put out workers under water and take them aboard again. "Lock-out" submarines that have followed include a new 1971 one designed by Link—the 5-man, 23-foot *Johnson-Sea-Link* (below). Four-inch-thick "bubble" of transparent plastic, forward, gives a pilot and observer a see-all-around view as pictured. Three divers leave and enter via a hatch in a diving compartment aft.

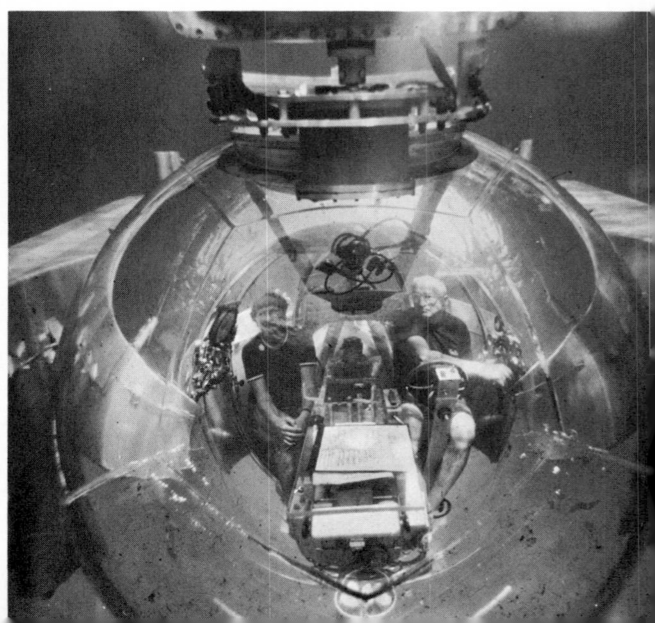

Navy's DSRV (Deep-Submergence Rescue Vehicle) is another new submarine of the seventies. She is flown to area of a submarine disaster. There she rides piggyback aboard nuclear sub to scene, then goes down to sunken submarine and joins hatches with it to rescue crew.

In 1969 Jacques Piccard and 5 others drifted 1,444 miles for 30 days in the submarine *Ben Franklin* (above)—man's longest trip in miles or in time in deep-diving sub. They went from Florida to Cape Cod area. In bathyscaphe *Trieste* in 1960, Piccard and Dan Walsh had made the longest trip to the sea floor—almost 7 miles straight down.

DRAG DREDGE, simplest device proposed for mining nodules, would gather scoopful at a time. It has an open mouth, sides of chains, curved bars at end that swing open for dumping into waiting barge.

Mining manganese nodules on sea floor is foreseen. This dredge would gather scoopful at a time. Bars at near end swing open for dumping into barge.

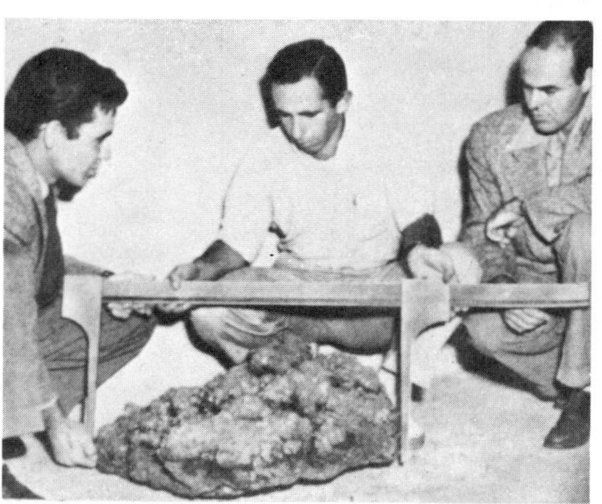

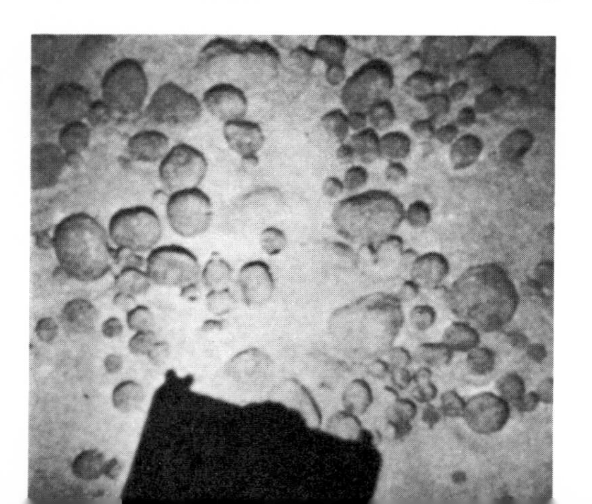

Biggest nodule yet—125 pounds—is seen above with collector Harris Stewart (center) and aides. Usual size is seen below. Engineer John Mero in 1959 *Popular Science* suggested dredging this wealth from sea.

Deep." These are forerunners of planned subs that will dive 20,000 feet deep (almost 4 miles). Such submarines will be able to reach all of the ocean floor except the extreme depths in the great trenches.

Says Steinbeck: ". . . all of these in addition to the pure clean wonder of increasing knowledge."

During the hundred years that *Popular Science* has existed, nine-tenths of man's exploring of the depths has been done, much of it since World War II. The 1970s will be the Internationl Decade of Ocean Exploration. Forty countries are exploring the depths. One important project: producing the first detailed map of the sea floor.

The United States in the 1970s will have unmanned sea-watching satellites, now being developed. With cameras and sensors, they will show such things as the temperature of the sea; where the fish are; rocks and shoals and mud-colored shallow areas; approaching storms and other weather; green, biologically fertile, ocean areas; and the blue Gulf Stream. Navigation satellites will tell an oceanographic ship where she is, to within the distance of her own length.

On the surface, unmanned buoys up to 40 feet in diameter will collect and telecast weather data and water temperature at various depths and wave heights. Advance warnings of hurricanes from buoys and satellites, ships and weather planes, already have massively reduced casualties.

Without change, the words that *Popular Science* applied to the voyage of the first *Challenger* in the 1870s can be applied to the oceanography coming up in the 1970s:

"Results of the highest scientific importance can be expected."

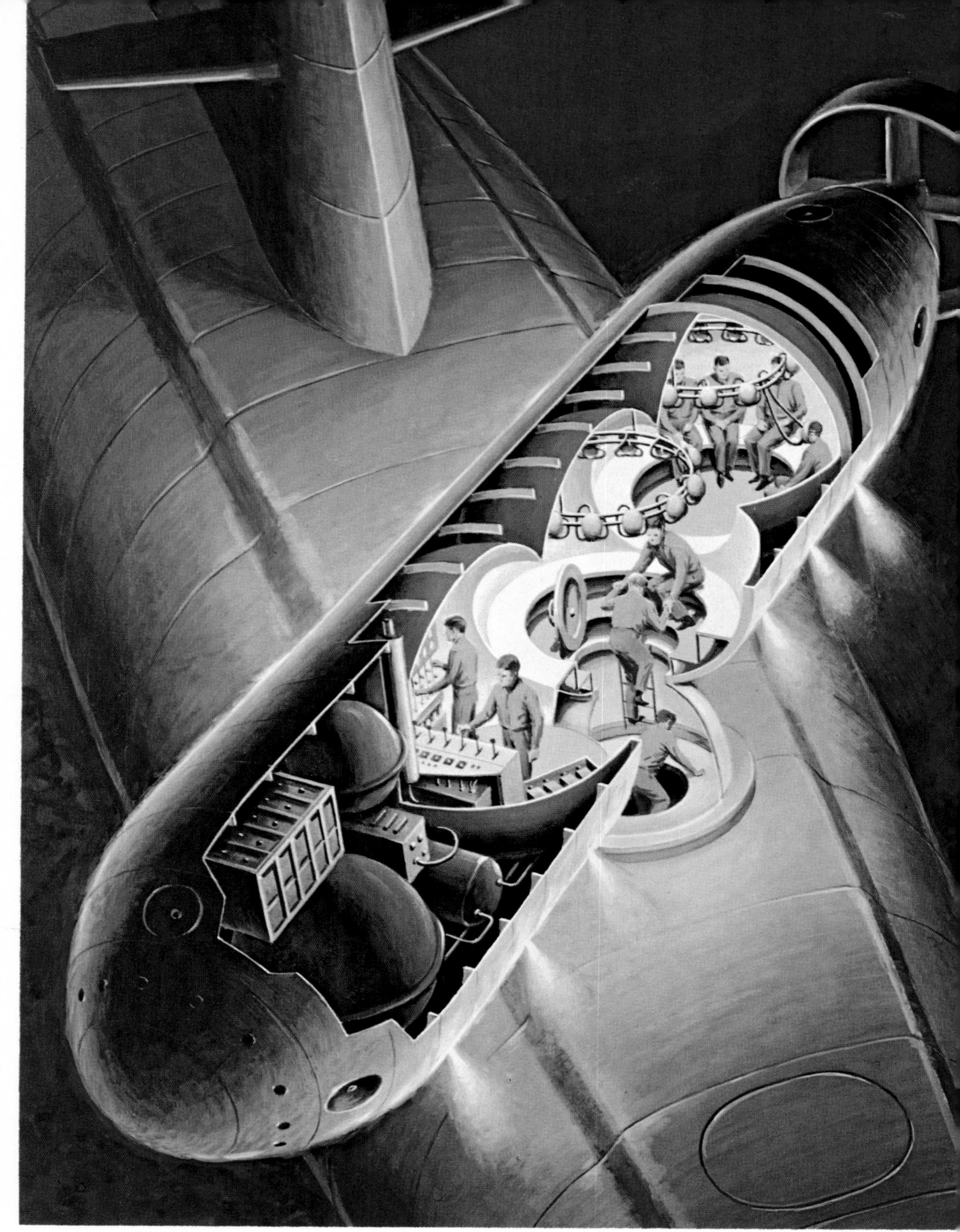

Saving a sunken submarine's crew is the mission of the U. S. Navy's fantastic new Deep Submergence Rescue Vehicle, a 44-foot undersea craft able to dive 3,500 feet. Flown to nearest port, it rides piggyback on a nuclear submarine to vicinity of a stricken one. Three-man crew maneuvers it to dock with the vessel's escape hatch—and the survivors climb aboard as pictured, 24 at a time. First of the type was launched in 1970. Advance 1969 story with this painting began *Popular Science* series on "inner space" by Cmdr. Scott Carpenter, famed aquanaut and astronaut.

PAINTING BY RAY PIOCH

1918: Aerial combat was new when a cover artist portrayed this action scene of World War I.

1922: Fiftieth-anniversary issue featured headphone-days radio—and what Edison saw ahead.

1927: Building model of Columbus' *Santa Maria* was a *Popular Science* do-it-yourself project.

1933: "Flying-wing" airplane looked far out—but in later years some big ones actually flew.

1956: Expected to be the world's first satellite: Vanguard, modeled by magazine. Then came Sputnik.

1957: Transistor radios arrived—for addicts to carry along with them wherever they went.

1960: Revolutionary Wankel engine was introduced to readers in first of many stories on its progress.

1962: A proposed 30-foot space station of inflated rubberized fabric was built by Goodyear.

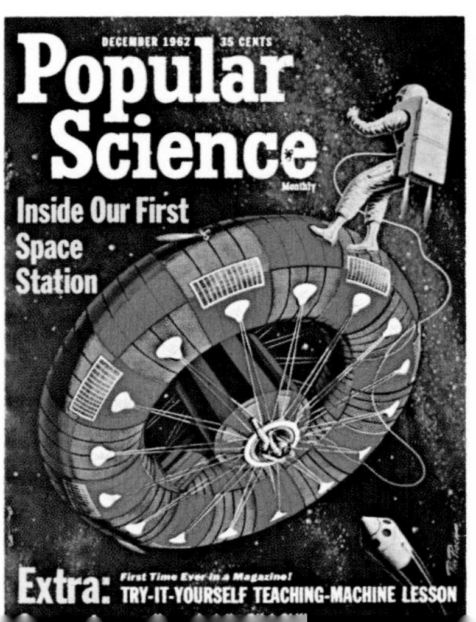

1965: Television sets, like radios, shrank to portable size with the aid of transistors.

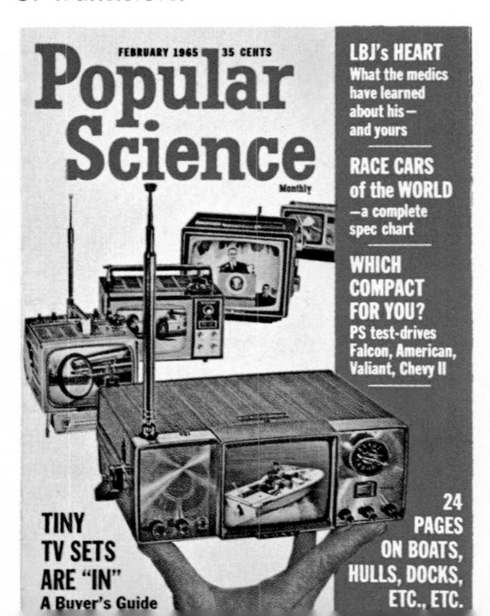

Windows on a Changing World

A selection of POPULAR SCIENCE covers

1928: Big airships looked promising then as passenger airliners—but disasters made them extinct.

1959: Air-cushion vehicles came on the scene—strange new machines that skimmed above land or water.

1967: Submarine *Deep Diver* was first deep-sea one to put out and retrieve divers on ocean floor.

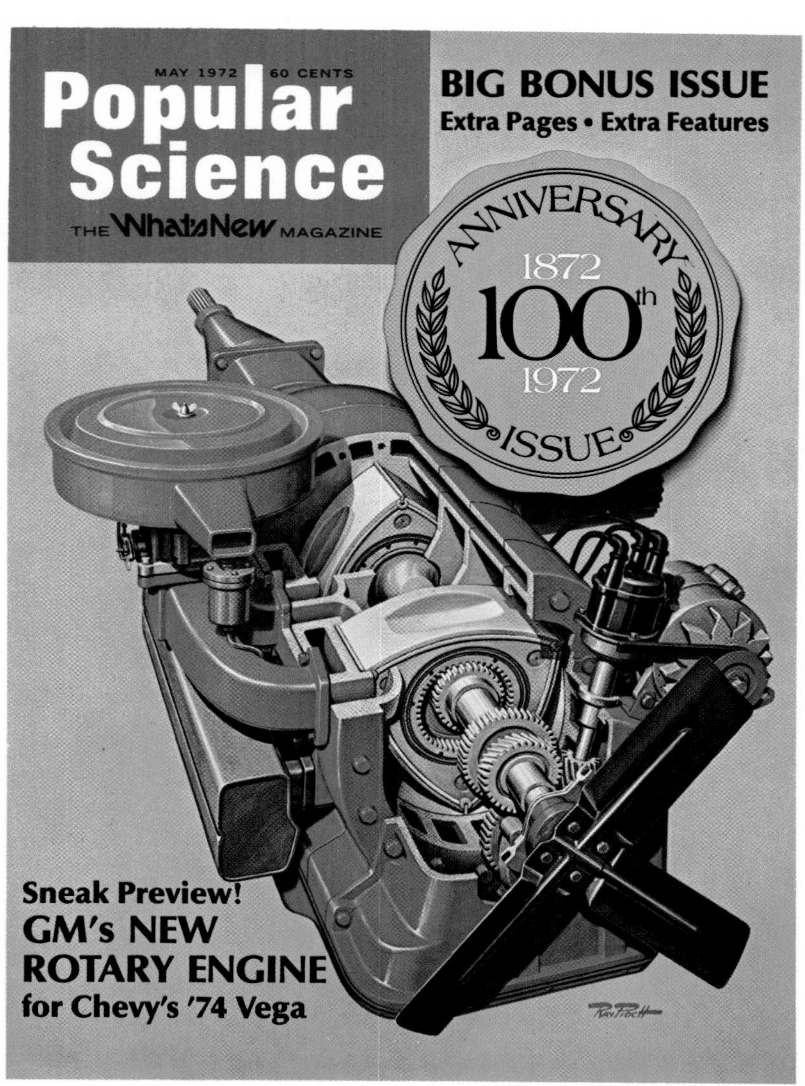

1972: 100th Anniversary cover story previews the first U.S.-built car with a Wankel-type engine: an optional version of Chevrolet's 1974 Vega. Car's debut will cap success story of radical rotary engine first described in magazine twelve years before (1960 cover, opposite page).

Famous cover artist Norman Rockwell, then twenty-seven, painted this do-it-yourself scene for April 1921 *Popular Science* cover. Artist's subject is mending a teapot with a novel 2-fuel soldering torch (described in magazine item on a following page), while kitty-cat superintends operation.

CHAPTER 13

A Century of Do-It-Yourself

PAPA'S in the barn putting a new wheel on the buggy, Sonny's currycombing the mare, Mama is baking bread, and the young ones are weeding the lettuce and picking bugs off the potato plants. . . .

Do-it-yourself was indeed the way of life a century ago.

At the time *Popular Science* was founded, this activity took in all the daily chores required to keep a household running and the homestead spruced up. The whole family turned to, if it had no servants. The U.S. was mainly agricultural. The horse carried or hauled all who didn't walk. The milk supply often came from the family cow—even in villages. Behind the picket fences, gardens grew much of the family food, and a hen coop furnished fresh eggs and mouth-watering fried chicken.

Many of the tools required to handle all the chores are now all but forgotten. Out in the barn or stable, you found a wagon or buggy jack, a wheel-nut wrench, a leather punch and brass rivets for keeping harness in repair; and for yard and garden, a dung fork, a scythe, and sickle. There'd be an iron crowbar, a "twist-stick" bucksaw, and maybe a cant hook for rolling over logs felled for firewood. A hatchet, a pond-ice saw, and even a shoe last were commonplace.

The housewife, busy from before dawn until after dusk, had a butter churn, washboard and tubs, flatirons, dough kneader, rug beater, treadle sewing machine, and the Sunday ice-cream freezer.

Leafing through old catalogues, seeing everyday needs, you sense the atmosphere of that long-gone era. You find a surprising number of saws: bone-cutting saws for cutting up meat bought by the side, pond-ice cutters, and a variety of saws for cutting up wood for building or warming homes.

A $7.50 sawing machine promised great things. The man who bought this stood up, working a lever up and down like a pump handle to work a linkage that pulled a 1-man blade back and forth across a log on the ground. Said the 1897 advertisement: "The uncomfortable bent position when sawing in the usual way is overcome and a natural upright position secured, enabling the full force and weight of the body to be thrown on the saw."

A little oddity of the day was a hand saw only 12 inches long called a "gentleman's panel saw." Presumably this shorty was for occasional repair jobs made by the elite!

In his 1879 catalogue, Henry Disston reported "over one hundred hands [employees] daily employed in filing saws, each hand wearing out from six to twelve files per day." For just this reason, Disston began to make its own files and in the process managed to improve the breed.

Many of the tools in the old catalogues had a long history even then.

Both the Romans and the Egyptians used double saws tightened with a cord. These were the ancestors of the twist-stick bucksaw widely used a century ago for cutting firewood. It had an H-frame with a metal blade across the bottom and a tightening connection across the top. In early versions, the tightener was often just a doubled cord with a stick through it, used tourniquet fashion. Hence, the name.

Handles for the finest hand saws a century ago were carved from applewood.

Marking gauges then were prized tools. Beechwood, applewood, rosewood, and boxwood were found in gauges of ascending fineness (and price) in that order. A marking gauge made of temperature-stable boxwood with brass parts was top of the line in 1879.

Circular wood saws, invented in England, were common by 1872. They were turned by steam, water, or horsepower. In 1879 you could also buy dished circular saws. These were of special interest to sawyers of chair or wheelwright lumber and makers of barrels and kegs. A 20-inch concave saw blade was priced at $11.25.

The pit saws that first converted trees into lumber ranged in length from 5 to 8 feet. Handles were sold separately—a down-curving tiller on one end, a straight-across handle on the other. It took 2 men to use a pit saw, one man in the pit that the huge saw log spanned and one dancing along the log above. In the basement of any house at least a hundred years old, if the undersides of the first-floor floorboards are bare, you can see the straight-across marks left by the pit saws that cut them. A circular saw leaves a curving mark.

Fencing of the wide open West was just beginning and the Samson tree planter and post-hole digger, patented in 1870, was probably a big seller. This had 2 spade blades, facing each other, on long handles. A pivot handle brought the blades together to lift dirt from the hole. Fence builders still use a tool just like this.

Turn-of-the-century catalogues listed sets of molding planes for cutting wood to all kinds of shapes, but Stanley had already brought out its "Patent Universal Plane" to take the place of the set. Jack planes, once called jackass planes because of their workhorse use in smoothing floorboards, were made in both metal and wood. You could also get Bailey's block plane with the new adjustable mouth. A block plane has the plane iron mounted bevel up for end-grain planing, which once was commonly termed "blocking in."

Farmers and gardeners were already facing a pest problem. From the 1897 Sears catalogue you could buy a wheeled sprinkler that applied the pest killer you favored "on two rows at once." Or a Paris-green sprinkler—"everyone who raises potatoes should have one," said the catalogue.

By 1902 tool sets sold at bargain prices. One catalogue's do-it-yourself advice was to use "our $5.28 Wood Butcher's Set" to "do your own carpenter work and save five times the cost of this outfit by keeping your property in perfect order." The set included Stanley's folding rule, a block plane, and a try square with a rosewood handle.

Modern do-it-yourselfers could pick a lot of useful tools from the antique catalogues—an alligator wrench with wide-opening jaws like those of its namesake, a monkey wrench, a solid adjustable-jaw wrench, a hand-crank drill, a breast drill, and an automatic push drill.

They could also have bought what sounds like a useful attachment for a bit brace—a 45-cent washer cutter that cut "any size washer with any size hole up to 5½-inch diameter." Apparently that product didn't make it into our times.

Also listed were "drive screws," said to be "preferred by some mechanics as they can be driven with a hammer without breaking the hold in the wood." Many old-time carpenters even today commonly resort to a hammer instead of a screwdriver to set screws in heavy work.

The twentieth century brought many new wonders and conveniences—and changed home chores. The hours once spent washing the buggy and greasing its wheels were given over to the new automobile. Owners of a "machine" had to learn new ways—changing a tire, adjust ing the carburetor, cranking the engine. But many an old device was still in use.

On farms, horse power remained supreme. You could buy one contraption for running, say, a grain grinder. A horse walked around a pole to turn it and transmit power, through gears overhead, to a fast-turning drive wheel you could belt to all sorts of machines. Or you could buy a horse-driven treadmill, an endless belt that revolved a drive shaft. With a third humdinger, a team of horses walked in a circle, turning a set of gears to saw wood or make cider.

Many tools were foot-powered. Sears' 1902 catalogue showed a foot-power bracket saw with a tilting table for making angle cuts with the thin reciprocating blade.

The first electric drills were developed in Europe early in the twentieth century.

This Soldering-Torch Uses Two Fuels

THE steady hand of a good mechanic is needed to apply solder. Many people find this out after they make several attempts to mend the teapot or the wash-boiler. It is indeed aggravating to see the solder roll off in small balls as fast as it is applied. After a little experience, the would-be mechanic generally ends up by taking it for granted that soldering is a difficult job that only a mechanic can accomplish.

An inventor has done a great service and filled a long-felt want by developing a blow-torch that any one can use in soldering. No soldering copper is needed. The most inexperienced person can make solder stick with this little torch. The flame is applied directly to the piece to be soldered. When it has reached the proper temperature—hot enough to melt the solder—the solder is brought in contact with it.

Two separate compartments make up the torch. They contain different fuels. One fuel feeds the burner while the inflammable vapor of the other is blown into the flame. This is done with a little hose that the user blows through. A very hot flame is produced. The flame is hot enough to bring metal to a red heat. The little torch may be used for light brazing as well as soldering.

Many household objects may be rescued from the family scrap-heap by the use of the little soldering-torch. Mother's favorite kettle may be saved, also umbrellas and saucepans.

Anybody can use this blow-torch to mend the family teapot or a favorite box; no soldering copper is needed

Do-it-yourselfers of half a century ago would have found it quite natural to buy this blow-torch to mend the pots and pans. Salvage kept the trash heap to a minimum. Note attempt in article reproduced above to make a difficult job seem easy. New blowtorch was basis for the 1921 Norman Rockwell painting at opening of this chapter.

They were unwieldy affairs. It took 2 ingenious young Yankees to set the stage for the drill to become the most popular power tool in history by adding essential improvements—the pistol grip and the trigger switch. They were Black and Decker. They filed claims for the added features in 1914 and obtained the patents in 1917. In 1961 they introduced a cordless drill. Nickel-cadmium batteries powered it. Variable-speed and reversing switches brought further advantages to the tool.

Spade bits and brad-point auger bits were especially developed for electric drills. Shockproof insulation made power tools safer. Unbreakable plastic casings and tungsten-carbide tool tips made tools more durable.

Do-it-yourselfers flocked to buy portable electric saws, too. *Popular Science* gave its readers their first look at one in April, 1925. An early model produced by Michel, forerunner of the Skil Corporation, it's shown here on page 264. Skil says it was originally intended for cutting sugar cane. But the saw's principal use soon changed to wood cutting.

Since a primary goal of do-it-yourselfers is to save money, making one tool do several jobs was bound to have strong appeal. As early as 1902 Sears was selling a treadle-powered wood lathe with a scroll-saw attachment. Delta's Handi-Shop, one of the first powered multipurpose tools, combined a lathe, a scroll saw, and a circular saw. It came along in the twenties and helped Delta pay dividends during the Depression. Many buyers used the scroll saw to make the jigsaw puzzles then popular and salable.

In 1947 appeared Shopsmith, a combination tool invented by an industrial engineer, Dr. Hans Goldschmidt, who had fled Germany in 1937. In 1950, *Popular Science* reported that some 50,000 had been sold in a little over two years to the tune of $10,000,000. The basic tool sold for $189.50; a heavy-duty motor for $34.50. The outfit provided 5 tools: a circular saw, a disk sander, a wood lathe, a vertical drill press, and a horizontal drill press.

The years between the 2 great wars brought improvements in many old tools and development of several new ones.

Tilting tables had been common for years for making angle cuts with shop saws. A tilting arbor, for tilting the circular-saw blade as it projected through a slot in the table, added versatility and convenience to table saws.

What a home workshop might have looked like in 1872 is shown above in this woodcut from a book published that year. Actually, they're apprentices working in cabinet shop.

Handsaw of many uses, this 1874 Disston has a 24-inch square, rule and straightedge, scratch awl, and plumb-and-level device.

Adjustable clamps like this were among the tools owned by artisans in the early days.

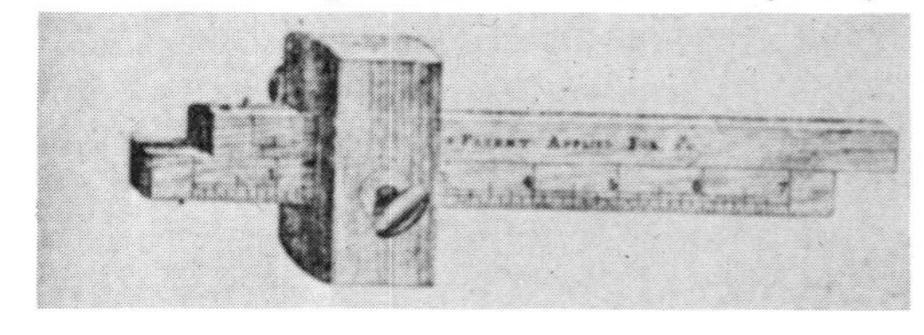

Stanley's combination marking and mortise gauge of boxwood had brass-faced bars and head, brass thumbscrews. It cost $1.50.

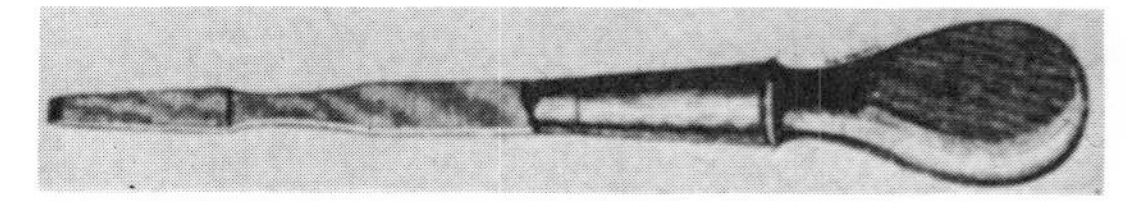

1870s' style for a screwdriver is shown by this illustration in old Stanley catalogue.

Foot-operated scroll saw first maae by W. F. and J. Barnes Co., Rockford, Ill., in 1872. Pedal power was common for shop tools.

Foot-power wood lathe in late nineteenth century turned spindles and rungs for furniture.

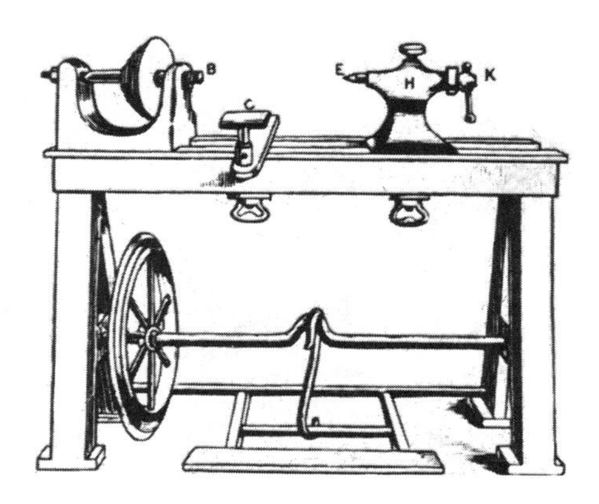

Home workshoppers these days can choose between a table saw and a radial-arm saw as their basic wood-sawing tool. Raymond E. DeWalt invented the radial arm in 1922. It evolved from the need to turn out more work at his plant without increasing the payroll.

"I rigged up a yoke," DeWalt said, "and hitched it directly to a motor and saw. Then I mounted it on a standard and arm so the saw could be raised, lowered, slid back and forth, or tilted at any angle; in fact, could do everything but a handspring."

The saw cut costs beyond DeWalt's fondest dreams. His shop caught up and took on a lot more work. Requests to buy his saw came so fast, DeWalt resigned his job and organized a company in 1923—now Black & Decker's DeWalt Division.

The useful belt sander invented by Arthur N. Emmons was developed for industrial metal polishing and patented in 1928. It took only a few years for Emmons, and others, to realize that the belt sander was a good one for wood-smoothing, too.

Old-time woodworkers spent hours pushing around a sanding block for fine finishing. Power to speed up sanding now leaves no excuse for skimping on the job. You can choose from straight-line reciprocating sanders, orbital sanders, a combination of both, belt sanders (portable and stationary), disk sanders mounted on a bench arbor or a drill, and drum sanders.

Man-made abrasives for use with the new power sanders have all but outmoded the old flint and garnet sandpapers. Inventors went nature one better by giving the grains (aluminum oxide or silicon carbide) a longish-shape—then making each grain stand on end to cut better. Applying an electrical charge causes the grains to repel each other and stand upright while they are bonded to the backing sheet.

Portable electric routers grew out of a makeshift machine created by R. L. Carter to solve a problem during World War I. Carter's pattern works had an order for a pattern and 16 core boxes for a large sectional boiler. Specifications called for radii to be cut in the core box, all with matching edges. Somehow the radii were overlooked and the casting was not accepted. The radii had to be put on the core boxes. How to do it?

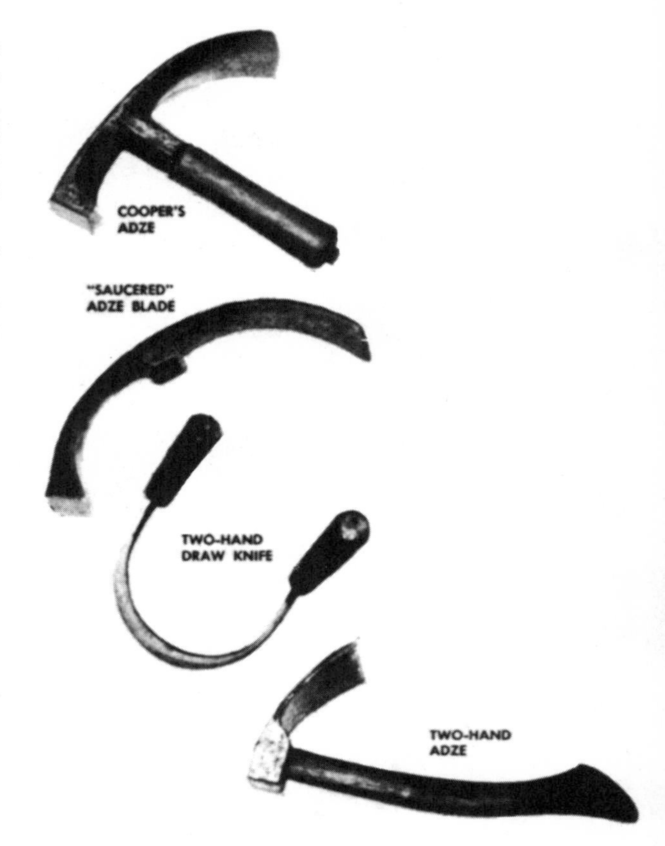

Before age of motor-driven shop tools, early woodworkers painstakingly shaped wood with primitive hand tools. Cabinetmaker apprentices had to master each of them. Adze at top was important to barrelmakers.

The cut made so easily nowadays with machine power was a job for a plow plane such as this a century ago. The antique tool still cuts a clean dado. Wooden screws are of boxwood; fence and blade are adjustable.

Developing art of making holes is chronicled on these pages. Woodworking shop's spiral augers once included this 3½-inch monster.

Pushing on handle of early push drill revolved chuck and bit at high speed.

Electric drill arrives! In 1917 *Popular Science* carried this photo of "shooting a hole with an electrically-operated pistol-shaped drill," and close-up of tool, from Black & Decker.

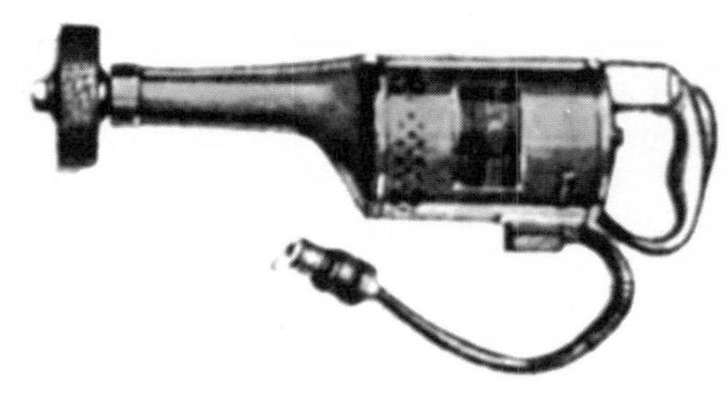

Accessories for electric drill appeared soon. This 1921 photo showed a grinder on one. Caption noted that drill stopped when you relaxed your grip on the switch in the handle.

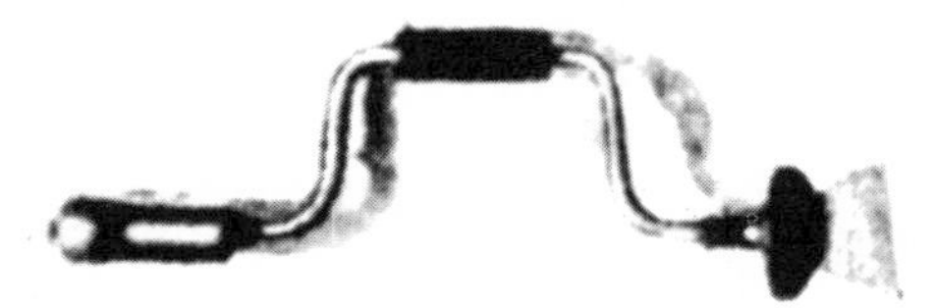

Hand-power hole maker that never wears out is a brace and bit. The owner of this one said in a 1923 advertisement that he had been using it happily since 1882.

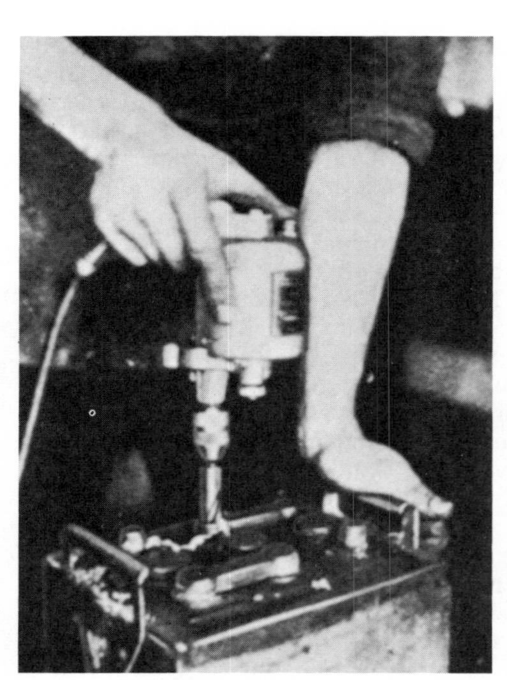

New one-hand drill was what this tool was called when the picture appeared in 1926.

Hole saw for an electric drill, still a good tool today, made news in 1927.

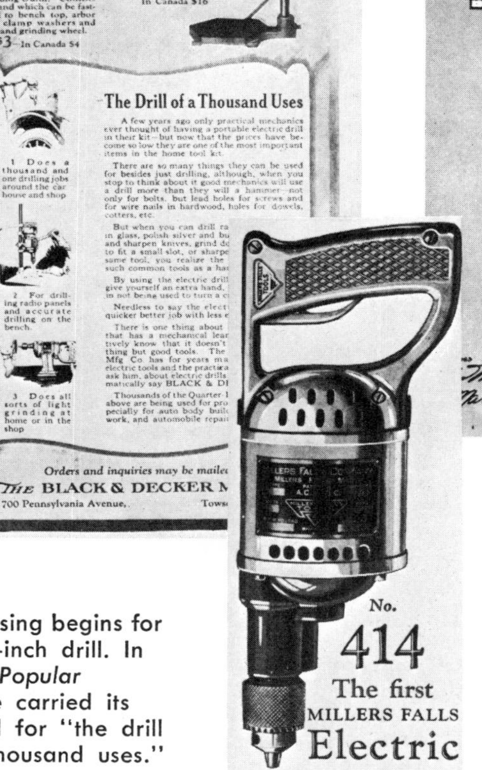

Advertising begins for the ¼-inch drill. In 1924, *Popular Science* carried its first ad for "the drill of a thousand uses." That boast soon proved no exaggeration.

No.
414
The first MILLERS FALLS Electric Drill

FOR fifty years Millers Falls Company have made hand, breast, chain and bench drills. To take up the manufacture of portable electric drills was a natural step forward. It depended on one thing—the development of an electric tool of true Millers Falls quality. Here it is—the first number of a series. Look for it in hardware and automobile supply stores. It's a fine tool—what more can we say than that it *belongs* to the Millers Falls group?

MILLERS FALLS COMPANY
Millers Falls, Mass.
28 Warren Street, New York — 9 So. Clinton Street, Chicago

"The first number of a series" of portable drills, Millers Falls called this one. The advertisement appeared in 1926. Stanley advertised its first ½-inch drill in 1928. Universal motors now ran all drills, furthering tool's popularity.

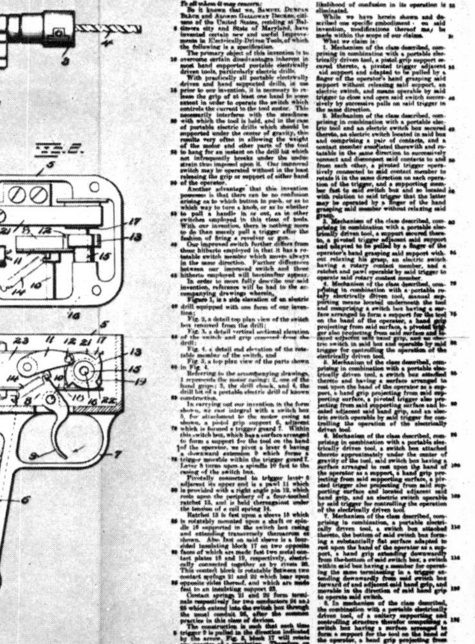

Reproduction of the original patent for the pistol grip and trigger switch drill.
Perhaps the most significant patent in Black & Decker history. The principles embodied in this patent are used universally today.

UNITED STATES PATENT OFFICE.

Patented Nov. 6, 1917.

BLACK & DECKER

Original patent granted in 1917 to Black and Decker for 2 electric-drill innovations—pistol grip and trigger switch. These 2 concepts and universal motor made tool popular with do-it-yourselfers of the day.

This heavily engraved drill was a memento of the twentieth anniversary of Portable Electric Tools in 1966. No shooting iron seen in the old West was more handsomely decorated than this creation.

Twist-stick bucksaw, oldest wood-cutting tool on these pages, was used for cutting firewood. Tourniquet at top tightened blade at bottom.

Radial-arm saw was invented in 1922 by Raymond E. DeWalt. It was the direct result of a need to turn out more work in his shop.

Early jigsaw was turned by hand crank on the large pulley wheel.

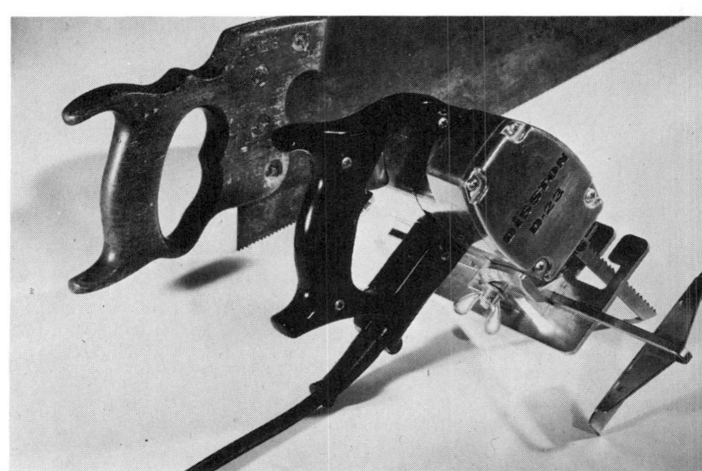

Saber saw at right above, a 1958 Disston, had same handle as original 1850 handsaw at left. Early chain saw seen in 1932 photo, below, rolled to the job on pneumatic tires.

Electric handsaw was first shown in April 1925. It was made by the Michel Company.

Chain saw was lumberman's tool when this ad came to *Popular Science* in 1948. But suburbanites and vacation-home owners soon found it handy.

Weight had dropped to less than 25 pounds by 1949. Since then McCulloch and Homelite have competed to make the tool still lighter.

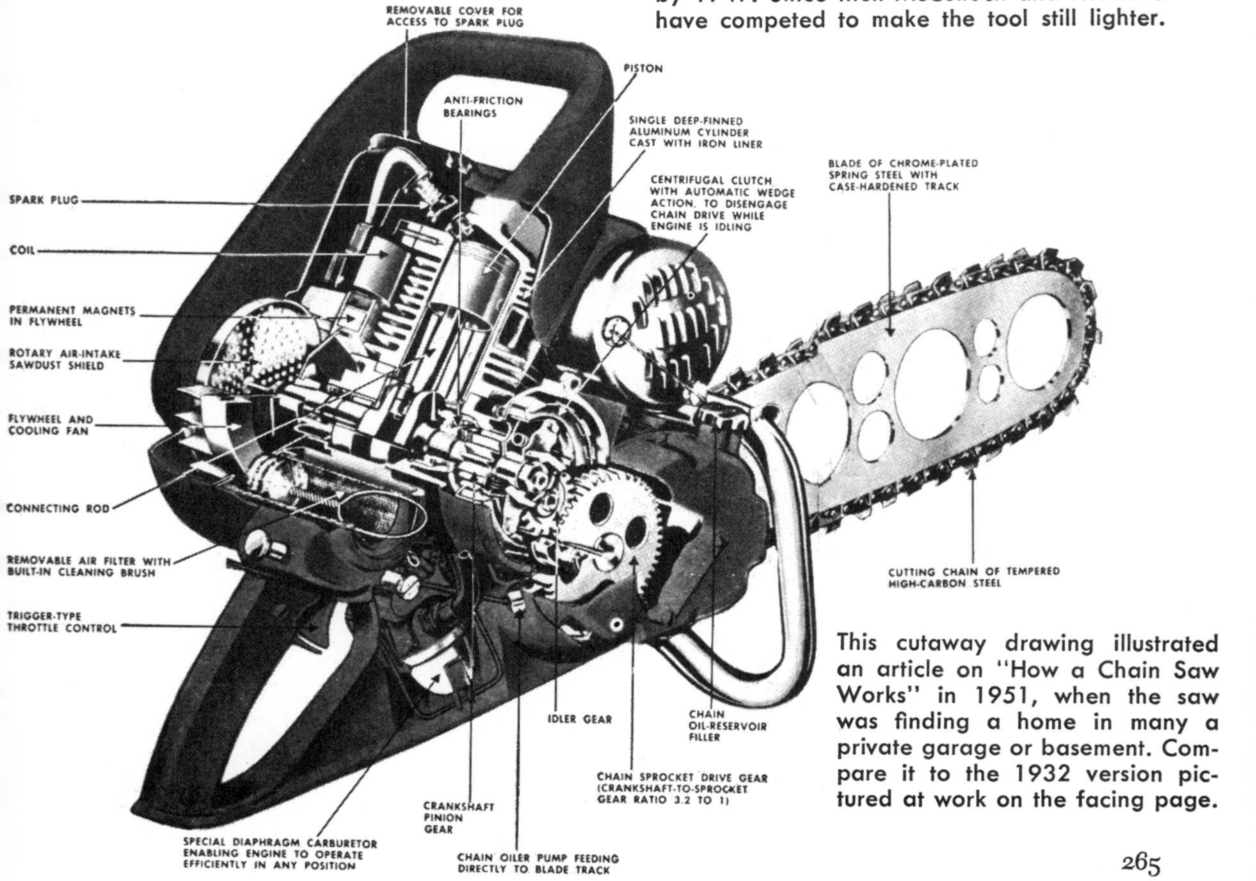

This cutaway drawing illustrated an article on "How a Chain Saw Works" in 1951, when the saw was finding a home in many a private garage or basement. Compare it to the 1932 version pictured at work on the facing page.

With a spokeshave it would mean long tedious hand-carving. Instead, Carter took the worm gear from the motor of a barber's electric clipper and ground a radius into it. To guide his newborn cutter for an accurate cut, he made 2 bevel guides, put them on a shaft with the cutter between them, and plugged in the motor. Carter and Junius A. Yates, a nephew, went to work with the improvised hand shaper and did the job on the 16 boxes in about 2 hours.

More than a year later, Yates visited a local cabinetmaker and found him laboriously cutting curves on walnut frames for the ornate sofas popular then. He showed the cabinetmaker how easily the curves could be shaped with the still unheard-of electric tool. The cabinetmaker immediately wanted to buy it.

After the armistice, Carter and Yates went into business making the new tools. Ten years later, the R. L. Carter Company was doing fine—more than 100,000 of the hand shapers were in use.

To handle another problem, Carter conceived the idea of mounting the shaper in a base and giving it a chuck to hold a cutting tool, such as a rotary file. Using the shaper motor, he designed a chuck to take cutter bits with quarter-inch shanks. Thus, the router was created. Stanley acquired Carter's business in 1929 and introduced the tool to do-it-yourselfers.

A completely new tool of the 1940s, the Weller soldering gun has been one of the most important of the electronics age. Radio and television workers took to it in great numbers. Builders of electronics kits agree that they could not have done without it. A *Popular Science* article gave this account of how Carl Weller invented it:

Back in the thirties, Weller was a radio repairman. His constantly used soldering iron was a time waster. In the shop, it was kept hot all day to avoid waiting when a solder joint had to be made. This meant frequent redressing or replacement of the tip. On house calls, you twiddled your thumbs while you waited for it to heat; and when you finished the job, it was too hot to put in the toolbox.

The trouble was, current flowing through a coil of resistance wire had to heat the wire first; then the hot wires had to heat the copper tip by thermal conduction. A lot of heat got soaked up before the tip reached soldering temperature. More than one would-be inventor asked the right question: Why not use the electric current to heat the tip directly? But Weller got to the Patent Office first with the answer on *how*.

His knack for ignoring traditional solutions paid off. Instead of starting from the assumption that resistance wire was necessary to convert electricity into heat, he took off from the premise that a good heat conductor was necessary for the soldering tip. This meant using copper.

But no engineer would consider connecting a chunk of copper across a power line; it would be a dead short. The copper would get hot all right, but so would the connecting wires—hot enough to melt. Two problems had to be solved: how to concentrate heat at the soldering tip and how to limit current flow to practical value.

The answers were in the textbooks. The resistance of copper is quite low, but it does have some. And its resistance increases with temperature. If a circuit is entirely copper wire of uniform size, the whole length of the wire will be uniformly hot. But if part of the circuit is much smaller in cross section, that part will have more resistance and will start to heat faster. As it heats, resistance goes up still more and it heats still faster. The heat, then, tends to concentrate at that point. These facts gave Weller his clue.

Limiting the current flow was simply a matter of applying Ohm's law. If you are committed to a circuit of low resistance, then you have to use a low voltage. A step-down transformer, converting 110-volt power to much lower voltage than had been considered before, was the answer. Weller patented a pistol-shaped soldering gun using these ideas in 1941.

The Moto-Tool—a little grinder that puts power at your fingertips—was invented in the thirties by Albert J. Dremel. Its busy little spindle runs at 27,000 rpm. Fit it with one of its almost endless variety of bits, cutters, and abrasive wheels, and it will tackle a drilling, cutting, smoothing, polishing, or machining job on any material. In 1967 the little electric handful was fitted with a constant-torque, permanent-magnet motor. A *Popular Science* writer reported that trying it out was "like finding that the house cat has turned into a tiger cub." He had to work hard to stall it. Dremel fol-

Designer of *Popular Science* ship models, Capt. Armitage McCann puts finishing touches on his *Santa Maria,* flagship of Columbus (starboard side is seen at right). Building instructions were so detailed, they ran 3 months in successive issues.

Four-foot model of Savannah, first nuclear merchant ship, was designed and built by Henry B. Comstock, former staff member of *Popular Science*. Floyd McGuckin built steam turbine; Howard McEntee provided radio control. Project appeared in 1960.

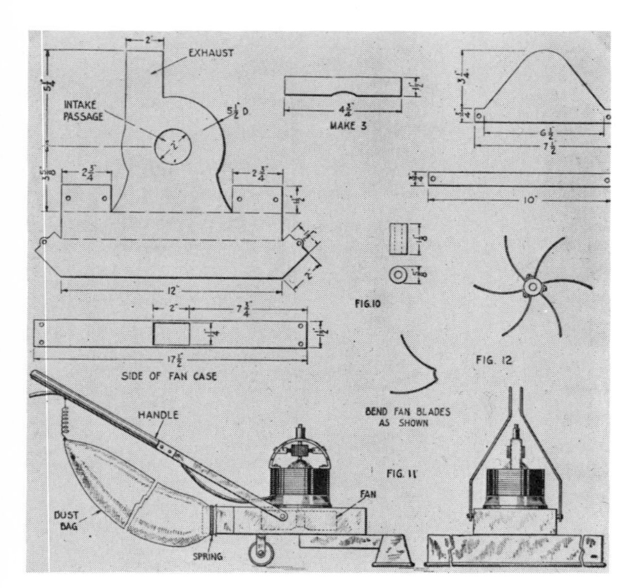

Make a home vacuum cleaner? Sketches above illustrated a how-to article way back in October 1917. There were even instructions for winding your own motor. Make-it-from-scratch was the rule in those days.

Plans for building an air-driven sled in 1921 foreshadowed snowmobiles, still a quarter century in the future. Air prop drove sled.

Electric hot foot for yowling cats was presented in 1921. Two bare wires ran along backyard fence from secondary of a Ford spark coil. Touching both induced urge to go.

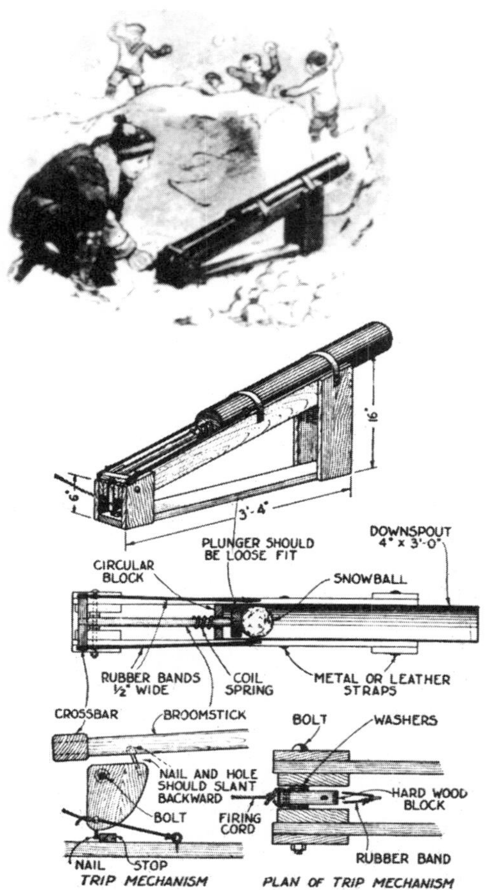

Cannon for shooting snowballs was detailed in February 1929. A coil spring over a broomstick shot balls over walls of snow fort.

Anton Flettner's rotor yacht was scaled down for modelmakers in 1928 drawings below.

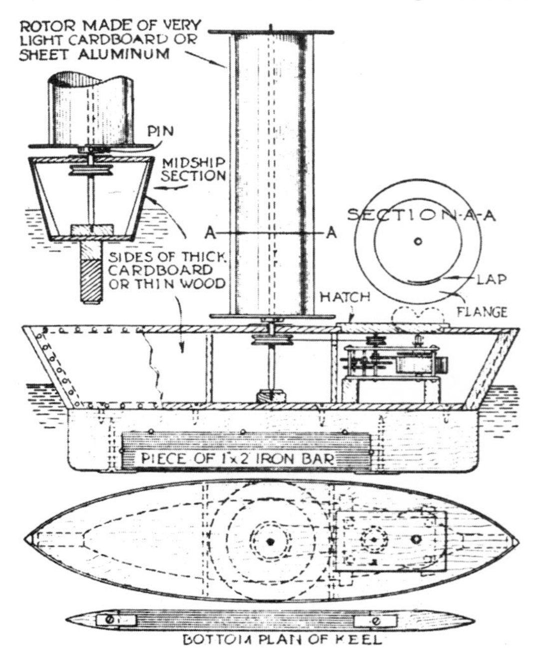

lowed this up with a new accessory that turns the hand grinder into a miniature router-shaper that can do surprising things for a woodworker.

One of the few non-electric hand tools invented in the twentieth century was described in the September 1956 *Popular Science.* Actually, it was a brand-new family of tools: Stanley's remarkable Surform tools. A perforated steel strip—the blade—fits a rasplike tool or a special block plane. The blade shaves, shapes, and smooths wood, plastic, metal—anything you want up to the hardness of mild steel. You can use it to plane a board, trim-fit an asphalt tile, smooth auto-body filler, shave an aluminum storm window to size, or plane the bottom of a door without splintering the grain at the ends.

Hundreds of tiny sharp knives "skim the surface like a food grater." The cutting is done in small bites, so you get a smooth, even finish. The blade doesn't tilt, skip, gouge, or dig in because the teeth dip only slightly below the surrounding metal, which rides on the surface to control depth. Soft materials that clog ordinary shaping tools are easy to work—the shavings pass out through the blade's perforations. The blade is held in the metal frame of either tool like a jigsaw blade. When one becomes dull, you just insert a new one.

Over the next decade, Stanley improved the tool. Blades now come in both flat and half-round forms and there are round-file holders and electric-drill drum holders.

Surform tools brought a brand-new way of shaping materials and they turned up in many places besides home workshops. They have been used by dentists to shape plaster molds of teeth, by craftsmen who make marionettes, and even in one circus—to manicure elephants' hoofs. One workshopper reported he uses his Surform to grate Parmesan cheese for spaghetti dinners! "Perhaps this is why Stanley has come up with their slogan ('it shaves everything but your beard')," speculated *Popular Science.*

An entirely new type of power saw showed up in the mid-fifties. Some makers called it a portable jigsaw; others, a portable saber saw, because its short blades resembled those in a bench-type saber saw.

The saw originated in Switzerland in the late forties, a product of Scintilla S.A. Later the Forsberg Manufacturing Com-

Do-it-yourself TV set was a popular project in January 1929. You can still build this antique from the wiring diagram below.

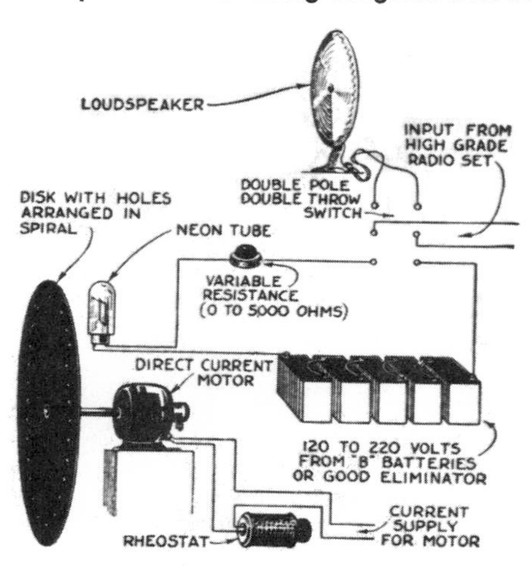

Home-built ruby laser (below) was detailed in November 1964, only four years after the first successful laser was reported.

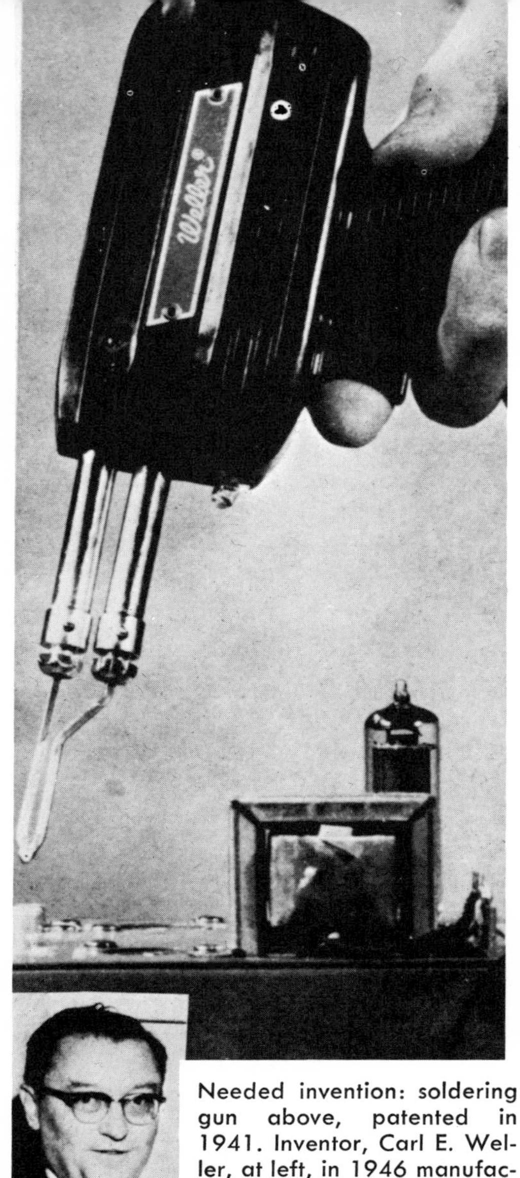

Needed invention: soldering gun above, patented in 1941. Inventor, Carl E. Weller, at left, in 1946 manufactured it himself; major companies had rejected it as an impractical tool.

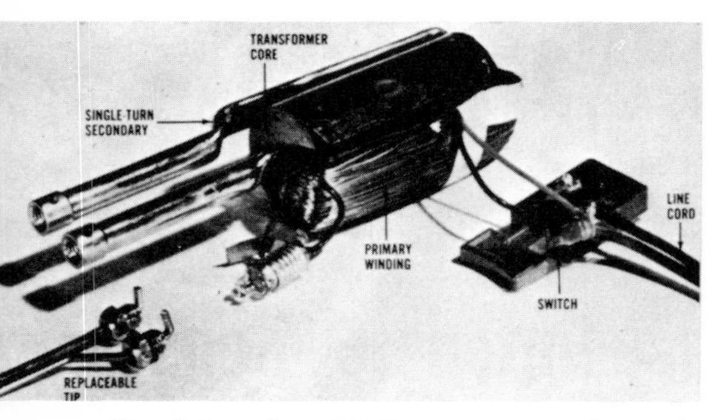

Simplicity of a Weller gun was shown in photo with 1963 article about invention.

pany, Bridgeport, Connecticut, brought out its version. Both tools were available to U.S. buyers. The early ones had no handles. You clasped the body of the tool within your palm.

Even though some people still call it a saber saw and some a jigsaw, the tool now has made a fine name for itself. For one thing, it will make its own starting hole for a cut in the middle of a wide expanse of wood. You simply rest the rear edge of the guide shoe against the surface of the wood, tilt the saw forward, and up and down action of the blade carries the tip down through the wood. For curved cuts, you can't beat a portable saber saw.

New wonder tools for outdoor chores have come thick and fast, too, in recent years. Homeowners have been praising the Lord for some years now for electric hedge clippers. Just lately their thanks could be doubled—for a cutter bar with blades on both sides. You can now buy a clipper that trims whichever way you swing it.

As early as 1872, inventors were trying to improve matters for the unhappy man whacking off grass with a sickle or a scythe. About midway in the nineteenth century, an English engineer patterned a grass cutter after a machine used in his textile mill for shearing cloth! Things improved little until about 1890, when lawn cutting with a cylinder rotating against a stationary-bed knife came into use. An English inventor even tried to propel such a cutter with a bulky steam engine. Then in 1919 a smart Yankee hooked up the gasoline engine from a washing machine to a reel mower.

Soon sulkies were hitched on and, in only a few years, riding mowers could trim as much as 7 acres in a day. After World War II, the rotary mower was heard throughout the land. Other outdoor machines were also saving the homeowner's back: snow throwers in winter, leaf vacuums in fall, garden tractors the year round.

It was 1936 before the young 'uns got out of the backbreaking work in the garden patch. A bright engineer came along with a cultivator to do the hoeing for them. Howard C. Ober was hoeing a strawberry patch when he got the idea that a small engine ought to be able to swing a hoe much better than human biceps. Why not put blades or tines on a revolving shaft that would both cultivate the soil and pull the machine? But would it be maneuverable and light enough, he wondered, to hop over strawberry runners without harming them?

In his first machine, the tines were bolts with the ends hammered over to give a hook shape. A $\frac{6}{10}$-horsepower engine from a washing-machine motor was mounted on the frame. "The thing stalled as soon as the tines hit the ground," Ober says. A bigger engine pulled the rotor around, but the bolts soon wore to stubs.

Ober changed to tines of malleable cast iron. In 1938 he got his first patent on the hoe. The cast tines were still too soft, and it was hard to keep them tight on the shaft. He substituted a hexagonal shaft for his first round one and made the tines with matching hexagonal holes.

With $5,000 and the machinery from his home workshop, he made and sold a few hoes. Then along came the war and almost before Ober was in business he was out of it. But the midget power plants he needed became available with war's end.

Not content with his tiller-cultivator, Ober designed attachments to convert the basic machine into a sprayer, lawn mower, edger, and rotary snowplow. By 1955 he'd sold more than $2 million worth of the machines.

It was do-it-yourselfers who helped bring the chain saw out of the woods. Old records indicate that a Disston representative put on a demonstration as far back as 1905. But it was many years before the saw we now know felled a tree.

A chain saw powered by an 11-horsepower outboard motor was produced in 1943 by Carl Kiekhaefer. It weighed almost 100 pounds. A man at each end supported it and walked it right through the tree. But lighter, more compact engines cut down weight. By 1963 Homelite had a direct-drive 12-pounder. By 1968 McCulloch had the "world's lightest"—a saw weighing only 6½ pounds.

All sorts of help for the handyman came out of the labs over the years. What would the new breed of "wood butchers" have done without wood putty? Especially that old and common kind with the familiar brand name: Plastic Wood? Among today's great aids can be numbered pressure-sensitive tapes and urethane foam.

In *Popular Science,* do-it-yourselfers followed the development of DuPont's Teflon S. First cousin to the Teflon that banished grease from frying pans, the S breed reduces friction on the moving surfaces of all sorts of tools and sets up protection against rust.

A guiding premise of do-it-yourselfers is that anyone can do anything. Well, you can paint a house, inside or out, with high hope, thanks to today's water-clean-up paints—casein, resin, latex, and others. When the job is done, you just wash the brush, and yourself, in water.

The paint rollers that have taken a load off the old paintbrush also promise more success. They roll on a fast and streak-free coat, especially on interior walls and on ceilings. For small jobs, good paints in aerosol cans let you spray on a finish, without waste.

In the hands of both pros and amateurs, staplers now do some of the work of hammers and other tools. And the roster of tricky fasteners and connectors for all kinds of building materials is longer than your arm.

Plywood has grown up since the do-it-yourself phrase first became common. Fir plywood led the way. Later, fine hardwood veneers were bonded to inexpensive backing. Eventually, plywood came with the final finish baked on—all you had to do to panel a room was cut the sheets to size and attach them to the walls.

Planer chips, a waste product at sawmills in the old days, are now used to make particle board and the fibers to make hardboard. Like plywood, such panels also come surfaced with native wood veneers or plastics.

A patent lawyer, writing in *Popular Science* in 1933, told of the invention by accident of another modern building material —the pressed-wood-fiber panels (hardboard) that many still call Masonite, after the fortunate inventor, William H. Mason.

In 1925 Mason was attempting to make paper out of wood fibers. Sheets of the fibers were placed, for thorough drying, between the steam-heated plates of a power press.

One day while running a test, Mason shut off the steam to go to lunch—or, thought he had. When he returned, he discovered that the valve he'd turned was out of order and had failed to shut off the

For the Handy Man and Every Man—

PLASTIC WOOD

Reg U S Pat Off

Handles Like Putty . . . Hardens Into Wood

HERE is wood that you can mould—a compound packed in air-tight tins so soft that it can be used like putty, but which hardens by action of the air into solid wood, waterproof and grease-proof, and adheres firmly to wood, metal, fabric, plaster or stone. It is invaluable in the home, for the automobile, and with boats. A few household uses are shown here, but their number is unlimited and each day brings reports of new successes.

Loose Rungs

Pull out the rung. Scrape hole to remove dried glue or dirt. Half fill hole with Plastic Wood and force rung back in. Clean off loose particles. Allow rung to set overnight. It is often advisable to hold rung fast with string while Plastic Wood sets.

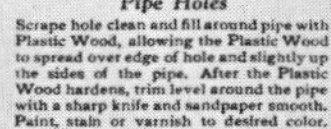

Pipe Holes

Scrape hole clean and fill around pipe with Plastic Wood, allowing the Plastic Wood to spread over edge of hole and slightly up the sides of the pipe. After the Plastic Wood hardens, trim level around the pipe with a sharp knife and sandpaper smooth. Paint, stain or varnish to desired color.

Baseboard Cracks

When the floor settles away from the baseboard Plastic Wood can be worked into the crack with the fingers or a knife and when hard, is easily sandpapered or pared smooth with a sharp knife, and painted or varnished to the color desired. Scrape crack clean before filling.

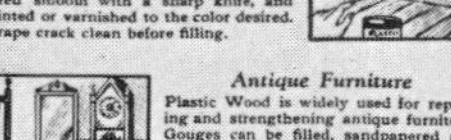

Loose Casters

Remove caster and fill hole half full Plastic Wood. Force caster back in hole. Remove Plastic Wood that squeezed out, and allow chair to stand overnight before using.

Antique Furniture

Plastic Wood is widely used for repairing and strengthening antique furniture. Gouges can be filled, sandpapered and colored, missing decorations can be modelled: loose mouldings and pieces can be affixed; rungs and joints permanently strengthened; drawer pulls made secure.

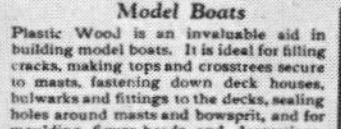

Model Boats

Plastic Wood is an invaluable aid in building model boats. It is ideal for filling cracks, making tops and crosstrees secure to masts, fastening down deck houses, bulwarks and fittings to the decks, sealing holes around masts and bowsprit, and for moulding figure-heads and decorations.

Wood Working

Plastic Wood is the one substance to satisfactorily repair mistakes or accidents in wood working, and its use will often simplify many tedious jobs. It can be used for strengthening joints and corners, and when hard, it can be chiseled, planed, sawed, turned in a lathe and worked like natural wood.

Loose Handles

On brooms, mops, carpet sweepers brushes toilet articles and hand tools Remove handle and insert sufficient Plastic Wood into hole so that when handle is forced back some Plastic Wood will be squeezed out. Allow handle to set well before using.

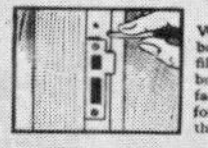

Locks, Latches, etc.

When it is necessary to move door locks, bolts or latches, Plastic Wood not only fills up the old holes with permanent wood, but can be used to cover any scarred surfaces around the hardware. Sandpaper for smoothness, and paint or varnish to the color of the surrounding surface.

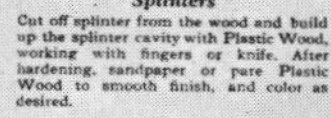

Splinters

Cut off splinter from the wood and build up the splinter cavity with Plastic Wood, working with fingers or knife. After hardening, sandpaper or pare Plastic Wood to smooth finish, and color as desired.

Plastic Wood is carried at Hardware and Paint Stores, and by Ship Chandlers. It comes in natural wood color in 1 lb. cans at $1.00, and in ¼ lb. cans at 35 cents. For use in replacing loose tiles, and for cracked porcelain and enamel there is a special preparation, Plastic Wood White Waterproof Tile Cement, in ¼ lb. cans at 35 cents.

Manufactured by The Addison-Leslie Company, 301 Bolivar Street, Canton, Mass.

PLASTIC WOOD

Trademark Reg. U. S. Pat. Off.

This "wood butcher's" friend has been around for years, as shown in 1928 ad.
First hardboard (below) was made by accident.

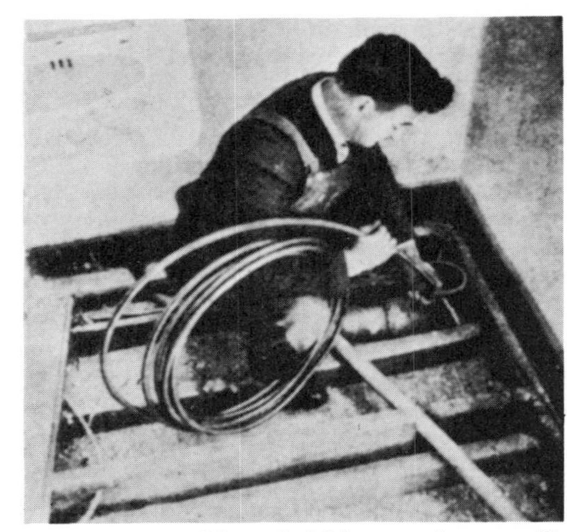

Soft copper tubing "which can be readily bent around corners is the latest thing in water piping," said *Popular Science* in 1930.

Making plywood: steam-operated "peeler lathe" keeps log spinning evenly against long knife that slices off veneer in continuous strip.

May, 1928 POPULAR SCIENCE MONTHLY

Uses for this *grainless* wood are steadily increasing from New York to Hollywood!

Possesses uniform strength and remarkable workability. Highly resistive to moisture. Very tough and dense. Has a smooth, attractive surface on the face side, and requires no paint for protection. Also takes any finish beautifully. Send for large, free sample.

TWO years ago, after a long series of tests and experiments, a modern inventive genius succeeded in producing a wood better, in several ways, than Nature's own material. It was much denser, far tougher, had greater moisture resistance. And it was *grainless!*

This product, known as Masonite Presdwood, has already shown almost unbelievable adaptability in many industries, and the number of its uses is steadily increasing throughout America.

In fact, the trails of Presdwood cross and re-cross endlessly—from breakfast nooks to motor boats, from cooling trays for hot castings to the paneling of beautiful homes, from the packing cases in a New York warehouse to the movie scenery of Hollywood!

This is happening right along

Step into the offices of a great manufacturing plant located in Camden, New Jersey, and you will find that all of its telephone booths are lined with Presdwood.

Out along the highways of Pennsylvania you will again see Presdwood—in the form of advertising signs.

In Pittsburgh, more Presdwood—a large department store ordered 14,000 feet of it just a few weeks ago.

In Cleveland, and other Ohio cities, candy manufacturers are making use of Presdwood for starch trays.

A Wisconsin manufacturer has also standardized on Presdwood—for incubators. A Minneapolis concern is using it in making safety wheels for bathing beaches and swimming pools. Out in Los Angeles, as well as in the East, it is going into concrete forms.

Presdwood is used for bed-room screens and for barbecue stands, for invalid trays and for theatre props, for artists' boards and for display booths. It will give as dependable service for garage doors in a Chicago suburb as for china closets in a Boston residence or for fixtures in a Seattle delicatessen store!

Only a short time ago, the head of a fast-growing American industry sat in his office running an eager hand over the fine, smooth surface of Presdwood. Quickly he decided that it was the very thing he needed, and it is now standard material for the manufacture of one of his leading products.

Similar incidents are happening right along. The simple truth is that you have no idea of what you can really do with Presdwood until you have seen a sample and made a few experiments with it.

The advancement of Presdwood has been especially rapid in planing mills and wood-working plants. Kitchen cabinets, show cases, counters, radio boxes and tension boards, work-bench tops, cupboards, office partitions and shelving—these are but a few of the many additional things which are made of Presdwood. And new uses are being discovered almost every day!

Remember that Presdwood is not only *grainless*, but that it has greater moisture resistance than Nature's own material, is much denser, and is far tougher. Yet it contains no chemical binder; no foreign substance of any kind. It is simply genuine wood torn apart and put together again—genuine wood, and nothing else!

Presdwood is manufactured by exploding fresh, clean chips from guns. This is done at a velocity of about 4,000 feet per second. The long fibres thus produced are then packed into hydraulic flat bed presses and subjected to hundreds of tons of pressure. The finished product is cut by automatic machines into broad smooth boards of uniform thickness.

Advantages of Masonite Presdwood

Masonite Presdwood has an exceptionally smooth, attractive surface on the face side, and requires no paint for protection. It also takes any finish beautifully. It won't crack, check, split or splinter. It shows minimum contraction and expansion.

It contains no grit, and can not damage tools. It can be used on saw, planer, sander, shaper. And it comes four feet wide by twelve feet long.

We urge you to make your own tests and experiments with Masonite Presdwood. Only in this way can you find out what a really wonderful product it is. We will gladly forward you a large free sample promptly on request. Send for it to-day.

MASON FIBRE COMPANY

Sales Offices: Dept. 1258, 111 W. Washington St., Chicago, Ill.

Mills: Laurel, Mississippi

© 1928, M. F. Co.

Plastic pipe so light that youngsters could tote it came in 1953. Flexible for bending, it cut with saw or knife, needed no reaming.

steam. As a result, the thick sheet of fiber, instead of merely drying, had been baked for more than an hour by hot steam. To his astonishment and delight, he found he had accidentally invented a fine, strong, and almost waterproof grainless wood. Now it comes with all sorts of convenient finishes.

And what a debt amateur builders owe to the man who first gave them plaster board in smooth 4-foot-by-8-foot panels! In 1923 a *Popular Science* writer proclaimed that another new building material made of "asbestos and cement may save our forests." The new material went into shingles, too, adding the advantage of their fireproofing.

In 1939 *Popular Science* told its readers of a new material that combined wood with a plastic that maintained the natural beauty of wood. The material was natural wood "impregnated and specially treated so that it has a permanent, glass-like surface that is impervious to all types of liquids and yet will not crack, chip, or break." It was rapidly followed by other stainproof and burnproof wall paneling and counter topping.

Do-it-yourselfers were guinea pigs for a succession of new and improved adhesives. They put their seal of approval on quick-grab contact cement, developed for sticking down counter tops when modernizing the kitchen was a favorite home improvement. They figuratively gave whoops of delight and rushed to buy, when *Popular Science* told them in 1959 that they could "bond almost anything to anything with those amazing epoxy adhesives."

A *Popular Science* article in 1953, told readers how "plastic pipe makes anyone a plumber." But for many years before that, mechanically adept men had been doing a mighty good job of home plumbing with a material they'd heard about earlier. In 1930 Roger Whitman, nationally known writer on home maintenance, had reported in the magazine on an "improved method of piping that eliminates most of the tedious process of cutting, threading, and fitting [and] greatly simplifies the laying of water pipes. The pipe used is copper, and instead of being in stiff, short lengths, it comes in 60-foot coils." This tubing could be bent around corners.

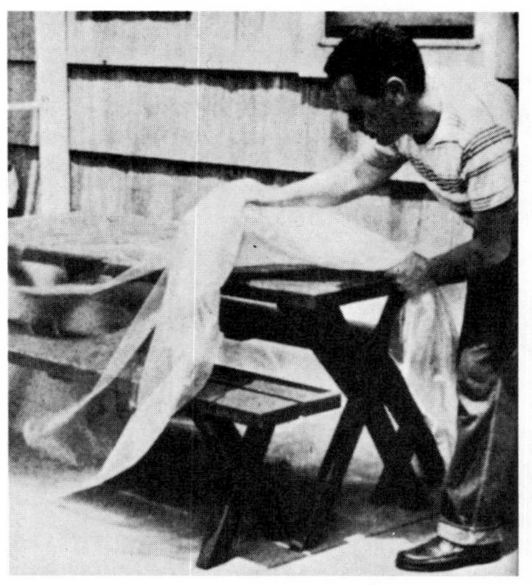

Polyethylene was new only a few years ago. In 1959 readers learned some of its uses. **Above, covering yard furniture for the winter.**

Contact cement became a staple item on home-workshop shelves in the fifties. Above, **spreading it on surface with applicator.**

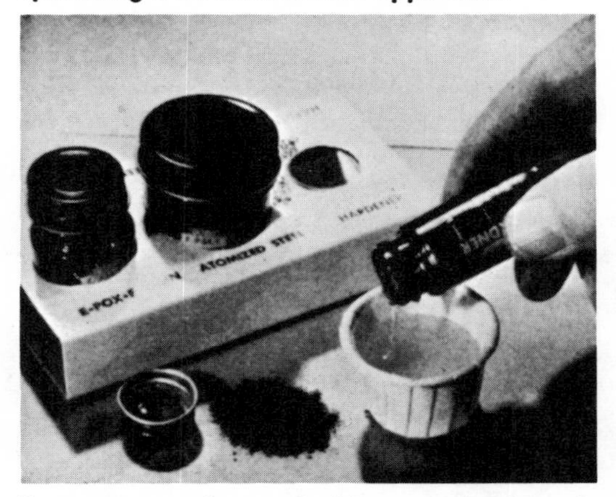

Two-part epoxies made big news in the adhesive world in 1959. Photo shows mixing of the 2 separate components.

Big new world of plastics opened up in the thirties—and *Popular Science* told readers how to work with the stuff. Above, sawing it with a bandsaw, and homemade castings.

For half a century *Popular Science* has encouraged readers to work with their hands. This was not true in its early days. A 1942 article marking the magazine's seventieth anniversary told why. The period just after the founding "was not a day of leisure, of preoccupation with hobbies and pleasurable activities in spare hours. There were few spare hours. Life was real; life was earnest."

The turning point came at about the time of the Norman Rockwell cover painting which opens this chapter—the beginning of the twenties. Rockwell portrayed his model at work on a task any homeowner might have encountered—the repair of a silver teapot. The elderly gentlemen with sideburns is shown using a new kind of blowtorch.

The need and desire to work at home in spare time spiraled after World War I and promoted a demand for tools and a convenient place to use them. Home workshops were being established.

In 1932 an editor of *Popular Science* wrote an editorial entitled "The Need to Make Things":

"Modern civilization has made it difficult for many of us to satisfy our inherited skill hunger to a normal extent. Mass production, desirable because it makes it possible for each of us to live better than the kings of old, unfortunately does not satisfy our skill hunger . . .

"This growing desire to satisfy skill hunger undoubtedly is the basic reason for the perfectly astounding increase in interest

"Cast Resins . . . marvelous new material for amateur handicraft," said 1935 article telling how to make jewelry with home workshop tools. At lower left, turning a bracelet on a tapered wooden mandrel; center, carving with a high-speed spindle; right, beveling on a grinding wheel. Readers were informed new plastic was "not the same as celluloid."

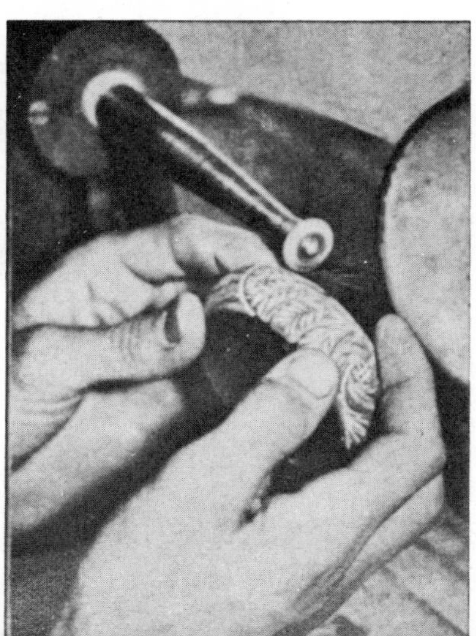

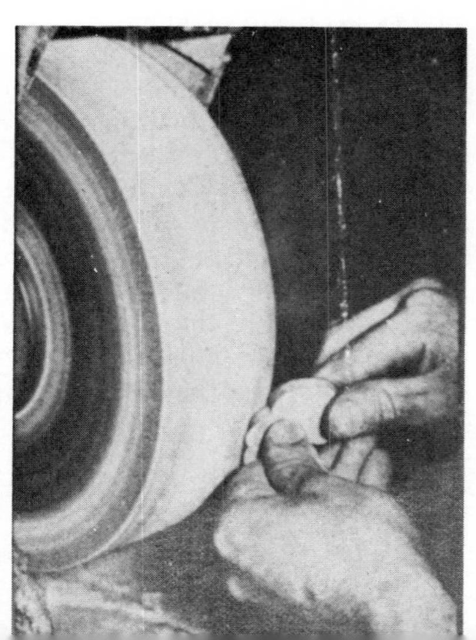

Shopsmith was a big success as a combination tool. In 1950, *Popular Science* detailed the invention of this remarkable combination tool by Hans Goldschmidt. Photo above accompanied that article. It shows how the tool was shifted from one use to another. It could also be used as a disk sander, a wood lathe, and a horizontal drill press.

Routers came to home workshops in the 1950s after long use in professional shops. The photos below illustrated an article in September 1955 describing how to use the new Porter Cable Routo-Jig. Advertising described it as an "amazing 3-in-1 tool"—a router for mortising, grooving, rabbeting; a rotary jigsaw for cutting intricate contours; and a shaper for accurate moulding, beading, and joining work.

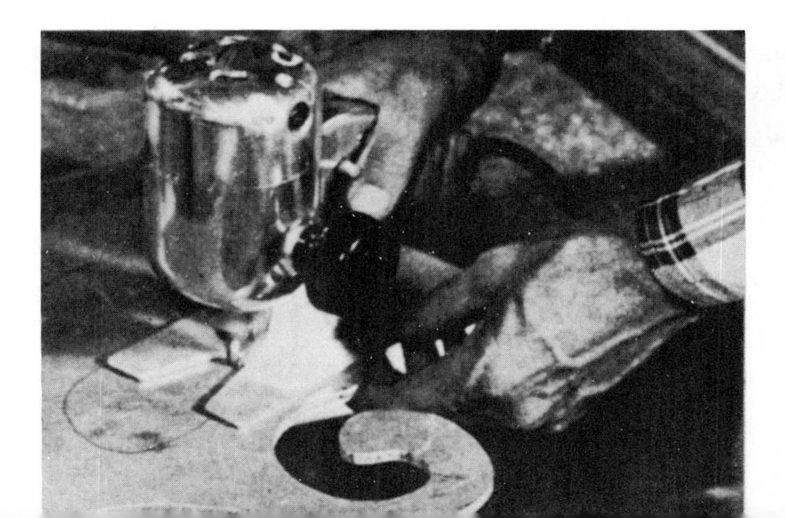

people are taking in all forms of hobbies that revolve around the use of various tools in the home workshop."

The Depression years of the thirties brought enforced leisure to many—and the need to economize. And the 5-day week gave workers spare time they never had before.

In 1933 the National Home Workshop Guild was organized. Sponsored by *Popular Science,* it gave great impetus to the home and shop movement. Its aim was to bring together men of like hobbies, help develop handicraft projects, and "make the most of increased leisure under the NRA." Many hundreds of chapters were organized all over the country and a monthly newsletter and a speakers' bureau brought the local clubs news and information from experts on tools and shop techniques. Home workshops took over the basement and the hunt for worthwhile projects was on.

Blueprints for ship models that appeared monthly in *Popular Science* sold by the thousands. The *Santa Maria* of Columbus and old sailing ships like the *Cutty Sark* set sail on mantels from Maine to California.

Radio was new and dime stores stocked the parts. Articles on building your own remained common reading fare in *Popular Science* well into the forties.

A list of *Popular Science* do-it-yourself projects tells a story of the decades. In 1917 it was building a home vacuum cleaner, including winding the motor; in 1923, building that wonderful appliance—an iceless icebox. In 1964 it was how to make a laser; in 1966, putting up a home version of Buckminster Fuller's geodesic dome. In the years between, you could build an early form of snowmobile (1921); a television disk, away back when television receivers had spinning disks instead of picture tubes (1928); a portable tape recorder or a Geiger counter (both in 1955); and in 1960, a GEM (ground-effect machine).

And the seventies? A stack-away swimming pool, the privacy meditator—a place to get away from it all—and the no-hands picture clock. And just to show that things really do come full circle, instructions on how to make your own stained-glass windows.

New and better paints have been common in the do-it-yourself era. Scene above illustrated 1926 article on brushing lacquers.

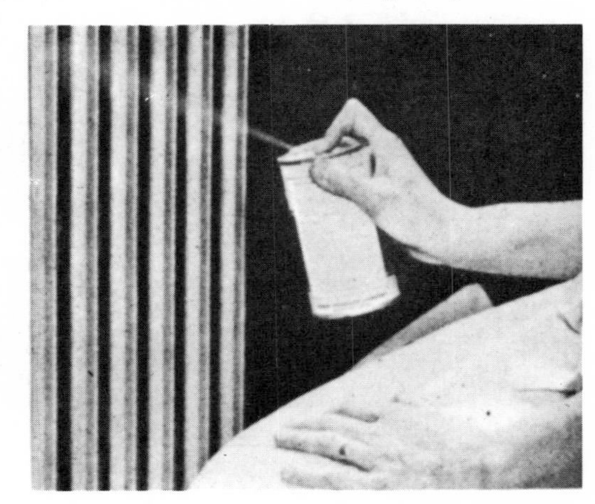

Remember first paint "bombs"? At right, article details advantages of the spray cans.

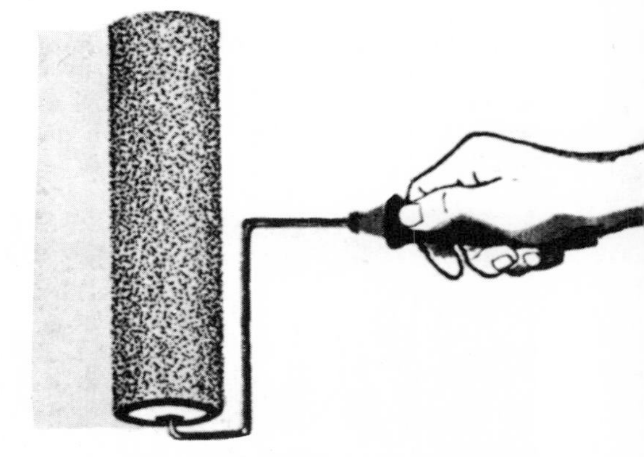

Paint rollers improved steadily in the fifties. Some pros didn't like them at first, but homeowners found them fast and streak-free.

WATER PAINTS AND THEIR USES

[PAINTING]

Type of Paint	Where Used	Kinds of Surfaces	Preparation of Surfaces
Calcimine, whitewash, other glue-bound water paints	Walls and ceilings of living rooms, dining rooms, bedrooms, halls, basements	Use only on surfaces not subject to dampness. Plaster, wall board, wall paper, primed wood, old painted surfaces	(a) Remove old calcimime. (b) Ordinarily no primer is necessary. (c) Surface should be clean and dry. (d) Plaster of uneven porosity requires oil- or water-type priming coat
Casein and casein-soybean paints	As above	Where normal conditions prevail, but where maximum washability is not expected	As above
Resin paints, interior type	As above	Surfaces subject to moisture or that must be washed frequently.	As above
Resin paints, exterior type	Exterior walls	Where surfaces are subjected to dampness, rain or snow. Brick, stone, concrete, concrete block, asbestos shingles, and old painted surfaces	See that surface is clean and free of dust
Cement powder paints	Interior or exterior walls	As above	Surface should be damp during application

Note: All types of paint are reduced with water according to the manufacturer's directions. They are applied with a brush, but may be sprayed if suitable equipment is available.

POPULAR SCIENCE MONTHLY SHOP DATA FILE

Data sheet designed to be clipped out and glued to file card accompanied April 1942 article on new water-based paints. Note instructions to brush or spray. Rollers had not yet appeared.

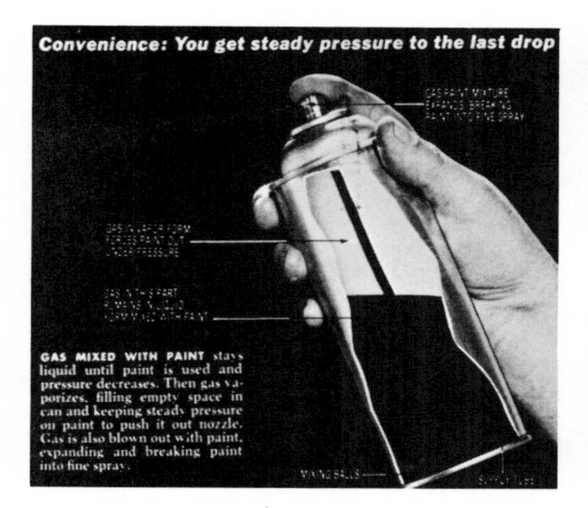

Convenience: You get steady pressure to the last drop
GAS MIXED WITH PAINT stays liquid until paint is used and pressure decreases. Then gas vaporizes, filling empty space in can and keeping steady pressure on paint to push it out nozzle. Gas is also blown out with paint, expanding and breaking paint into fine spray.

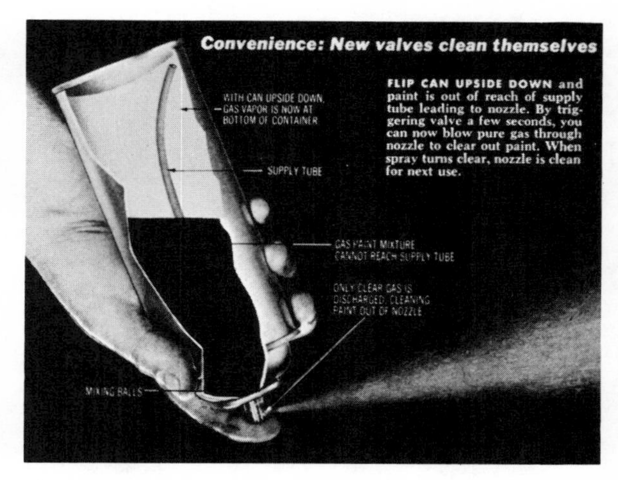

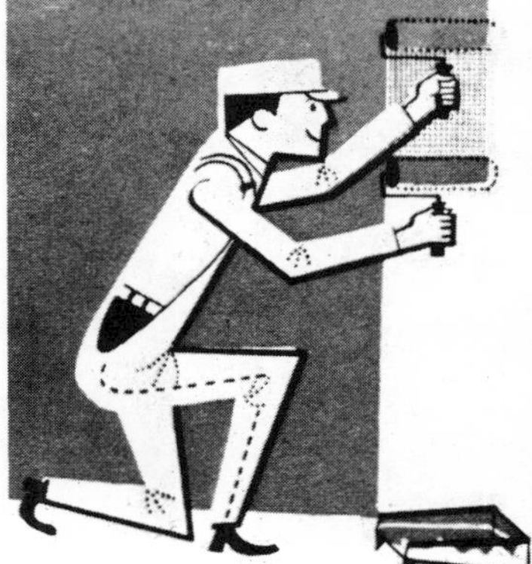

Convenience: New valves clean themselves

SUPPLY TUBE

FLIP CAN UPSIDE DOWN and paint is out of reach of supply tube leading to nozzle. By triggering valve a few seconds, you can now blow pure gas through nozzle to clear out paint. When spray turns clear, nozzle is clean for next use.

Cartoon panel made important points about latex paints when they still were new. Illustrating an article in May 1957, panel showed that you could paint on rainy days, apply 2 coats the same day, and when the job was done, wash the rollers clean in water.

Do-it-yourself projects of recent years included those on these 2 pages. Above, a convertible rocker-table for the patio created by Les Walker, a talented young designer and architect who has become well known for such multiple-use "living machines."

Strips of redwood are the major materials for a handsome 16-foot canoe, weighing less than 90 pounds. At least 10,000 builders are now paddling on U.S. waters in this craft.

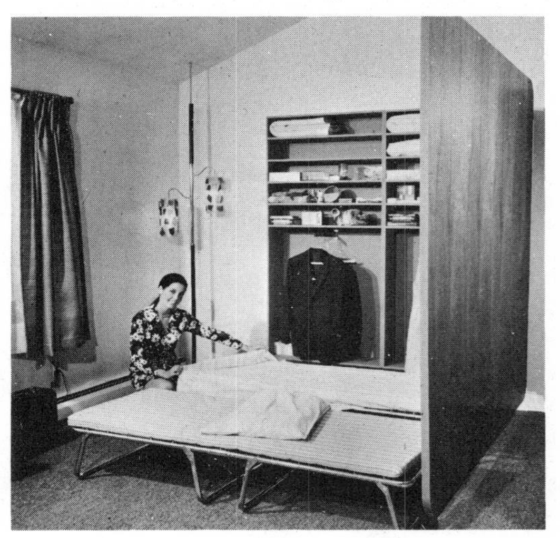

Guest room that folds into a 10-inch space against a wall was idea of Ken Isaacs, design consultant of *Popular Science*. Shelves store 2 cots, everything else needed to sleep 2.

This is the Superchair! If someone will bring you your lunch, you can settle in for a full day of reading and relaxing, or even drop the back to enjoy a nap. The chair was another unusual design from Ken Isaacs, whose invention of the Living Structure and the Microhouse attracted worldwide attention several years ago.

Buckminster Fuller's famous geodesic dome was inspiration for this low-cost pool cover **or greenhouse. Under license from Fuller,** thousands of suburbanites have built the *Popular Science* Sun Dome. Wood strips stapled **into a frame are covered with sheet plastic.**

Bulky 1919 power mower had a chain-driven 27-inch reel. Photo appeared in a looking-back article in 1959.

The old handyman who had held sway so long, making a good living for himself, was a mythical figure by the forties. His kind were generally older men—and gradually they disappeared. In their place grew up—out of necessity and often just for the fun of it—a new class of home handymen and Saturday mechanics.

In the postwar years, weekend home improvement meant installing new flooring and ceiling materials that came in easy-to-handle squares. Installation of drop ceilings was popular, too. And miles of counter topping modernized older kitchens.

Upstairs in the unfinished floor of many a small postwar house the owners turned to and finished off the space that had been a shell when they signed the mortgage. Garages and carports were added.

As growing families continued to find living space tight, do-it-yourselfers flocked to their basements to build a recreation room. Both before and after such improvements, dismayed homeowners often called for help in choosing sealers for curbing the water that seeped, or flowed, through the basement walls and floors.

Weekend workers set up their aluminum ladders and climbed all over the outside of their homes, applying new paints and surfacings and making the repairs they read about. Inside, in the late fifties, colorful ceramic tiles were laid on thousands of bathroom walls and floors. In 1954 new adhesives made it possible even for duffers to attempt this job.

The old sociable front porch had long since disappeared with the horse and buggy. A patio now appeared at every back door in the land, and the outdoor fireplace brought back a piece of the kitchen coal stove.

But the bricklayer and patio spreader had something to call down blessings on: ready-mixed concrete that came in bags. And in 1964 *Popular Science* reported on the new shrinkless concrete and in 1970 told about the next development: a concrete that needs no joints.

Putting in a driveway and properly winterizing it are still weekend chores of many homeowners. And as new products become simpler and simpler to use, new projects will proliferate. For the do-it-yourselfer, there seems no end of good news in sight.

Outdoor power equipment makes life a lot easier for do-it-yourselfers today. Above, snow thrower; below, modern garden tractor with blade used for grading or snow removal.

CHAPTER 14

The Unpredictable Marvels of Electronics

ELECTRONICS is the common denominator of twentieth-century science . . . the keystone that supports an enormous array of other modern-day marvels . . . the hinge pin on which much of our life today turns. Try to imagine our vast space program without giant electronic computers, our fine medical care without X-ray machines, or the safety of air transportation without radar and radio navigation.

Electronics is a vigorous, versatile giant you find hard at work in scores of occupations: in the entertainment field (radio, television, and high-fidelity sound), in industrial production (automation control systems), in communications (worldwide radio and telephone service), and even in banking (automatic check handling). And that's just scratching the surface.

Today's world relies heavily on electronics. Yet, strangely, electronics has been a virtually unheralded technology. Men dreamed about taking flight for centuries before the first airplane took off. But no citizen of the Middle Ages ever had visions of an electronic computer, or a television set, or a microwave oven. Perhaps this is one reason why every breakthrough in electronics seems so incredible.

Our electronic age began with a kind of scientific detective story. During the last half of the nineteenth century, the leading scientists of the day were struggling to explain a curious discovery. Unable to pin down what they had found, they called it "radiant matter." Sir William Crookes, a resourceful British physicist, had developed a series of demonstration devices—we call them "Crookes tubes" today—that illustrated the puzzling phenomenon.

One of the gadgets consisted of a bulbous glass tube that housed a diamond gemstone sitting on a pedestal. All the air was pumped out of the tube to create an internal vacuum, and there were two aluminum disks mounted inside the tube, facing the diamond. When the disks were connected to a source of negative high voltage, the diamond began to glow with a strange green light.

"On darkening the room you see the diamond shines with as much light as a candle," reported Crookes in the November 1879 issue of *Popular Science.*

The conclusion was unmistakable: under the influence of the high voltage, the aluminum disks were emitting some kind of "radiant matter" that made the diamond glow.

Another Crookes tube was a pear-shaped glass bulb with a metal cross mounted inside it. When an aluminum plate in the narrow end was connected to a negative high voltage, the glass on the wide end would glow softly, except for a well-defined cross-shaped area. Could it be that the metal cross was casting a shadow and preventing radiant matter from reaching the glass and making it glow?

Yet another Crookes tube contained a little paddle wheel resting on a track. When voltage was applied to an aluminum plate in the left end of the tube, the paddle wheel spun off to the right. When voltage was applied to an aluminum plate at the right side of the tube, the paddle wheel

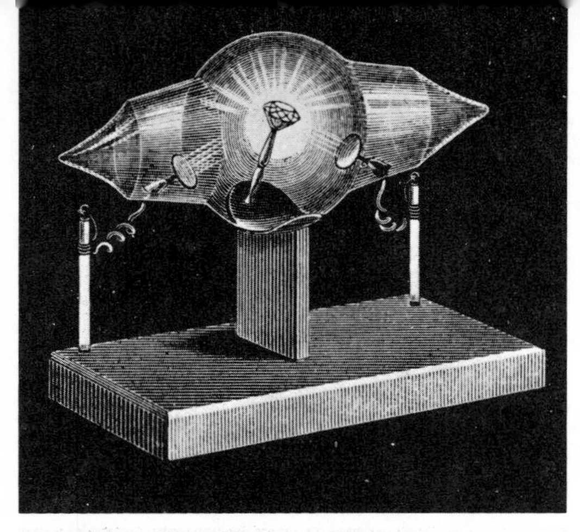

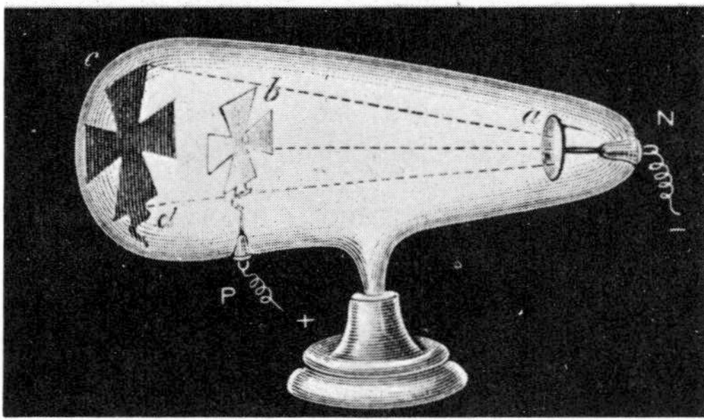

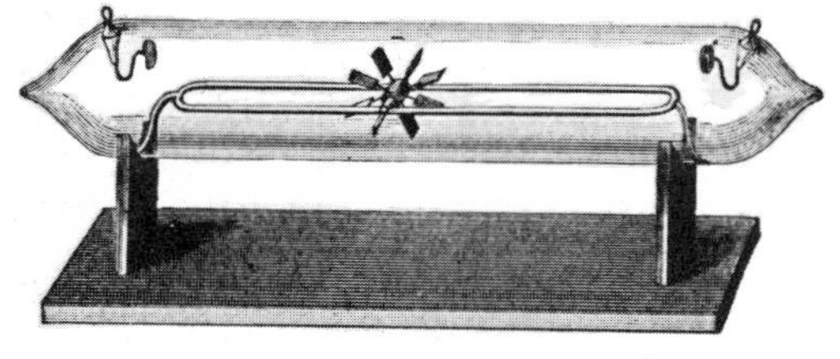

Crookes tubes demonstrated cathode rays, found to consist of electrons. Engravings are from 1879 *Popular Science*.

moved left. It was clear that the force of radiant matter striking the paddle wheel was making it turn.

But what *was* this radiant matter being given off by the metal plates?

Some researchers held the theory that it was a kind of radiation—or cathode ray (so called because the negatively charged metal plates were called cathodes). But Crookes suspected that the cathode rays were really a stream of tiny bits of matter . . . and he was correct.

In 1897 J. J. Thomson, another Britisher, proved that radiant matter was a flow of minute electrically charged particles. He called the particles "electrons." They were part of the atoms that make up all matter—including the aluminum disks in the Crookes tubes. The high negative voltage applied to the plates forced electrons out of the metal, and propelled them at high speed toward the diamond, glass, and paddle-wheel targets.

Curiously, Crookes tubes were to surface again decades later, when they would become the prototypes of the cathode-ray tubes (or CRTs) that are used as picture tubes in television sets and as display screens in radar sets. Inside a CRT, a cathode assembly at one end generates a narrow beam of electrons that soars through the tube and strikes a phosphor screen on the faceplate, making the phosphors glow. As the beam sweeps across the phosphor screen, pulled back and forth by a changing magnetic field, it "paints" the picture you see.

It soon became apparent that electrons

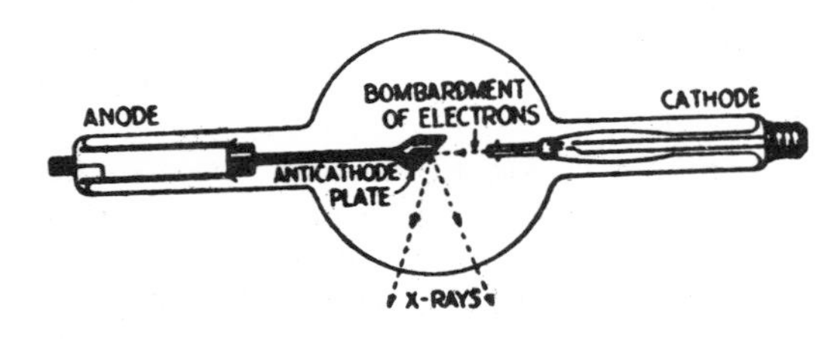

First X-rays came from Crookes tubes like those below. They were ancestors of 1927 X-ray tube (above) in which X-rays are emitted as cathode rays bombard metal plate.

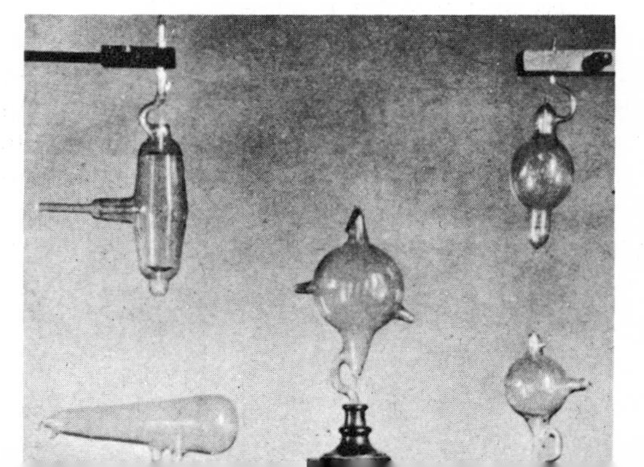

Roentgen's X-ray of a human hand, described as "weird" in 1896 when picture was made, was soon world-renowned. Making "radiographs" became a sort of parlor game. The dangers of injudicious exposure were not understood. Many pioneer experimenters died of burns and ray-induced cancer.

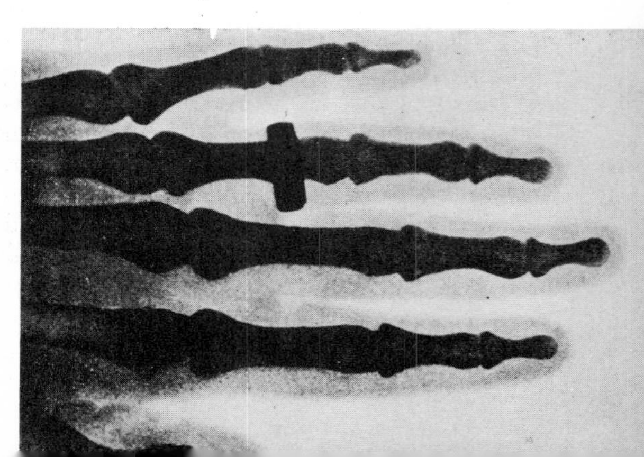

are the "stuff" that electricity is made of, and electric current is actually a stream of negatively charged electrons flowing through a metal conductor. And the science of *electronics* was born: "that branch of physics that treats of the emission, behavior, and effects of electrons."

The first truly significant electronic device was discovered by accident . . . a bit of serendipity reported in the December 1908 *Popular Science* as an aside to history. One morning in 1893 Wilhelm Roentgen, the German physicist, was experimenting with a Crookes tube when his wife called him to lunch. He laid down the glowing tube (still connected to its high-voltage source) on top of a book. Unbeknown to Roentgen, there was a metal key inserted between the book's pages and a piece of photographic film in a film holder underneath the book.

Some time later that day, he used the piece of film and developed it. To his surprise, he found a shadow picture of a key on the film. Enormously intrigued, Roentgen duplicated the "experiment." When he produced a second image of a key on another piece of film, he quickly discovered the key lying inside the book.

But, what caused the shadow picture on the developed film?

Eventually, Roentgen deduced that the glowing Crookes tube was emitting some kind of radiation—Roentgen called it X-rays—that could penetrate some materials better than others. The X-rays could pass through the book and film holder without difficulty, but were blocked by the metal key. The X-rays that reached the film fogged it to create the key's shadow image.

Roentgen announced his discovery of X-rays in 1895, but for several years there was controversy about the find. Not until early in the twentieth century was it understood that X-rays are produced when fast-moving electrons smash into metal objects. The collisions between electrons and atoms of the metal surfaces inside the Crookes tube emitted bundles of electromagnetic energy—a kind of ultra-short light waves—that are the X-rays.

Early experimenters soon discovered that X-rays could be used to photograph the bones of the human body. It became a kind of scientific parlor game to make "radiographs" of hands and other areas, and

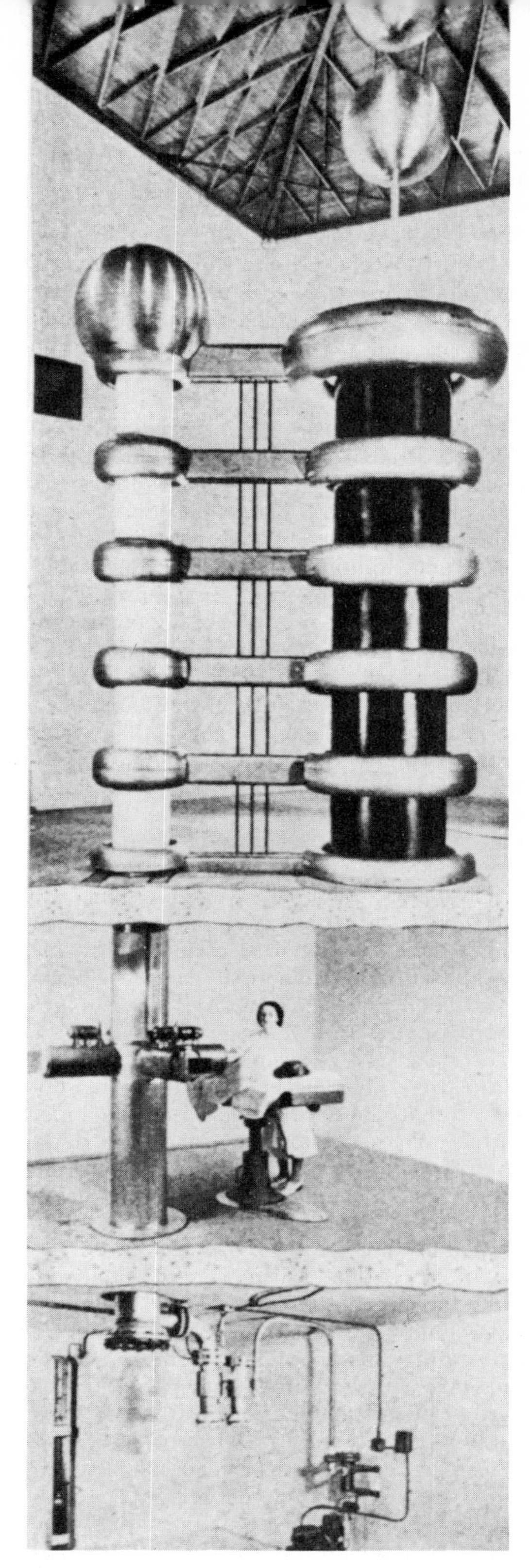

Four decades after discovery of X-rays, ever bigger sources of rays were wanted. This 30-foot monster of 1938 used a million-volt power supply to speed electrons on collision course with metal target.

Roentgen's hand X-ray became world-renowned; "The realism of this weird picture simply fascinated all who beheld it," wrote Professor D. W. Hering in *Popular Science* of March 1897.

Unfortunately, no one then knew of the dangers involved in the indiscriminate use of X-rays. Many of the X-ray pioneers died of burns and ray-induced cancers due to heedless overexposure.

Practical X-ray tubes look a lot like Crookes tubes. Inside them, a cathode produces a stream of electrons that bombards a copper or tungsten "target" and causes the target to emit X-rays. Modern X-ray tubes are often very large to permit them to operate at millions of volts. The huge voltages are needed to produce intense beams of high-energy X-rays of great penetrating power for use in medicine and industry.

The gadgetry we associate with modern electronics—such as radar, TV, computers—owes its existence to a family of devices that can precisely control the flow of electric currents. They are the building blocks of electronics. The vacuum tube, the transistor, and, more recently, the integrated circuit, are all members of this family.

The first vacuum tube—the Fleming vacuum valve—was invented in 1904. But it might have been created a full twenty-one years earlier by that great American tinkerer, Thomas A. Edison. And therein lies another fascinating tale:

In 1883, when Edison was hard at work perfecting the electric light bulb, he observed that as a light bulb burned, the inner surface of the glass bulb gradually darkened. Today we know that this happens because the glowing filament slowly evaporates and settles on the glass as a thin, dark film.

One way that Edison investigated the bulb-darkening process was by sealing a pair of metal wires inside a test bulb and connecting them to a current-measuring meter. When the bulb was switched on, he was surprised to find that the meter indicated a flow of electric current.

Unfortunately, Edison completely ignored the incredible significance of the "Edison effect" he discovered. Presumably he failed to understand what it meant: that electrically charged particles (electrons) were being "boiled" off the hot filament and that their movement to the cool metal wires constituted the current measured.

Twenty years later, J. Ambrose Fleming, an Englishman, resurrected the Edison effect during his search for a "one-way valve" for electric current.

A device that would pass current flow in one direction, but block flow in the reverse direction, was necessary to advance the fledgling science of radio communications. It would serve as a "detector" to capture the information broadcast on a radio signal.

Fleming's diode (2-element) valve was surprisingly similar to Edison's experimental set up. It consisted of a heated cathode (negative terminal), which boiled off electrons, and a cool metal anode (positive terminal), both mounted inside an evacuated glass bulb. Electrons could flow from the cathode to the anode, but *not* from the anode to the cathode.

In 1933, fifty years after Edison's work, *Popular Science* reported that a replica of Edison's experimental "tube" was going to receive a radio broadcast. It worked just as well as a Fleming valve.

From Fleming's diode, it's a short hop to the first vacuum tube that could amplify electric signals: the De Forest Audion tube. In 1907 Lee De Forest, an American electrical engineer, added a third electrode—a meshlike grid—between the anode and cathode of the vacuum diode and turned it into the triode (3-element) tube. In effect, De Forest added the "control handle" to the electric valve.

De Forest discovered that a weak electrical signal applied to the grid would control the flow of a large electric current between cathode and anode. If the grid was made slightly negative, the stream of electrons would be repelled back to the cathode, cutting down current flow through the tube. A more positive grid signal would permit a greater current flow.

Now, for the first time, it was possible to amplify a weak signal into a much stronger replica of itself. Consider a few of the possibilities:

The tiny electrical signal produced by a phonograph cartridge as it coasted along a record groove could be made strong enough to drive a loudspeaker, making the electronic phonograph possible.

Or, a low-power radio signal could be boosted into a powerful signal capable of

First, Thomas A. Edison, while experimenting with an incandescent electric light bulb, placed a metal plate inside the bulb near the filament and discovered the so-called "Edison effect," later applied by others to radio

Next, Prof. J. A. Fleming, seeking to improve the wireless detector, produced the first vacuum tube, known as the "Fleming valve," using the "Edison effect"

Finally, Dr. Lee de Forest added to the Fleming valve a third element known as the "grid," producing the vacuum tube as it is in use today

Evolution of vacuum tube is shown by 1922 picture group from *Popular Science*, above.

being transmitted over long distances.

Or, a weak radio signal pulled in by an antenna could be detected and amplified easily, making low-cost home radio receivers practical.

And telephone signals could be amplified by vacuum-tube "repeater" amplifiers, making coast-to-coast long-distance telephony easy and low-cost.

For fifty years following the triode's invention, the vacuum tube reigned supreme in electronic product design. More electrodes were added inside the tube to refine its performance; tubes were made tiny enough to fit inside hearing aids and inside the nose of an artillery shell; tubes were developed to handle huge amounts of power for the control of welding machines; tubes were built to work at the high frequencies demanded by radar. In concept, though, the vacuum tubes of the 1950s closely followed the basic designs of Fleming and De Forest.

And, during that fifty-year period, the uses of vacuum tubes ranged from the practical (radio and television) to the warlike (radar and the firing circuitry inside atom bombs) and on to the sublime (electronic organs and musical instruments).

But, versatile as the vacuum tube is, it has a few annoying characteristics. Heating the cathode hot enough to emit electrons takes a lot of electrical power, so portable electronic gear using tubes requires heavy, bulky batteries. Because of this great quantity of generated heat, a tube-equipped chassis needs ventilation. Most vacuum tubes require relatively high voltages to operate efficiently. Tubes eventually burn out. And tubes tend to be bulky,

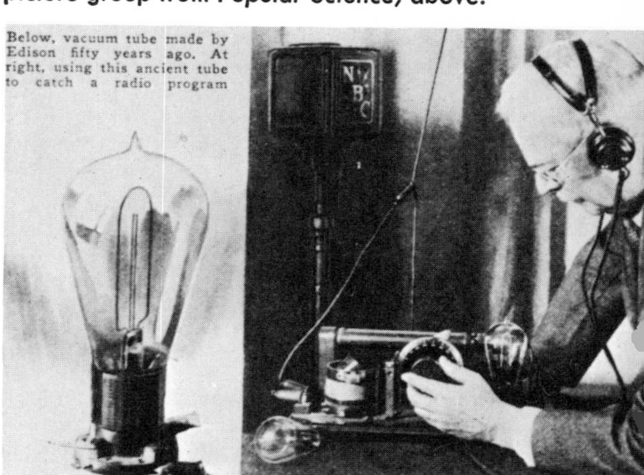
Below, vacuum tube made by Edison fifty years ago. At right, using this ancient tube to catch a radio program

Edison built better than he knew. In 1933, 50 years after it was developed, replica of his "vacuum tube" was used to receive a radio program. Edison built the tube to investigate age-darkening of light bulbs.

Vacuum tubes like these are still used in many television sets because they're less expensive than equivalent solid-state devices. The 2 taller units at the far right are logically called "Compactrons." Their single glass envelopes house a number of tubes' components to save space.

DATE Dec 24 1947
CASE No. 38139-7

We obtained the following A.C. values at 1000 cycles

Eg = .015 R.M.S. volts   Ep = 1.5 R.M.S volts

Pg = ~~6x10-5~~ 5.4x10-7 watts   Pp = 2.25x10-5

Voltage gain 100   Power gain 40

Current loss 1/2.5

This unit was then connected in the following circuit.

Audio signal   90 volts   Scope
261B   261B
125,000:1000   125,000:1000

This circuit was actually spoken over and by switching the device in and out a distinct gain in speech level could be heard and seen on the scope presentation with no noticeable change in quality. By measurements at a fixed frequency

8 DATE Dec 24 1947
CASE No. 38139-7

in it was determined that the power gain was the order of a factor of 18 or greater. Various people witnessed this test and listened (were present) of whom some were the following R.B. Gibney, H.R. Moore, J. Bardeen, G.L. Pearson, W Shockley, H. Fletcher, R. Bown. Mr. H.R. Moore assisted in setting up the circuit and the demonstration occurred on the afternoon of Dec 23 1947

Read & understood by G.L. Pearson Dec 24
H.R. Moore Dec 24, 1

Notebook pages of Dr. Walter Brattain record the events of December 23, 1947, the date when a transistor was first used to amplify a speech signal about 40 times. Brattain was one of the co-discoverers of the transistor. The others: Dr. William Shockley, leader of the Bell Telephone Laboratories research team, and Dr. John Bardeen.

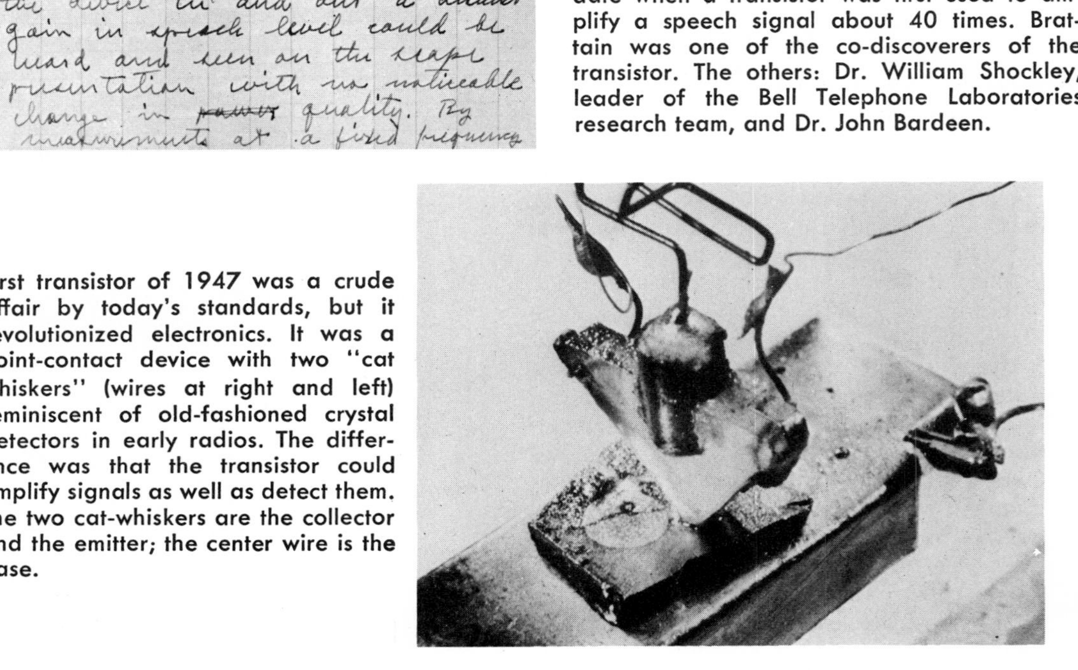

First transistor of 1947 was a crude affair by today's standards, but it revolutionized electronics. It was a point-contact device with two "cat whiskers" (wires at right and left) reminiscent of old-fashioned crystal detectors in early radios. The difference was that the transistor could amplify signals as well as detect them. The two cat-whiskers are the collector and the emitter; the center wire is the base.

especially those that handle high power.

In 1948, *Popular Science* reported on a new device, a "capsule," which, it said, "challenges the vacuum tube." Three scientists at the Bell Telephone Laboratories, Drs. John Bardeen, Walter Brattain, and William Shockley, that year developed a revolutionary solid-state amplifying device they called a "transistor."

"Solid-state" simply means that the amplifying action takes place inside a piece of solid material—a tiny crystal of germanium or silicon. This is in contrast to vacuum tubes, in which electrons flow through an open space between the electrodes.

Silicon and germanium are both considered semiconductor materials because in their pure state they are neither good electrical conductors nor good insulators; they are somewhere in between.

A transistor is built much like a miniature sandwich made of three layers of crys-

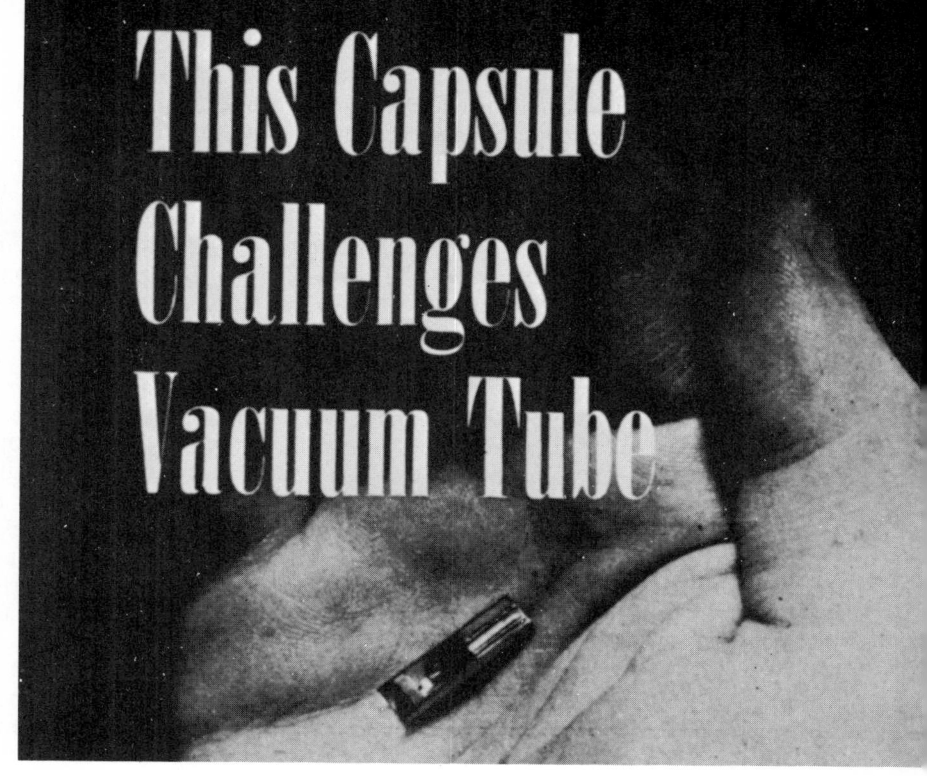

Early transistor at right, case cut away, was pictured in 1948 *Popular Science* article hailing discovery. Below: enlarged view of transistor ready for use (left) and breakaway (right) to show key parts—2 tiny wires and block of metal under the pencil point. A third wire is out of sight under the block. The transistor is about ⅓ the size shown here. One of the first consumer electronic products using transistors was the hearing aid in photo at lower right of group.

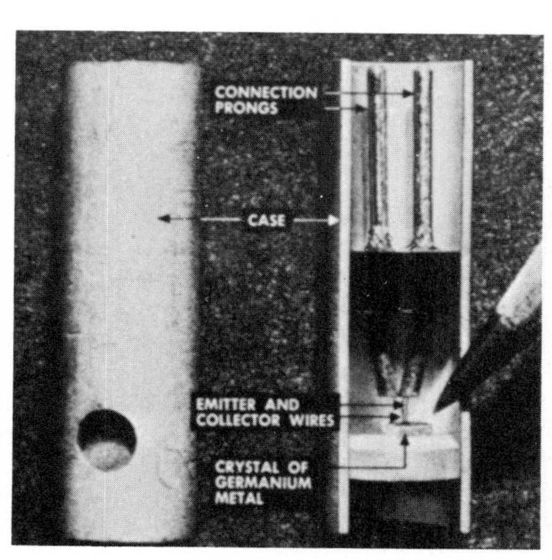

talline material. A small current fed into the central layer (the "filling" of the sandwich) causes a much larger current to flow between the two outer layers (the slices of "bread"). This is how the transistor amplifies.

"Transistor," incidentally, is a contraction of the name "transfer resistor," in recognition of the fact that the device behaves like a variable electric resistance that is controlled by the input signal.

The development of the transistor (and other solid-state devices) created a chain reaction that soon made the vacuum tube obsolete in many kinds of electronic circuitry. Semiconductor devices don't waste power heating cathodes; they are tiny and rugged; many types have essentially unlimited operating lifetimes.

An immediate by-product of the transistor's small size was the accelerating trend toward miniaturization of electronic circuitry. The designers shoehorned transistors and tiny electronic components into ever smaller packages. The results were button-size hearing aids, palm-size TVs, and shoe-box-size digital computers.

Then in the early 1960s, a new technique of circuit miniaturization evolved. The integrated circuit (or IC) came along, proving, said *Popular Science*, that "electronically speaking it *is* the little things that count."

An IC is a tiny chip of silicon—about the size of this letter "o"—that has gone through a complex fabrication process. As a result, a complete electronic circuit (composed of several transistors and other components) has been created on and within the chip.

An IC can be more reliable than the conventional circuit it replaces, since it behaves like a single component rather than a collection of dozens of parts, any of which can go bad. Also, there are no interconnections to short-circuit, and no solder joints.

Today, ICs are at work in computers, space satellites, military electronics gear, and every conceivable kind of consumer electronics product—from electric organs, to color TVs, to pocket-sized electronic calculating machines.

From the earliest days of the electronic age, designers have been fascinated by devices that respond to light. The electric eye has a long and honorable pedigree.

Nineteenth-century scientists found that the chemical element selenium would respond to light by changing its electrical resistance. More current would flow through a chunk of selenium as the light shining on its surface became brighter.

Alexander Graham Bell noted this property of selenium and tried to use the substance to build a machine that would transmit spoken sounds over a light beam.

A clever gadget using selenium photoelectric cells was built in 1916 by B. F. Meissner and J. H. Hammond. It was a battery powered, self-propelled "dog" that followed a flashlight beam controlled by its "master."

An improved version, but this one an "electronic squirrel" named Squee, was re-invented in 1952 by Edmund Berkeley, a computer expert. Squee had phototube eyes. These are diodelike tubes equipped with large, unheated cathodes made of alkali metals. When light strikes the cathodes, they emit electrons that flow to the plate. The brighter the light, the stronger the current flow.

Over the years came dozens of other devices employing circuits that see. Electric-eye devices opened doors, sorted industrial products, and put out flash fires. Cameras had them built in to control exposure.

Modern photoelectric cells bear strong resemblance to their ancestors. The cadmium sulfide photocells used in photographic exposure meters and automatic cameras work much like the early selenium cells: current flow through the cell increases when it is exposed to light.

Next to the atom bomb, radar was probably the most important secret weapon of World War II. The British credit radar as being one of the deciding factors in the Battle of Britain and the air war with Germany. Happily, since the war, radar has become equally important as a navigation aid for aircraft and shipping and as a weather-forecasting tool.

Radar (*ra*dio *d*etecting *a*nd *r*anging) has its roots in a discovery made in 1922 by two scientists at the U. S. Naval Research Labs, in Washington, D.C. Dr. A. H. Taylor and Leo Young found that objects crossing the path of a radio signal interfered with the signal's reception. Early radar schemes used separate radio transmitters and receivers positioned widely apart to make use of this phenomenon, including a "secret" German system unveiled in 1935 in *Popular Science*.

During the next decade and a half, this fundamental property of radio waves was developed in the U.S., Britain, and Germany into the prototypes of practical radar systems. The culmination was the creation of radar sets that permitted the transmitter and receiver to be at the same location.

The key to modern radar is the fact that high-frequency radio waves will carom off the surfaces of dense objects (a discovery made in 1926). In effect, if the frequency is high enough, the radio waves behave like light waves striking a mirror.

Thus, the "detection" function of radar is performed by a high-frequency radio beam launched into space by a special vacuum tube called a "magnetron" (the British developed the first unit in 1940). Any object—ship, plane, cluster of raindrops—in

Actual size of 3 integrated circuits can be seen by comparing them with silver dollar on which they are resting. The inset picture shows 3 stages of electronics evolution: a vacuum tube (base shown), a transistor, and (lower right corner) a tiny integrated circuit.

front of the beam reflects part of it back to the radar set as an "echo" signal.

In 1934 the ranging (or distance-measuring) feature of radar was invented. The technique involves chopping the radio beam into a series of short pulses. Elaborate electronic circuitry measures the total time required for each pulse to travel from the transmitter to the target and back again to the receiver. The speed of radio waves is a constant 186,000 miles per second, so, by measuring the round-trip time of the reflected signal, distance to the target can be accurately determined.

A typical radar system consists of a high-frequency transmitter and a matching receiver, connected to a highly directional antenna that can produce a sharp, narrow beam of radio waves. The transmitter sends off short bursts of radio waves, and the receiver listens for echoes. The antenna rotates to provide 360-degree search coverage.

A modern automatic camera uses a light meter to measure light that is coming through the lens and to adjust the lens opening and the time of film exposure. Meter's active element is a cadmium sulfide photocell.

Robot Squirrel Hunts His Own Food

When you shine light in Squee's phototube eyes, this little electronic squirrel goes nuts.

PSM PHOTOS BY HUBERT LUCKET

EDMUND C. BERKELEY, an expert on electronic computing machinery, created Squee to "stir people's interest in the extraordinary possibilities of robots."

LIKE REAL SQUIRREL, Squee has sensing organs (phototubes, switches, feelers); acting organs (three motors to drive, steer and operate scoop); and thinking organs (12 relays, seven vacuum tubes). Powered by one 12-volt wet battery and two 67½-volt dry batteries, what Squee really eats is electricity.

THIS electronic animal behaves like a real squirrel. It looks for a nut, runs to it, picks it up, carries it to a nest, then hunts for another nut. The nuts that it collects, however, are usually golf balls, and the nest to which it takes them is an aluminum plate. The sensitivity of Squee's phototube eyes accounts for his action. They respond differently to a steady light than to a flickering light. Whoever runs the squirrel must place a steady light (in this case a flashlight) behind the nut. When the nut enters the scoop, the squirrel's circuits are switched to make him search for a flickering light placed over the nest. A gas-filled lamp that flashes on and off 120 times a second then lures Squee to his nest, where he deposits the nut—and starts all over again.

Reproduction of 2-page *Popular Science* item describes and illustrates mechanical "squirrel" of 1952 with phototube "eyes." Guided by lights, it automatically hunts and collects food (golf balls) and carries it to nest.

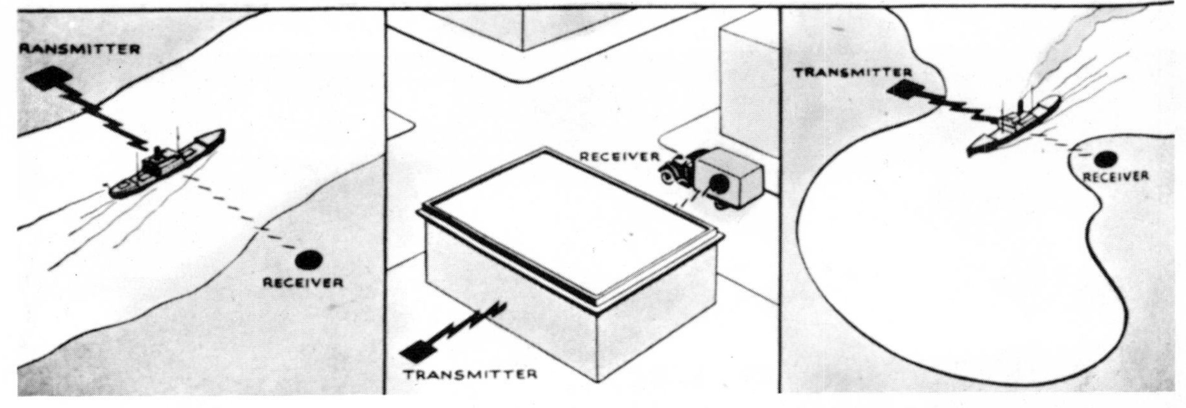

As early as 1922, U.S. scientists discovered that objects crossing path of radio signals hindered reception.

Similar effects were noted if a building stood between a truck-mounted receiver and a transmitter.

At harbor entrances, waves sent between transmitter and receiver on opposite shore would detect passing ships.

A vital element of any radar system is the display (or screen) that shows the operator the echo signals returning from objects in the path of the beam. Virtually all modern radars use a CRT display that looks somewhat like a television picture tube. Aircraft, ships, clouds, and land features appear as bright "blips" on the glowing screen.

In recent years, the magnetron tubes from radar sets have found a new application: they now serve as the heart of microwave ovens. The magnetron produces a stream of high-frequency waves that are absorbed by food placed in the oven. The microwaves cause internal heating of the food, and the food literally cooks itself, and does it in a fraction of the time required by conventional ovens. And because the microwaves bounce off the metal walls inside the oven, the oven itself stays cool.

During World War II the U. S. Navy found itself buried under a mountain of complex calculations required to produce ballistic tables for their big guns on the high seas. They needed machines that could "think," so they turned to electronics for help. In an air-conditioned basement at Harvard University, IBM built a 35-ton calculating machine, called the "automatic sequence-controlled calculator," under the direction of Commander Howard H. Aiken, a Harvard engineering professor. *Popular Science* in 1944 called it a "robot mathematician that knows all the answers." Actually, it was the first practical programmed computer.

Programming is the key to computer operation. A program is simply a detailed series of instructions that tells the computer what to do and when to do it. A vital part of a computer is a machine that will do basic arithmetic operations (addition, subtraction, multiplication, and division). Programmed operation sets the machine free to work at fantastic speed. Once the human operators have written the program of instructions and given the computer the numbers to work with, the machine goes off on its own to solve the most difficult problems.

The Aiken machine used a combination of electronic, electrical, and mechanical gear to do arithmetic. It could add or subtract 23-digit numbers in 3/10 of a second, multiply in 5.8 seconds, and divide in 14.7 seconds. The program of instructions was stored on a punched paper tape that unwound automatically, as the calculations progressed, and controlled circuitry that directed the computer's operation. Numbers for the machine to work with were entered by setting a bank with 1,440 dials.

Shortly after World War II, the first all-electronic computing machine was built at the University of Pennsylvania by Dr. John W. Mauchly and J. Presper Eckert, Jr. It was called the Eniac (for *e*lectronic *n*umerical *i*ntegrator *a*nd *c*omputer). It was 1,000 times faster than the Aiken computer.

Eniac filled a room 30 feet wide and 50 feet long. It was built of 18,000 vacuum tubes, contained about a half-million soldered connections, and cost $400,000.

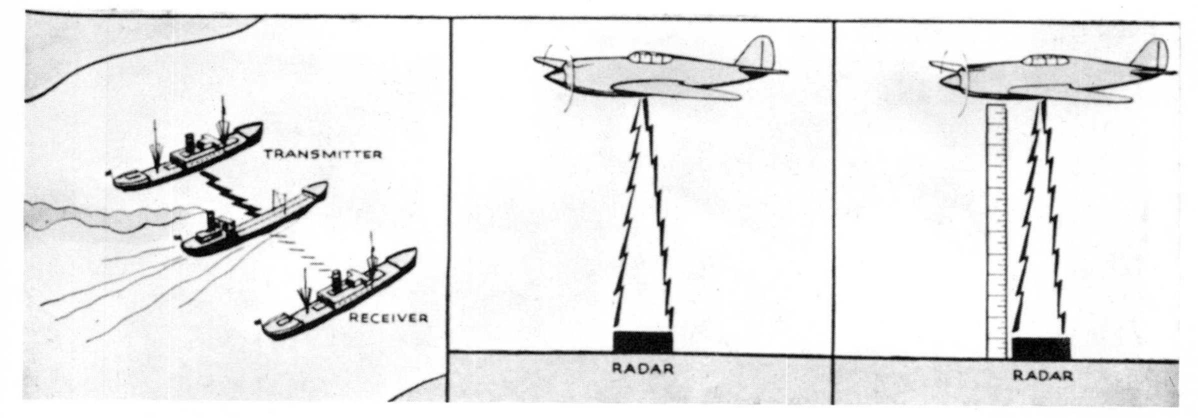

On land or sea—device was limited by need for gear on opposite sides of the target.

In 1926 it was found that surfaces would reflect radio waves—and transmitters and receivers got together.

Last step was measuring distance between radar and plane. In 1934 round trip of waves was first clocked.

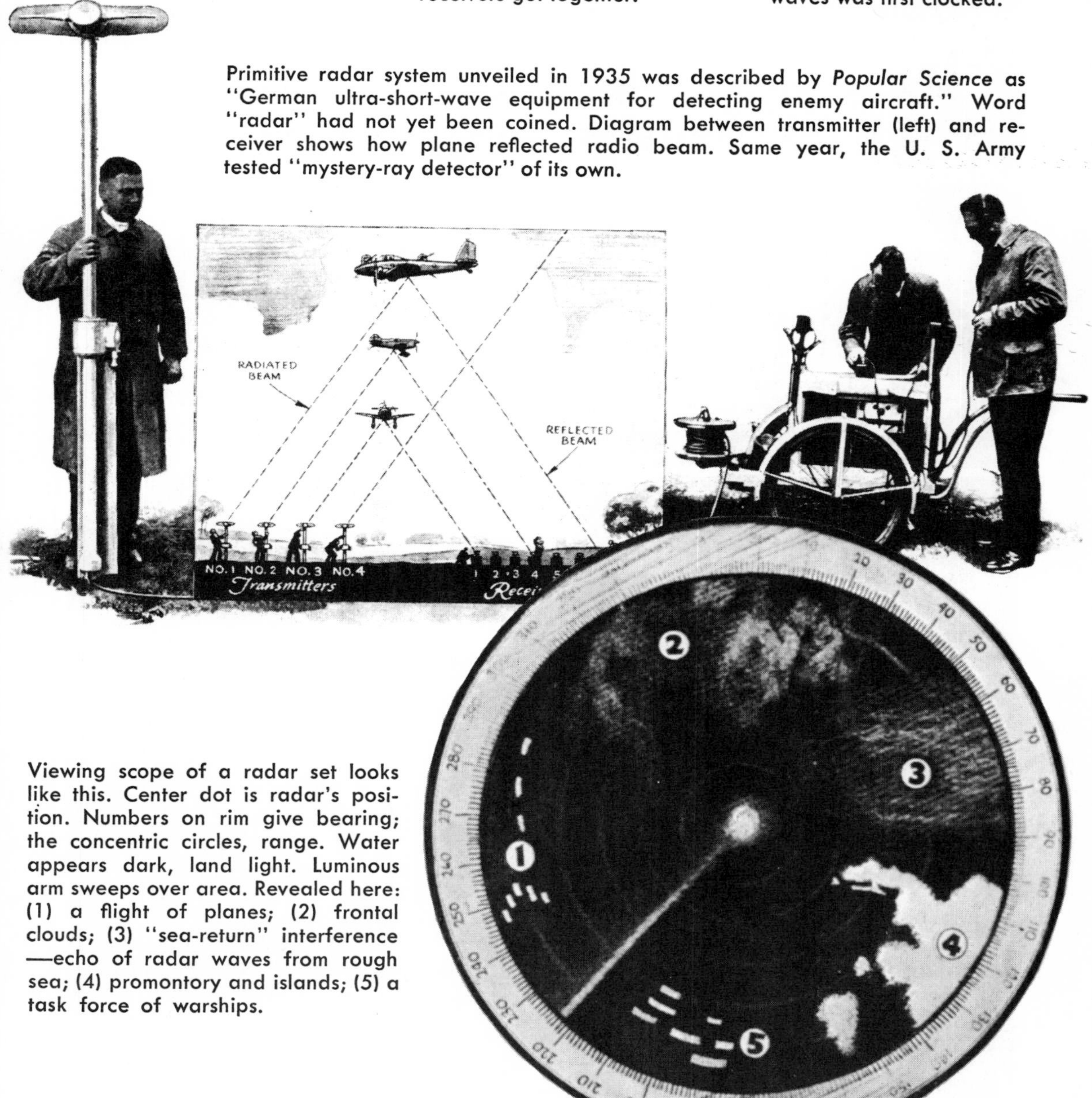

Primitive radar system unveiled in 1935 was described by *Popular Science* as "German ultra-short-wave equipment for detecting enemy aircraft." Word "radar" had not yet been coined. Diagram between transmitter (left) and receiver shows how plane reflected radio beam. Same year, the U. S. Army tested "mystery-ray detector" of its own.

Viewing scope of a radar set looks like this. Center dot is radar's position. Numbers on rim give bearing; the concentric circles, range. Water appears dark, land light. Luminous arm sweeps over area. Revealed here: (1) a flight of planes; (2) frontal clouds; (3) "sea-return" interference—echo of radar waves from rough sea; (4) promontory and islands; (5) a task force of warships.

The first all-electronic computer was the Eniac, built at the University of Pennsylvania shortly after World War II. It cost $400,000.

The first Univac computer used 1,500 vacuum tubes mounted on pull-out racks like those at left, *Popular Science* reported in 1949, the year it was built. It cost $200,000.

An electronic memory system temporarily remembered numbers the Univac was working with. Electrical signals produced sound waves, which traveled along columns of mercury arranged within a drum-shaped structure.

By 1949 the electronic computer as we know it today was taking shape. The British Edsac computer contributed the concept of internal program storage. Instructions for running the calculations were electronically "remembered" inside the machine, rather than externally on paper tape, significantly boosting computing speed.

The first Univac (for *univ*ersal *a*utomatic *c*omputer) built in 1949, for example, could perform 500 multiplications a second under program control. Program steps and problem numbers were stored on magnetic tape, and numbers produced while the program was in progress were stored in the form of sound waves moving back and forth along columns of liquid mercury.

In the two decades following Univac I, astounding progress has been made in boosting the operating speed and capacity of digital computers. Modern machines use integrated circuits instead of vacuum tubes, and they can perform millions of calculations per second.

Today computers are at work in business, law enforcement, and science. The big 7074 even watches over your income taxes, as *Popular Science* warned it would in 1963. Applications have gone far beyond the wildest dreams of the early pioneers, and no end in sight. It seems incredible that when Howard Aiken first conceived his computer, he believed that only one machine of its type would ever be needed. This year, tens of thousands of different kinds of electronic computers will be built.

Close to the turn of the century, one of the great schemes of the pioneers of electronic science was the concept of wireless power transmission. Heinrich Hertz, the discoverer of radio waves, was one of the first to suggest the feasibility of generating powerful radio beams that could be "broadcast" over long distances and converted into usable electric power at the "receiver."

Nikola Tesla, the famed electrical engineer who developed the induction motor (and other fundamental electrical devices) went one step further: He proposed transmitting high-frequency electric currents

Radar magnetron tubes are now being used in microwave ovens that cook a meal in 1/4 the usual time.

Giant tower was intended by Nikola Tesla to transmit power through earth. It was mysteriously destroyed in 1914.

Beam of microwave energy carried this miniature helicopter to a height of 50 feet. A bedspring-like blanket of solid-state diodes converted the beam into electrical power to work the craft's electric motor.

through the earth, from special transmitting towers to widespread receiving stations. One such tower was built on New York's Long Island.

But the radio-beam concept has stayed alive for seven decades because it seems like such a good idea. In theory, massive power-generating plants could be located far from populated areas (one recent proposal suggests on the moon!) and users of electric power could tap the power beam simply by erecting an antenna in its path.

Unhappily, in practice there are many thorny problems to be solved. Among them: the difficulty of converting electric power into radio waves and back again at tolerable efficiences; the difficulty of focusing the radio waves into a narrow enough beam to travel long distances without spreading; and the certainty that anyone or anything passing in front of a high-energy-density beam would be french-fried.

Where does the future of electronics lie? In an exciting parade of new products and devices. If we let the pages of *Popular Science* be our crystal ball, we can single out 3 developments as having special significance for the decade ahead:

Large-scale integration (LSI) is the next plateau of integrated-circuit technology. Whereas ordinary ICs pack a few dozen components into a small silicon wafer, LSI devices cram *thousands* of parts onto a single chip. Often, the component density of an LSI device is 500,000 parts per square inch of silicon surface.

A single LSI device can contain the complete circuit of a TV or stereo set; a dozen can implement a good-sized computer.

As the cost of producing LSI drops, we may see the long-awaited day of "throw-away" electronic circuitry that is simply discarded if it fails.

Minicomputers will revolutionize our production industries and bring computerized automation to diverse fields such as health care and supermarket merchandising. These compact (shoebox size), low-cost (some now cost about $2,000—with prices coming down every day) special-purpose electronic brains can be designed to perform specific control and computational tasks, such as monitoring the conditions of several patients in an intensive-care ward, and automatically signaling an alarm in case of trouble. Or controlling an automatic assembly line or keeping tabs on the inventory of a supermarket by automatically tallying the individual cans of soup, say, as they are rung up on the cash registers.

The stage beyond this will be the development of even smaller minicomputers that may some day take over the control of automatic transmissions and fuel-injection systems of automobiles. Or ones that will "run" a home of the future, automatically monitoring fire-alarm and burglar-alarm systems, automatically locking and unlocking doors and windows, turning lights on and off, keeping tabs on lawn sprinklers, and even warning of low level in the oil tank.

Light-emitting diodes (LEDs) will literally change the way we look at things. A LED is a solid-state source of light that has an almost unlimited life. Scientists predict models that will produce light for 1,000,000 hours before their brightness fades.

Right now, there are LEDs under development that will be bright enough to be used as illumination sources. Others will be used in numerical displays that may replace the tuning dial on radios, and, eventually, the speedometer needle in cars.

Still others will be used to create flat TV picture "screens" that you'll be able to mount on the wall as easily as a picture.

And given time, electronics will take over your car on the freeway while you simply sit back enjoying a TV program or telephone ahead over Bell System's new Picturephone.

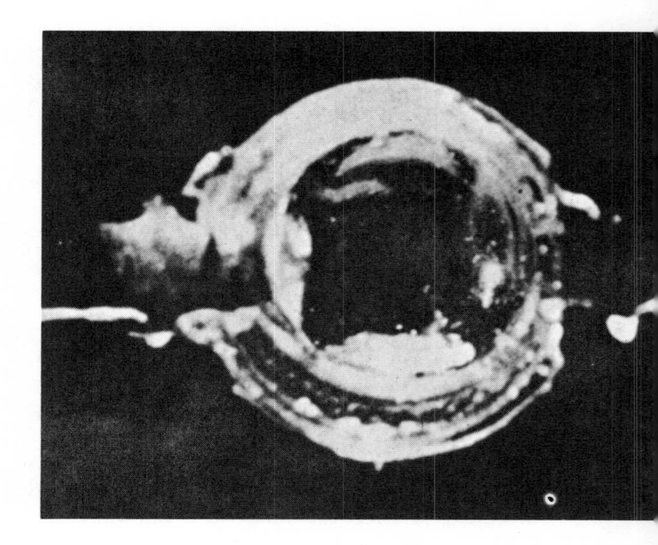

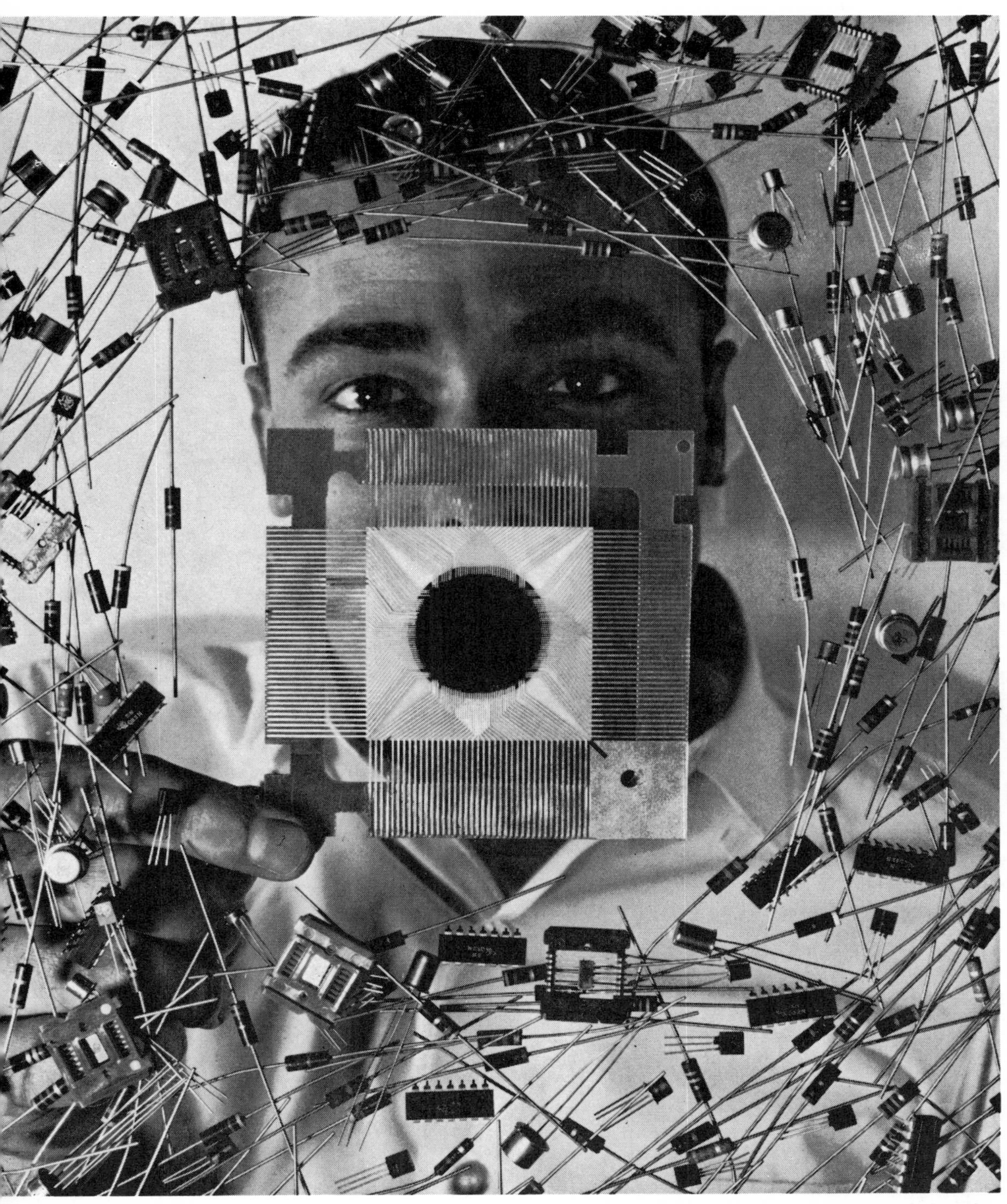

On facing page: enlarged view of light-emitting diode that may edge out conventional lamps for many lighting jobs.

Large-scale integration, represented by this Texas Instruments device, employs layer on layer of integrated circuits with hundreds of components. A coming thing, it promises "throwaway" electronic circuitry that will simply be replaced when it fails.

CHAPTER 15

So What Else Was New?

A pictorial scrapbook rounds out the story of popular science from 1872 to 1972 with assorted innovations, novelties, and adventures that have helped to make over our everyday life, our ways of travel, and our knowledge of the world about us—a miscellany of science highlights

The typewriter was invented not for business but to help the blind. Devised in 1868, it had assumed this marketable form by 1874. It had a lid, to be swung down over the keyboard and locked. Early typist is pictured at work.

"Giant" ocean-going steamship of 1876 weighed 5,000 to 6,000 tons. Sails helped propel liner. The *Germanic* (British), pictured, set a transatlantic record in that year, covering 2,780 nautical miles in 7 days, 11 hours, 37 minutes.

The standard passenger and express locomotive of that time developed 800 horsepower—which *Popular Science* said in a meticulous footnote was nearly equivalent to the actual power of 1,200 horses. It could pull a 150-ton train at a speed of 60 miles an hour.

The Mississippi River steamboat, driven by paddle wheel because water depth would not permit the use of screws, was in its full glory. Some could cover the distance from St. Louis to New Orleans—1,200 miles—in 4 days.

An electrical wonder of 1880 was the rein that administered a shock to fractious horses. The driver generated current by cranking a magneto. When a horse felt this, testified the inventor, he was so surprised he stopped short.

An electric burglar alarm with bell and pointer showed in which room of a house a miscreant was at his nefarious work. The device was powered by the Leclanché wet-cell battery at the right.

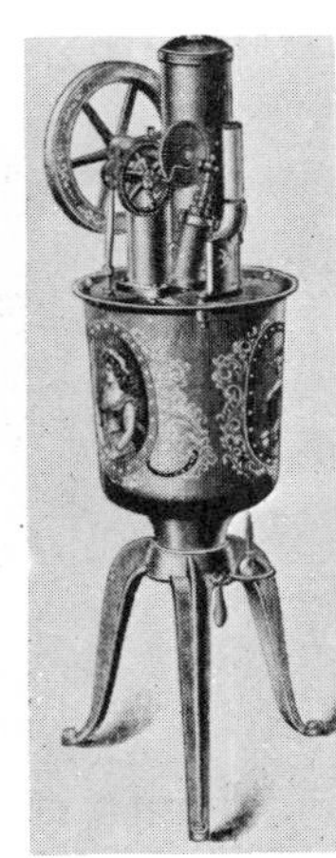

Home appliances were powered by gas or steam. Gas engine (above, left) produced "about one man-power." Tiny steam engine (right), burning gas, coal, oil, or wood, was "ornamental in design and handsomely finished, fitting it for any room of a dwelling . . ."

Solar energy printed a newspaper in 1882, in the garden of the Tuileries, Paris. A parabolic mirror directed the sun's rays at a boiler to generate the steam that ran the printing press.

Statue of Liberty went up in 1886, on site overlooking New York Harbor. Monumental gift of France to United States, statue by F. A. Bartholdi had crossed Atlantic dismantled, in 214 packing cases. Erected on granite-faced pedestal, it became cherished symbol of America.

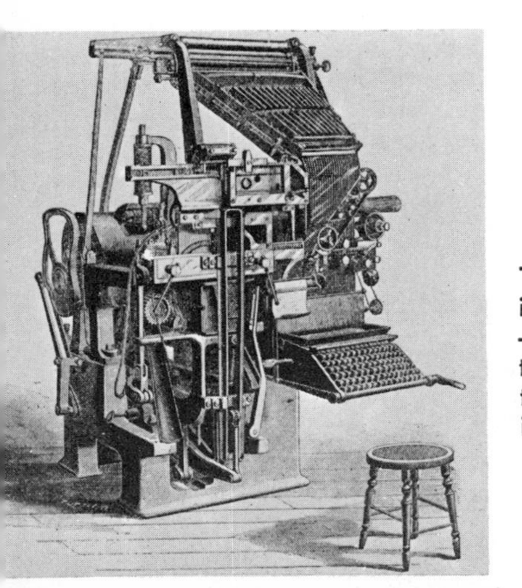

Type-casting machines gave the printing industry a rebirth. Typesetting—by hand—had not changed in 500 years. Linotype of 1885 (left) was 4 times faster than an expert hand compositor, counting the return of letters to their cases.

In 1887 France's Ferdinand de Lesseps was trying to dig a canal across the Isthmus of Panama to top his feat in building the Suez Canal. The chief machine used consisted of scoops on an endless belt (above). After seven years the project failed. U.S. tackled job in 1904. Panama Canal opened in 1914.

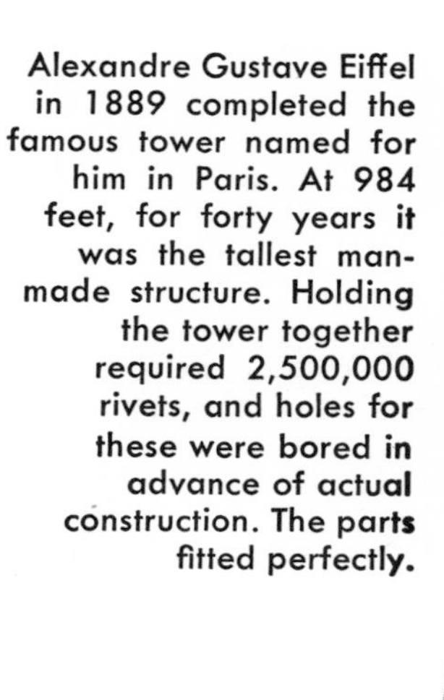

Alexandre Gustave Eiffel in 1889 completed the famous tower named for him in Paris. At 984 feet, for forty years it was the tallest man-made structure. Holding the tower together required 2,500,000 rivets, and holes for these were bored in advance of actual construction. The parts fitted perfectly.

This early bicycle drew from a doctor of philosophy an article of 9 ½ pages in *Popular Science* explaining how the thing remained upright. Picture showed how "lines of force" controlled the bike in motion.

King C. Gillette invented the safety razor five years before the turn of the century to render shaving harmless for the first time. He simply inserted a guard between the skin and the razor blade.

Marvels of the artisan's skill were mechanical music players employing all manner of "records," from perforated steel discs (below) to cylinders studded with projecting pins. The juke de luxe shown here was one of the finest instruments of the Gay **Nineties. Today it's a collector's item.**

The horse-drawn fire engine, with pumps operated by steam power, was common fixture in American cities in 1895. General adoption of self-propelled fire engines was still fifteen years away. Hand-pulled steam equipment was used to fight fires as early as 1840.

The world's first steam-turbine vessel, the *Turbinia,* was completed in 1897. Propelled by engines designed on the same principle as that adapted to aircraft decades later—using gas instead of steam—it achieved the previously unheard-of speed of more than 37 miles an hour. *Turbinia* was 100 feet long, with a displacement of 44½ tons.

Arrival of the electric trolley spelled the end of horse-drawn street cars. These, in Washington, D.C., took power from underground conduits.

The outboard motor was invented in first decade of the twentieth century. The 1909 Evinrude (above, left) had a knob on its horizontal flywheel for starting. At right is one of the same brand fifty-eight years later. A V4 design of 100 hp, it set a world outboard speed record of 130.9 mph.

Halley's comet, named for the astronomer who discovered it, coursed through earth's sky in 1910. It's due to appear again as soon as 1986.

Searchers for the early ancestors of modern man were led astray in 1908 when "Piltdown man," shown here restored from bone fragments, was unearthed in England. In mid-century he turned out to be a hoax—the evidence had been "planted." The jaw was from a modern ape, stained to look old, the teeth filed flat.

"Ice" skating in the summer without ice (left) was proposed in 1916. One artificial rink consisted of soluble glass, fluor-calcium, asbestos, ground glass, paraffin, and soapstone. It proved easier to manufacture ice.

The slide fastener, or Zipper, in 1917 began easing the job of getting into and out of galoshes and trousers. Actually patented about a quarter-century earlier, it needed only a slight improvement—a change in the shape of the teeth—to permit mass production.

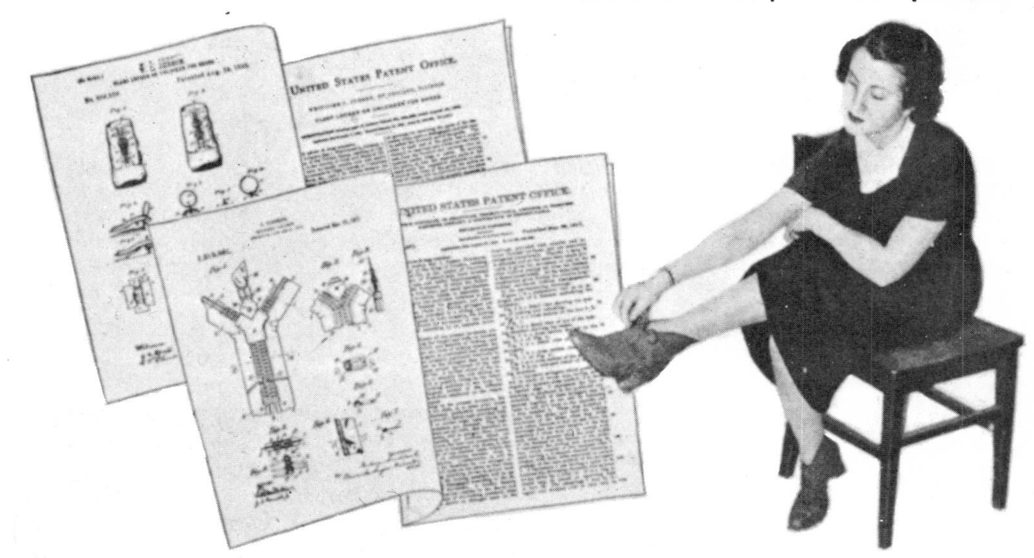

Collapsible aluminum tubes for items like shaving cream and toothpaste (above) arrived in 1920. The first advertisement in *Popular Science* picturing such a tube appeared that year. Before, soap sticks served for shaving and powder for teeth.

The voting machine (pictured on the cover of the magazine, right) began simplifying the job of counting election returns in 1920 too. Pulling levers instead of marking paper ballots—which could be tampered with—assured honest elections.

The hydrofoil boat dates from 1920. Co-experimenter on 70-mph craft in top photo was Alexander Graham Bell, the telephone inventor. A modern successor of the 1960s is shown below it.

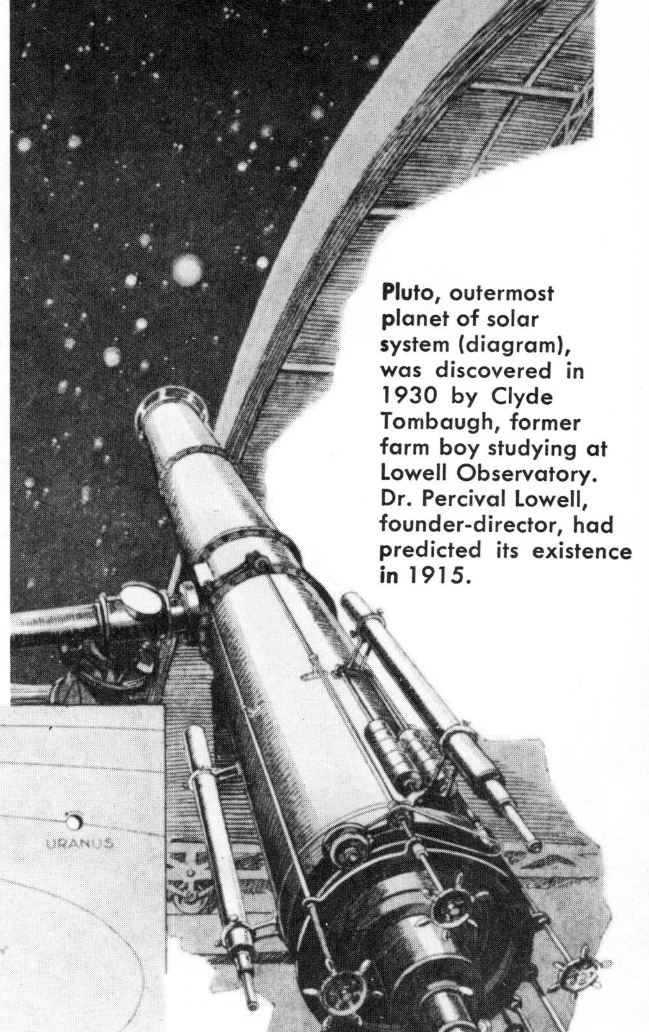

Pluto, outermost planet of solar system (diagram), was discovered in 1930 by Clyde Tombaugh, former farm boy studying at Lowell Observatory. Dr. Percival Lowell, founder-director, had predicted its existence in 1915.

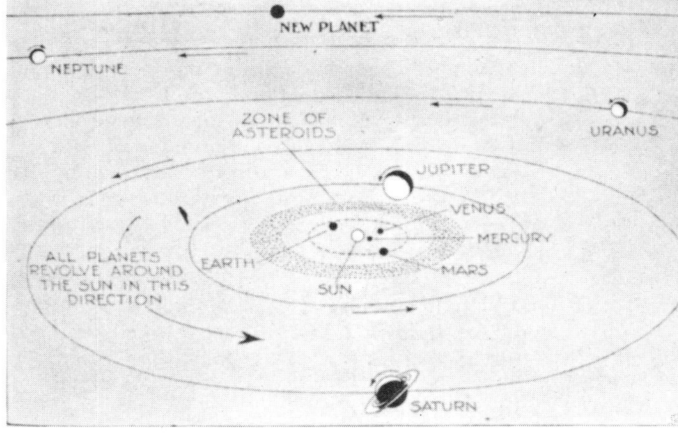

The coffin of Tutankhamen, Egyptian boy-king who ruled 400 years before Solomon, was archeology's biggest find in decades. It had been beaten out of $250,000 worth of gold bullion.

In a 1925 *Popular Science* issue: a sail-less sailing ship, the *Buckau* (right), propelled in wind by whirling "Flettner rotors." It worked, as its German inventor Anton Flettner foresaw, but it proved no competitor for steam or diesel ships.

Superfast streamlined trains made their debut in 1934. The Union Pacific and the Chicago, Burlington and Quincy put them into service to achieve new highs in sustained speed. The latter's "Zephyr," pictured above, made 104 mph where roadbed permitted, and engineers said it was capable of 125. All steel, nonetheless it was so light 10 men could pull it. The secret lay in new lightweight diesel-electric motive power. All this notwithstanding, in the following thirty years the airlines captured most long-distance passenger traffic.

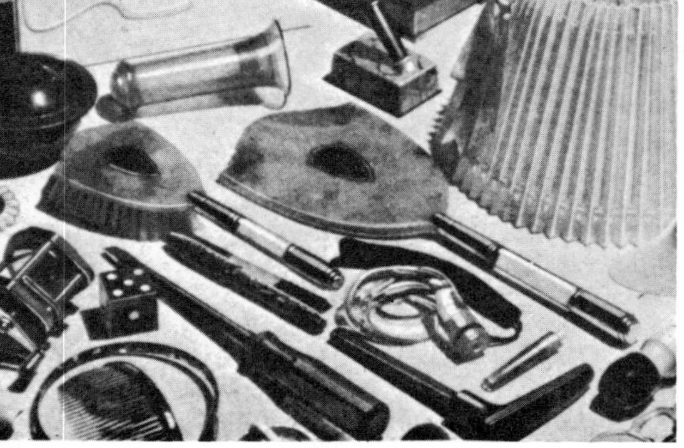

It was as the Decade of Plastics that the 1930s made their most conspicuous mark on the everyday world (above). Plastic materials invaded homes.

Superliners like these battled for speed supremacy on the North Atlantic. In *Popular Science,* a 1937 feature (right) pictured *Queen Mary* (top) and *Normandie*. *Queen Mary* topped the pack, covering 3,120 nautical miles from Ambrose Light to Bishop's Rock in August 1938 in 3 days, 20 hours, 42 minutes. U.S.'s *United States* beat that by 10 hours, fourteen years later. But the age of golden galleons was ending. Passengers preferred airliners. The vessels became cruise ships or were retired.

Battle
OF THE
Superliners

Contenders for the blue ribbon of the Atlantic. Note the larger

SULFANILAMIDE...
A Waste Dye Product
Is Medicine's Latest
Germ-Killing Weapon

"Wonder drugs" came out of the laboratory to combat man's infections. The first were sulfanilamides, waste dye products used against meningitis and 13 other diseases including scarlet fever. Three years after announcement, the drugs had saved the lives of 12,000 persons in the United States alone, it was estimated.

Penicillin was even more powerful against germs than the sulfas. An excretion from a common mold discovered in 1929, it was first produced in useful quantities in World War II. In photo above, a laboratory worker examines a culture of the mold on the surface of a sugar solution.

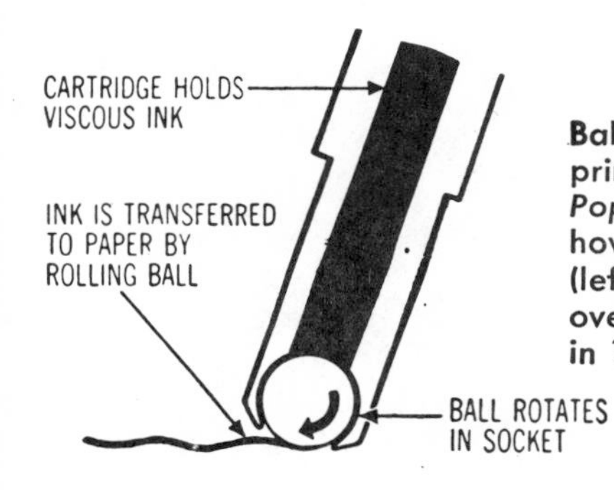
CARTRIDGE HOLDS
VISCOUS INK
INK IS TRANSFERRED
TO PAPER BY
ROLLING BALL
BALL ROTATES
IN SOCKET

Ballpoint pen, its principle explained by *Popular Science* in how-it-works diagram (left), began taking over the pen market in in 1945.

Slinky leaped into space like a striking serpent at a flick of the wrist. It flip-flopped over and over down a flight of steps. Merely a flat-coiled strip of Swedish blue steel, Slinky became a toy when an engineer saw spring roll off a workbench and do antics on floor.

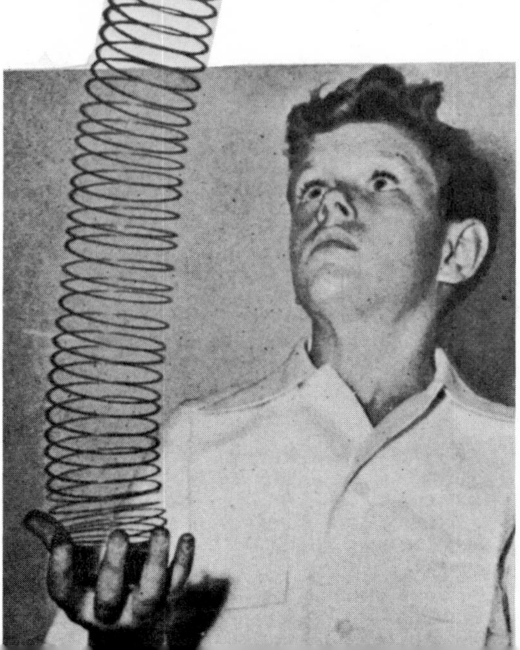

Two novelty toys appeared in 1947—the mysterious drinking bird and Slinky the Spring. Secret of bird's performance: methyl chloride changing back and forth from liquid to vapor.

The most fascinating scientific detective story of modern times began to unfold in 1947. Men came upon scrolls that were buried at the time of Christ. Known as the Dead Sea Scrolls, they were thin sheets of leather and copper. Their subject matter ranged from the Old Testament's Book of Isaiah (shown above, with an old print of the prophet inset at right) to prophecy, prayers, history, laws, and descriptions of places where treasure was hidden from the Romans. The period covered was from about 100 B.C. to A.D. 68. But who these people were, aside from being Jewish, and whether they had any direct contact with Christ, went unanswered. In A.D. 70 the Romans burned Qumran, home of the sect that produced the scrolls.

In 1947 the most powerful mirror telescope built to photograph the heavens (left) was put in operation on California's Mt. Palomar. A keen eye can spot the observer sitting (center) at the focus. It had taken twenty years to plan and build the 200-inch monster, twice as big as the one at Mt. Wilson Observatory. The mirror was cast of Pyrex, the same glass used in coffeemakers and ovenware. The 530-ton telescope is supported on a film of oil 3/1000 of an inch thick, forced into bearings at high pressure.

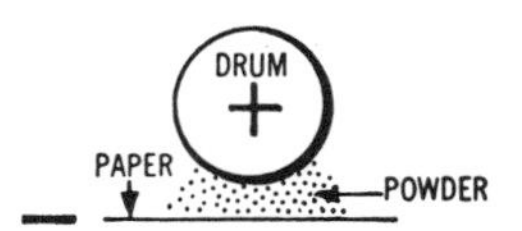

The Xerox copying machine did not become as usual as typewriters in offices until the 1950s, though it had been invented in 1938. Cutaway at right and diagram above shows how one kind works to produce duplicates for pennies. Key to its operation is electrified powder, fused on to any kind of paper positively charged in the process. Some copiers on the market use the blueprint principle—chemical action induced by light. Another type works by heat.

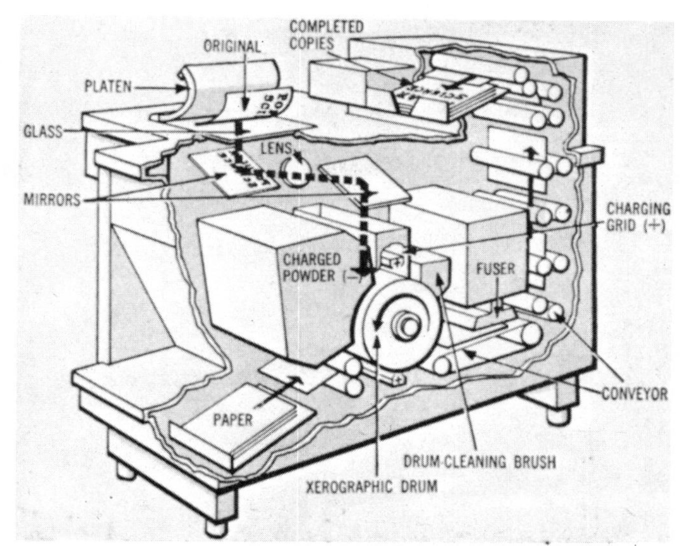

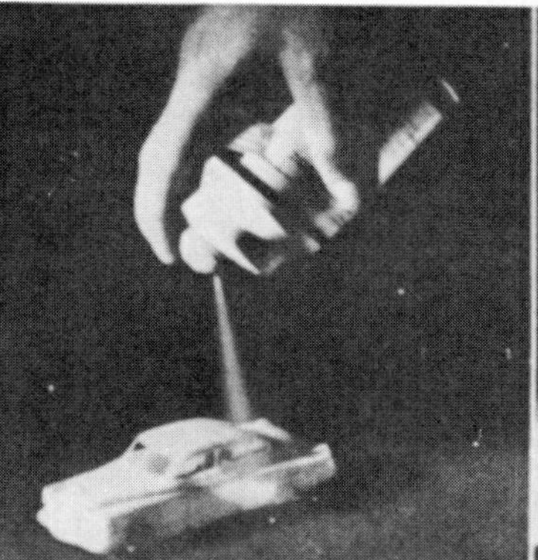

World War II's "bug bomb," dispensing DDT, led to the spray cans of the 1950s, above. Now they dispense everything from shaving soap to shoe polish, from paint to pancake batter. Light tinplate, they're essentially beefed-up beer cans. The propelling agent, an inert liquid, vaporizes as the contents are used.

In 1953 the war against "polio"—poliomyelitis, the dreaded killer and crippling disease—was won with March of Dimes research funds (right). Vaccines were developed for all 3 strains of the virus that caused it, so small that electron microscopes cannot detect them.

GET YOUR
POLIO
SHOTS
AND PLAY
SAFE
THE NATIONAL FOUNDATION FOR INFANTILE PARALYSIS

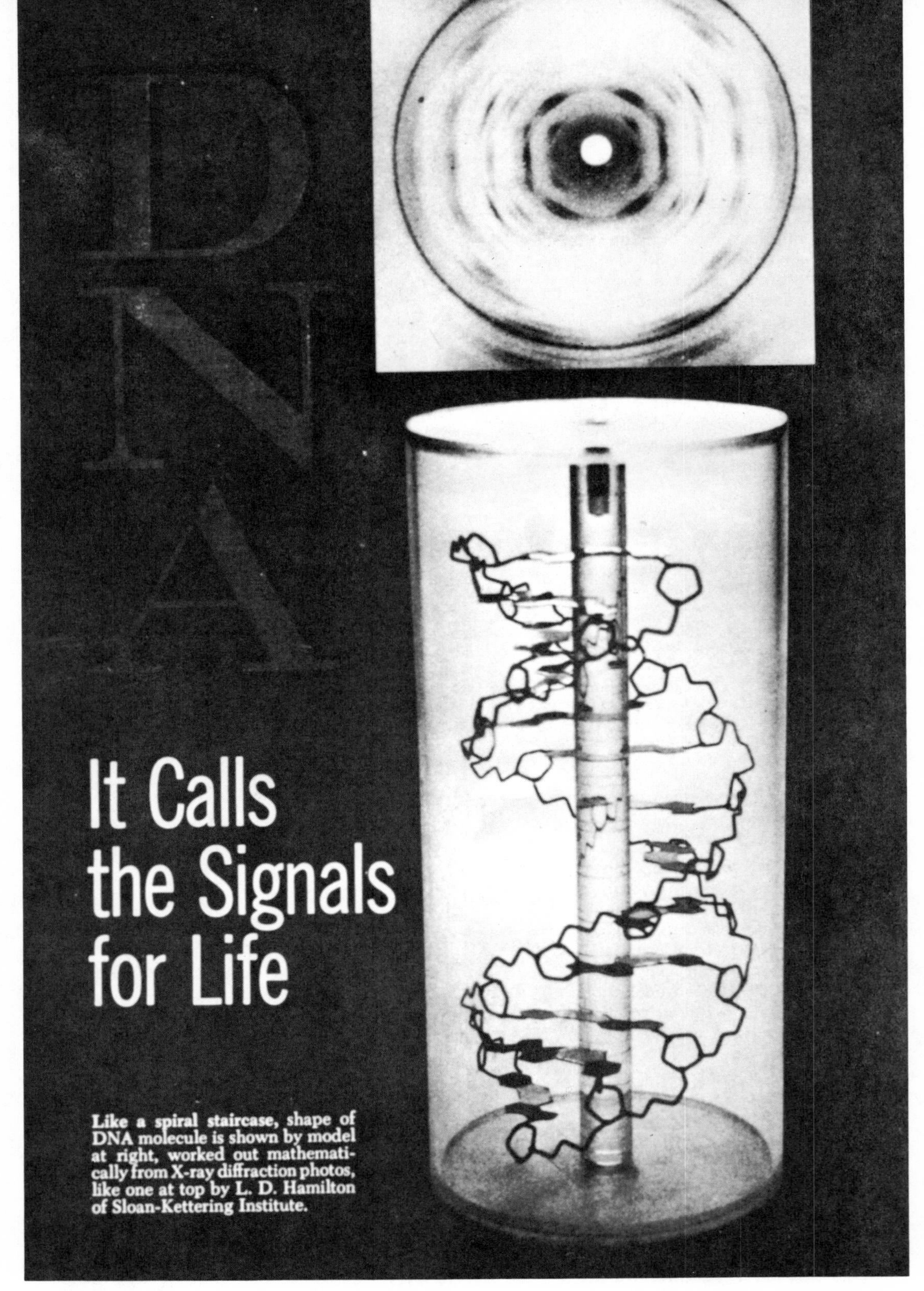

It Calls the Signals for Life

Like a spiral staircase, shape of DNA molecule is shown by model at right, worked out mathematically from X-ray diffraction photos, like one at top by L. D. Hamilton of Sloan-Kettering Institute.

In 1962 three scientists were awarded the Nobel prize for discovering, nine years before, the structure of life's most important molecule—deoxyribonucleic acid, known familiarly as DNA. They were Drs. James D. Watson, Francis H. C. Crick, and Maurice H. F. Wilkins. For more than a century scientists had been asking questions: How do egg cells "know" how to become plants, animals, people? How does a heart muscle "know" its role? Answer: DNA, the substance that genes are made of.

Eight times within the century men on foot tried to reach the top of the world, Mt. Everest, 29,002 feet high (foreground in air photo above). Eight times they failed. Several men perished in the attempts, succumbing to lack of oxygen, fatigue, glaciers, and sheer cliffs. Surrounded by formidable companion mountains of the Himalayas, on the border of Tibet and Nepal, Everest finally was conquered by New Zealand's Sir Edmund Hillary and a Nepalese Sherpa companion, Tenzing, on May 29, 1953. The 2 men are pictured at right as they neared summit. Hilary breathed oxygen from a flask. But, coming down, he was exhausted.

On a hot treasure hunt, a California utilities company struck steam. In a test of steam pressure in Sonoma County, workmen let wellheads (left) blow off for 3 weeks. Then, in 1960, they proceeded to harness the steam to generate electricity. Where in the world did the steam come from? From the earth's core. Only 20 miles of crust insulates mankind from a caldron of unthinkable energy. Where molten rock invades the crust, volcanoes and steam vents occur. Surface water, seeping down, becomes steam.

A new kind of mammoth transport without wings, wheels, or a conventional hull (above) had been invented—an air cushion vehicle, or Hovercraft. A 42-ton version built in 1964 could carry 150 persons. Later Hovercraft shuttled **across the English Channel.**

Japan led the world in high-speed trains. The one above, electrically driven, could make 126 mph, and in 1964 began averaging more than 100 over the 322 miles from Tokyo to Osaka.

The biggest bridge in the world, the Verrazano-Narrows suspension span leaping the entrance to New York Harbor, was opened in 1964. Sixty feet longer than San Francisco's Golden Gate Bridge, its main span measures 4,260 feet. Workmen are shown here in the midst of construction, attaching hangers to one of the main cables to carry the 13,700-foot roadway. Four such cables, each 35⅞ inches thick, had to be spun for the $325,000,000 bridge. Their 142,500 miles of steel wire would circle the globe 6 times.

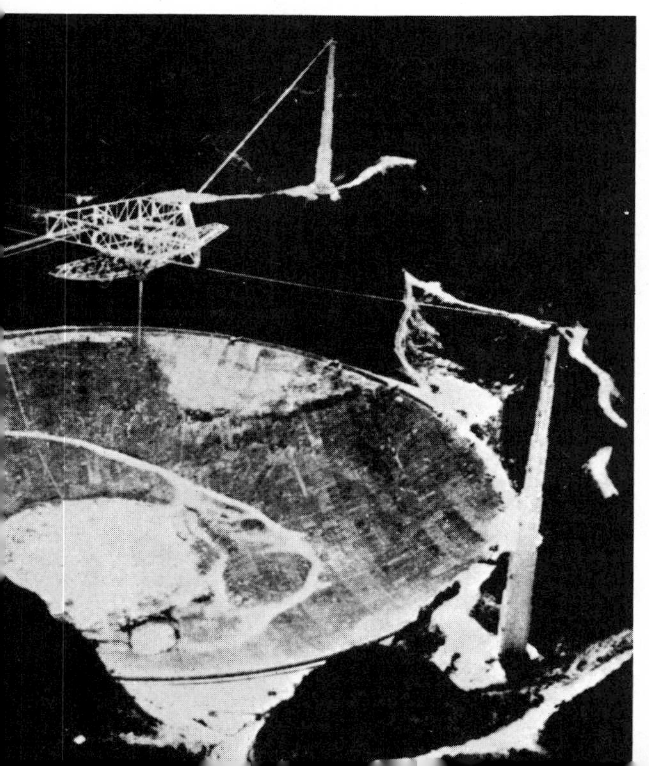

The world's largest telescope (left), 1,000 feet wide, for the new science of radio astronomy went into operation at Arecibo, Puerto Rico. The dish itself does not move. Three towers, 265 to 300 feet high, suspend a transmitter-receiver carriage above center. The 'scope is "aimed" by shifting the carriage.

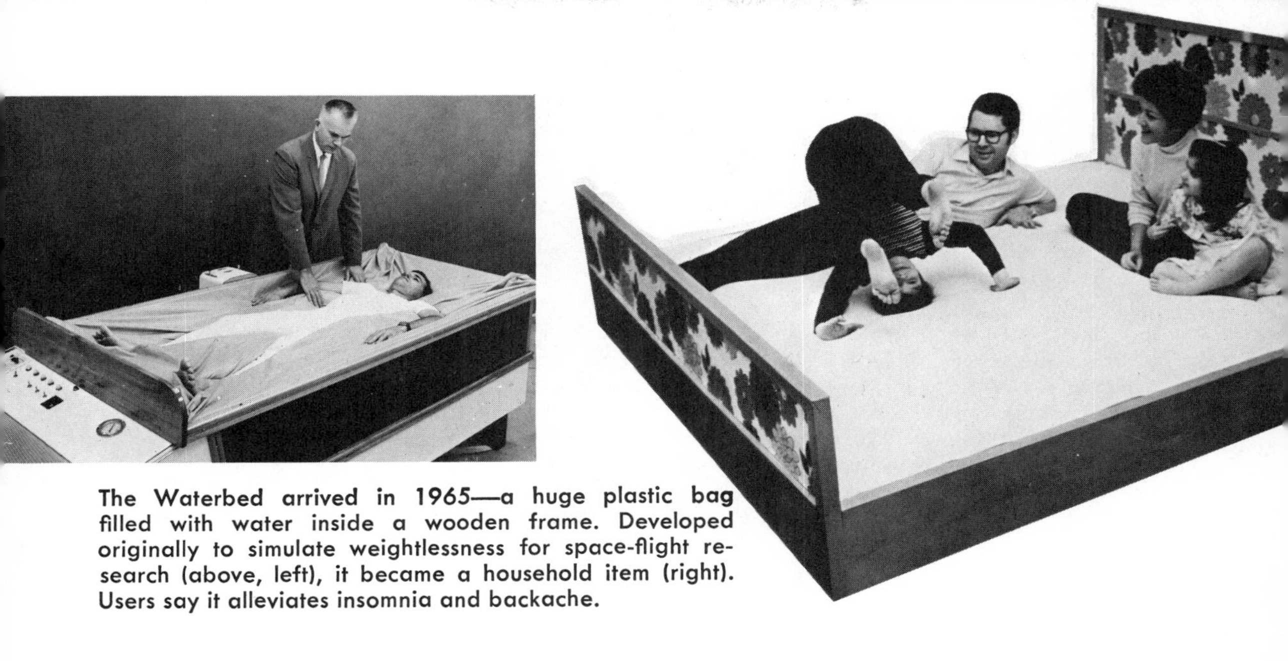

The Waterbed arrived in 1965—a huge plastic bag filled with water inside a wooden frame. Developed originally to simulate weightlessness for space-flight research (above, left), it became a household item (right). Users say it alleviates insomnia and backache.

The U.S. began emulating Japan in designing high-speed trains. Some were driven by gas turbines (below), some were electric. The turbotrains in use on the New York-to-New England run are built to do 100 mph when roadbeds are reconstructed to permit it. Cars hang from a frame overhead, swing outward as they take a turn. The other high-speed trains, the electrics, average close to 80 between New York and Washington, 20 miles faster than standard trains.

Human heart transplants were termed the medical miracle of 1967. Picture at right shows plastic model of heart. The first heart transplant was performed by Dr. Christiaan Barnard in Capetown, South Africa. In the next three years scores more followed throughout the world. Then the luster began wearing off the achievement. Only 1 in 7 patients survived very long. The big problem was the body's rejection of a foreign object. Drugs and serums helped. Research goes on.

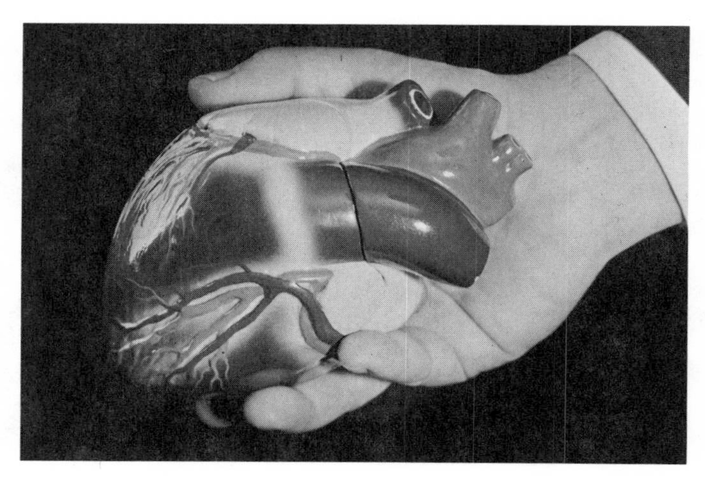

A mammoth oil tanker, the *Manhattan* (left), aided by an icebreaker, plowed from the Atlantic through ice-locked Northwest Passage to Point Barrow (see map) in 1969 to test feasibility of bringing out Alaska's North Slope crude oil riches by sea. Alternative was a trans-Alaska pipeline.

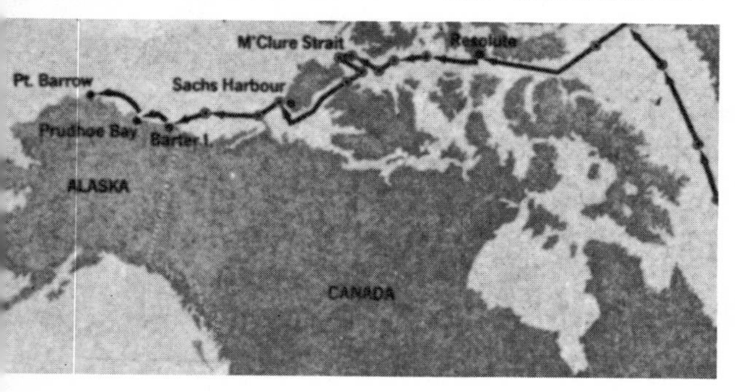

Popular Science reporter Robert Gannon (right, in rear) helps scientist to helicopter with an ice corer on the *Manhattan*'s historic trip.

Concluding nearly 2 centuries of debate, the Department of Commerce in 1971 recommended that the U.S. switch to the metric system in a ten-year period. Below, a lathe is converted from English to metric measurement—or vice versa—by an attachment (closeup at right). A full turn of this wheel advances a screw 0.02 of an inch or 5 mm.

Solid-state, light-emitting diodes were used for first time in a consumer product when wristwatch above was developed. Without hands or moving parts—apart from the oscillations of a quartz crystal—it flashed the time when a button was pushed. Power was from a rechargeable, silver-zinc battery. Price: $2,000.

The world's tallest building (right), supported as it rises by a diminishing number of modular steel tubes, was under construction in Chicago's Loop as this book went to press. The 109-story Sears Tower will stand 1,450 feet high. The method of construction will reduce weight and wind sway. There will be 102 elevators in 3 zones. Two express elevators will run to an observation level 1,350 feet above the street.

The U. S. Government was spending millions exploring substitutes for private automobiles in traffic-choked cities. One idea was PRT—personal rapid transit (left)—which has several variations. Computerized, automatic cars will carry 6 to 17 passengers. Get in, push a button for your destination. Most cars are electric-driven. PRTs hauled visitors at Transpo-72 at Dulles Airport near Washington, D.C. West Virginia University students will ride PRTs between campuses and into town.

CHAPTER 16

Afterword

MOST of the world's major inventions were made during the last hundred years, noted Chapter 1 of this book. Text and pictures in succeeding chapters have borne it out with example after practical example. By now it is time to turn from the trees to look at the forest:

What we have been seeing, during this Century of Wonders, is a technological speed-up unprecedented in history—dramatically narrowing the gap between "pure" and "applied" science until they have all but merged.

How the time span between discovery and application of scientific knowledge has been shrinking is pointed out by NASA's Dr. Ernst Stuhlinger:

While it took 3,500 years to get from simple lenses to telescopes and microscopes, the laser took only two years from theory to application. It took about 1,600 years to go from Egyptian steam toys to a real steam engine in which the steam did some useful work. Leonardo da Vinci made sketches of gliders and sailplanes four hundred years before we had real airplanes. The time lapse between discovery of principles and practical use, from Faraday to electric motors, was forty-five years; for radio, thirty-five years; and eight years for the transistor.

"In our modern times," said Stuhlinger, "engineers can hardly wait until scientists have made a new discovery. They build a gadget around it and they try to make it applicable to the solution of one or another of our problems here on earth."

As for new technology, he says dramatically, "we almost tear it out of the hands of the inventors."

* * *

This was not intended as a book to defend science against its critics, but it will not be complete until we address ourselves to what a recent article in *Fortune* called "The Senseless War on Science."

It may come as a surprise to many that there is such a war. Who are the antagonists? Many of the young are suspicious of science and technology and antagonistic to them. Many of their elders seem to feel they must make a choice between the world of science and the world of the humanities. A best seller warns that all of us are facing a nervous breakdown caused partly by the technological speed-up. Another book claims that the "myth of the machine" has obsessed our society and misdirected our energies.

The author of the latter book is the eminent architect and author Lewis Mumford, and is called *The Pentagon of Power* (*The Myth of the Machine*, vol. 2). In it Mumford declared that science and technology are bringing Western civilization to a breakdown. The force of what he calls "megatechnics" has gotten out of hand, putting mankind on the brink of disaster.

The popularizer of the idea that we face a general nervous collapse caused partly by the technological speed-up is Alvin Toffler, author, editor, and teacher. He calls both his best-selling book and the impending breakdown *Future Shock*. This, he says, is the disease of change, "a real sickness from which increasingly large numbers already suffer." What are the symptoms? They vary according to the stage and intensity of the disease. The author describes a range "from anxiety, hostility to helpful authority, and seemingly senseless violence, to physical illness, depression, and apathy." We are experiencing, he says, "the most rapid and deep-going technological revolutions in history," and then he declares, "In our haste to milk technology for immediate economic advantage, we have turned our environment into a physical and social tinderbox." Yet Toffler insists we can't and mustn't turn off the switch of technological progress, concluding that "we clearly need not less but more technology."

The war on science is confirmed by the editor-in-chief of the *Bulletin of the Atomic Scientists,* Eugene Rabinowitch. "The public," he writes in his introduction to the book, *Alamogordo Plus Twenty-Five Years,* "including much of the student youth, is beginning to see science and its child, technology, as enemies of mankind—creators of apocalyptic nuclear and space weapons and destroyers of nature. Science, they believe, is suppressing human individuality by making men servants of an immense soulless technological machine."

Rabinowitch, like Toffler, declares that we can't meet the challenge of our scientific age by deprecating science and destroying technology.

The consequences of the "senseless war" are summarized in the *Fortune* piece: "The immense prestige of U.S. science is being undermined by assaults from several different directions. If this wildly irrational campaign doesn't end soon, the U.S. can become a second-rate power and a third-rate place to live."

Its author, Lawrence Lessing, also points to the new youth culture, which turns away from science and toward the primitive, toward astrology and drugs. This group, he states, accepts the technology that pleases it, such as pharmaceutical chemistry for its drugs, power for its electric guitars, amplifiers, stereo, and motorcycles and motor cars, but wants to destroy the technology that does not please it.

Antagonism to science and scientists isn't new. In its early years (1872 and 1892) *Popular Science* devoted considerable space to one scientist's ancient struggle with the anti-science spirit of his age. His name—Galileo Galilei—appeared in the news recently when astronaut Colonel David R. Scott concluded Apollo 15's moon visit by dropping a hammer and falcon's feather simultaneously to the moon's surface to demonstrate Galileo's theory of falling bodies in a vacuum.

What got Galileo in trouble with the Church was his support of the then-heretical theory of Copernicus: that the earth orbited around the sun and was not, as theology held, the central heavenly body around which all others circled. Galileo was warned in 1616 to stop his heretical teachings or languish in its dungeons. Although he complied for a while, he published another pro-Copernican work in 1633 and was sentenced to imprisonment and commanded by Pope Urban to renounce his heresies. At the age of seventy, Galileo repudiated his writings in public as commanded and his sentence was commuted. When he died he was buried without benefit of clergy and his grave was denied a monument or epitaph.

(Ironically, Lewis Mumford, in the aforementioned *The Pentagon of Power,* blames Galileo and Copernicus for launching "a mechanistic and dehumanized view of the world" when they replaced the earth as center of the solar system with the sun—the "heliocentric" theory. When he dates modern man's fall from the discovery that the earth revolves around the sun, does Mumford mean that man should have continued ostrich-like to insist—against all rational evidence—that the earth was not only the center of his universe but also flat?)

A more recent demonstration of antagonism to science and technology occurred in England early in the last century. Bands of rioting Luddites destroyed factories' labor-saving machinery, which they blamed for reduced wages, unemployment, and workmanship inferior to their own. Supposedly they took their name from Ned Ludd, a half-wit who broke up stocking frames about 1779.

In our own century another anti-science event occurred—the famous 1925 trial of John Scopes, accused of violating a Tennessee law which forbade teaching evolution in public schools. This trial pitted the perennial presidential candidate, William Jennings Bryan (for the state) against the great trial lawyer Clarence Darrow (for the defense). Scopes was convicted and fined $100 (paid for him by the author H. L. Mencken). On appeal, the verdict

Galileo (above, left) got in trouble with the Church in the 1600s because his discovery, by telescope, of the moons of Jupiter supported the then-heretical theory of Copernicus (above, right) which held that the earth orbited around the sun.

Model of glider in 1924 *Popular Science* view above was based on sketch made four hundred years before airplanes, by Leonardo da Vinci.

Steam jets spun sphere of this Egyptian toy of 200 B.C., pictured in 1877 *Popular Science*.

was upset on a technicality by the Tennessee Supreme Court. In September 1929, *Popular Science* reported how teachers in Tennessee and Arkansas (which enacted a similar law, as did Mississippi) had to resort to subterfuge to tell their pupils the accepted scientific theory of man's descent. It was only as recently as 1967 that Tennessee repealed its law. In 1968 the Arkansas version was struck down by the U. S. Supreme Court. Mississippi did not wipe out its law until 1972!

From ancient to modern times, science and scientists have been in trouble. Is the new assault on scientific knowledge and technical knowhow any more valid than that of the ancient popes, the Luddites, or the modern anti-evolutionists?

For one answer, here is what Hubert P. Luckett, the present Editor-in-Chief of *Popular Science* wrote in a prospectus for the May 1972 issue. He said, in part:

"The wave of anti-technology hysteria sweeping the country grows out of fear of misunderstood forces and general confusion about technology's role in society. People see social ills around them—pollution, unemployment, war. They note that these unpleasant conditions are associated with machines—with the products of science and technology. So it is easy for the technically naive to conclude that the ills grow out of technology itself.

"They see that the way we have sometimes used—or misused—technology can be dehumanizing. But they fail to take into account that this is a *human*—not a technical—failure. Technology is the application of man's accumulated knowledge to the

Lately expanded ninefold, Dresden atom-power plant near Chicago helps meet energy crisis. Two 800,000-kilowatt reactors of 1970–71, in big building at left, now add output to that of plant's original 200,000-kilowatt reactor of 1959 within the dome at right.

This modular design was one of 22 winners in government-sponsored competition, Operation Breakthrough, to spur new building techniques.

Among the solutions to current transportation problems is this sleek prototype of a new transit car for BART—Bay Area Rapid Transit, the new 75-mile system in San Francisco. Program took ten years to develop.

problems of fulfilling some of his wants. Technology does not define wants; it simply supplies the means of fulfilling them.

"The two institutions we have evolved for defining wants are the market place and the government. If we have complaints about the pollution of the environment, the decay of the inner cities, the depletion of natural resources, they should be directed to the market place through our actions as customers and to the government through our actions as voters.

"The remedy [for the ills of society now being blamed on technology]," Luckett concludes, "is to make the social decision to apply even more advanced technology to solve the problems from pollution to transportation to housing to feeding the runaway population—and to making the benefits of technology available to all men everywhere."

Where do we go from here? What hopes do rational men have to look forward to? What are the specific challenges of the near future that only more and better science and technology can meet?

Described in a story in the New York *Times* as "a counterattack against those who increasingly criticize science and technology as disruptive and destructive forces in society," a New Technology Opportunities Program is being launched currently by the federal government. Basically, the program plans to establish priorities for research and development of technologies to solve domestic problems and provide incentives for industry to pursue the goals defined by the program. Dr. Lawrence Goldmuntz, who is co-ordinating a preparatory study of the program, cited the areas for which proposals will be evaluated and developed; they include solutions to such problems as transportation, communications, air quality, health care delivery, and productivity.

In the prospectus for the 100th Anniversary issue (May 1972), the present editors of *Popular Science* declared, "For the first time in human history, the tools are in hand for providing man's most pressing physical needs: food for all, decent housing, convenient mass and personal transportation, instant worldwide communications, a clean environment, abundant recreational resources and time to enjoy them."

Laser TV, recent development in area of new information systems and communications, was introduced in 1967 to *Popular Science* readers. Picture shows giant image projected on screen by pencil of ruby-red light from gas laser of experimental Zenith apparatus.

The editors then outlined 7 basic challenges that must be met—challenges that can be resolved only if we make the necessary social, political, and business decisions to tackle the problems with ever more advanced technology and science:

1. *Ample Energy.* Energy is the prime mover, the irreplaceable resource that runs our economy—the machines and tools of business and industry and the conveniences and amusements of our homes. But we are facing an energy crisis. To avoid running out of power we must use all energy resources such as coal, gas, and oil as carefully as possible. We must build nuclear capacity—lots of it. In the distant future we must work to produce cheap, abundant, clean power from the ultimate source: nuclear fusion.

2. *Healthy Environment.* To stop the pollution of our rivers and streams, the murk in our air, the litter in our countryside, we must recycle used aluminum, steel, and other metals into new products, turn waste

fibers into new paper, purify the sewage of cities, and scrub the smoke from factory stacks. For this we will need huge amount of clean, electrical power. The automobile, a major source of pollution, must be sanitized. Beyond catalytic converters and steam cars lie other technological solutions.

3. *Increased Recreational Resources.* Already there are undreamed-of recreational opportunities for the outdoors: campers, motor homes, new boats, snowmobiles, specialized vehicles, camping equipment. And in and around the house: television, radio, sophisticated high-fidelity music systems, an infinite variety of power tools. New developments in all these areas are visualized in the crystal ball. Vast new lakes for boating, swimming, fishing, and picnicking are created by giant dams built for power generating and flood control. Remote areas become accessible through new bridges and highways. The jet airplane has already opened the world for pleasure and education. Leisure time is one reward of our industrial society and more pleasurable recreation is a promise for the century ahead.

4. *Adequate Housing.* Technology has the answer to substandard and too-expensive housing: the factory-built house that can slash labor costs and produce individualistic high-quality dwellings at a price people can afford. Operation Breakthrough projects already demonstrate a galaxy of new techniques showing that housing factories can turn out a quality product at a bargain rate. The problems lie in unrealistic building codes, political squabbling, and bureaucratic foot dragging—but they can be solved, and better housing for all Americans lies ahead.

5. *Efficient Transportation.* We have fallen far behind in developing cheap, pleasant mass transit but the techniques are here: tracked air-cushion vehicles, linear-induction propulsion systems, BART (Bay Area Rapid Transit in San Francisco), individual people-moving networks. New high-speed trains are needed to solve the short-hop problem (such as Washington to Chicago). So are short takeoff and landing (STOL) planes for proposed trial city-center-to-city-center routes. Ahead lie the results of an advancing transportation technology: new rapid transit, better air service, improved automotive engineering.

6. *Effective Information Systems and Communications.* Satellite communications—with its vast potential—is just beginning to play a role in the global network. Worldwide satellite-TV broadcasts directly into the home are forecast. Other wonders: home shopping by Picturephone, direct information from the library through a home console, daily newspapers clicking out of home teleprinters, coat-pocket computers, and tiny transceivers—made possible by transistors and large-scale integrated circuits—that are forerunners of the universal, portable telephone we'll one day carry in our pockets. And last, the laser, playing a new part in the home through video-recording systems and large-screen projection TV.

7. *Expansion of the Frontiers of Knowledge.* The search for knowledge—another name for science—leads man toward space and rockets, the oceans, and the universe itself. The crystal ball shows fantastic new space ships to explore the vastness of the solar system and submarine devices to probe the alien world beneath the seas. It shows also a series of breathtaking discoveries about the nature of our universe which astronomers and astrophysicists have already begun. These and other probings will give man a far greater understanding of his world, his universe, and himself.

* * *

It is true, as many have said, that the predicaments in which we find ourselves are caused not by science but by man's misuse of science. Thus our blueprint of challenges to be met and the decisions man must make to meet them give us an optimistic note on which to close.

The crystal ball shows man knowing what he wants and what he needs. He has ways to make these wants and needs known—at the polling place as a voter and at the market place as a customer. Government and commerce must respond. By using—not misusing—science and technology—man will get what he wants and needs. And so, perhaps, another century of wonders lies ahead.